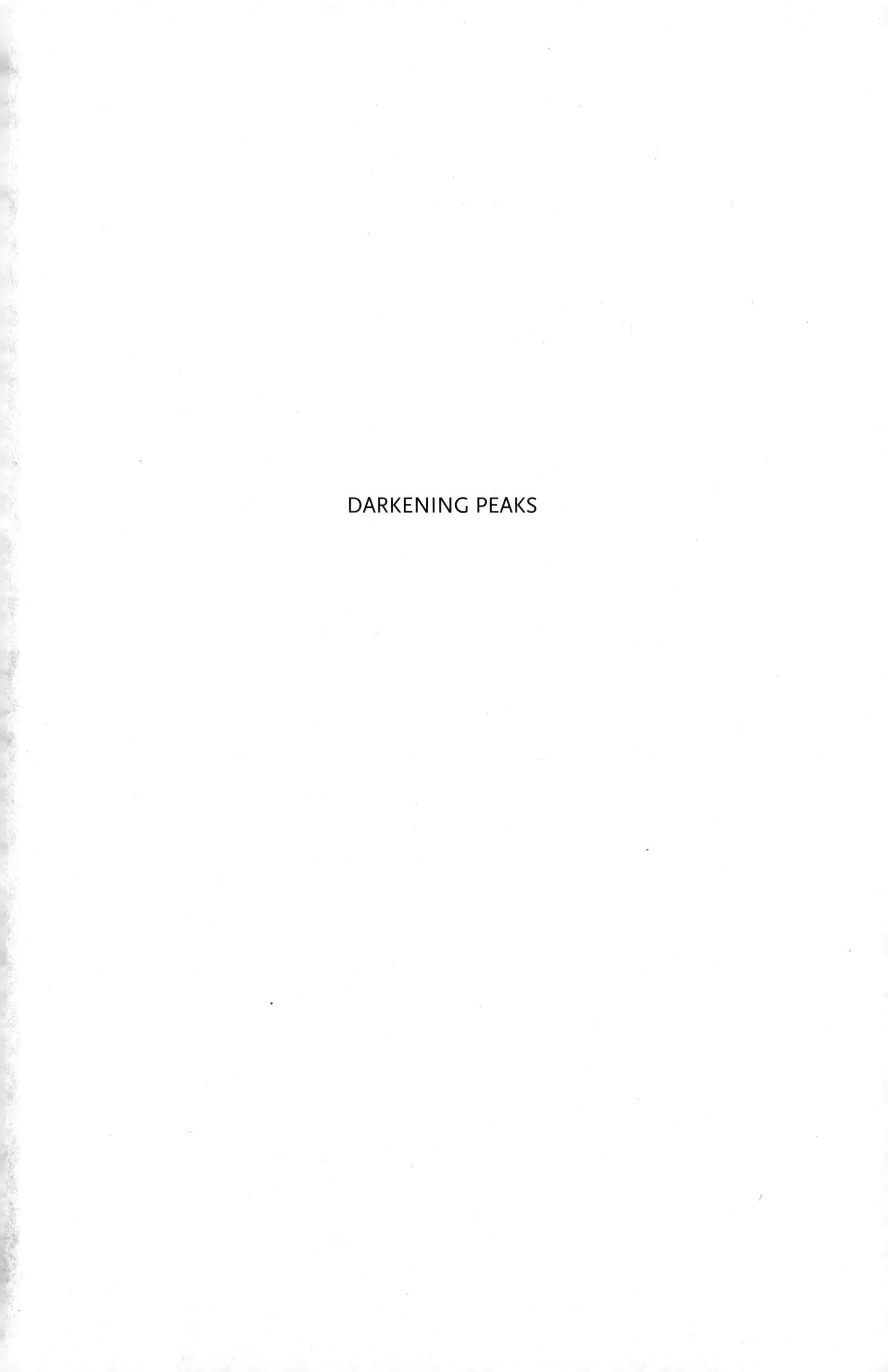

DARKENING PEAKS

DARKENING PEAKS

Glacier Retreat, Science, and Society

Edited by

Ben Orlove, Ellen Wiegandt,
and Brian H. Luckman

UNIVERSITY OF CALIFORNIA PRESS
Berkeley Los Angeles London

The publisher gratefully acknowledges the generous contribution to this book provided by the Ralph and Shirley Shapiro Endowment Fund in Environmental Studies of the University of California Press Foundation.

University of California Press, one of the most distinguished university presses in the United States, enriches lives around the world by advancing scholarship in the humanities, social sciences, and natural sciences. Its activities are supported by the UC Press Foundation and by philanthropic contributions from individuals and institutions. For more information, visit www.ucpress.edu.

University of California Press
Berkeley and Los Angeles, California

University of California Press, Ltd.
London, England

Library of Congress Cataloging-in-Publication Data

Darkening peaks : glacier retreat, science, and society / edited by Ben Orlove, Ellen Wiegandt, and Brian H. Luckman.
p. cm.
Includes bibliographical references and index.
ISBN 978-0-520-25305-6 (cloth : alk. paper) 1. Glaciology. 2. Glaciers. 3. Glaciers—Social aspects. 4. Glacial climates. 5. Climatic changes. 6. Glacial erosion. I. Orlove, Benjamin S. II. Wiegandt, Ellen. III. Luckman, Brian H.

GB2405.D37 2007
551.31—dc22 2007019338

Manufactured in the United States of America.
10 09 08
10 9 8 7 6 5 4 3 2 1

The paper used in this publication meets the minimum requirements of ANSI/NISO Z39.48-1992 (R 1997) *(Permanence of Paper)*.

Cover illustration: Piedras Blancas Glacier, Patagonia, Argentina. The glacier has diminished, its lower portion detached and calving into a newly formed lake. Photograph by Mariano Masiokas.

CONTENTS

Part 3 • Trends in Natural Landscapes: Climate Change

Part 4 • Impacts on Human Landscapes: Resources, Hazards, and Cultural Landscapes

Part 5 • Responses: Adaptation and Accommodation

ACKNOWLEDGMENTS

We have been fortunate to receive support and advice in every stage of the preparation of this volume. In fact, so many people have generously given their time that limitations of space preclude a complete list of names. We therefore mention only a few, trusting that the others who were involved will understand our appreciation.

The immediate ancestor of this book is the Wengen-2004 International and Interdisciplinary Workshop on Mountain Glaciers and Society, held in Wengen in the Berner Oberland of Switzerland in October 2004. Its genesis was a session on glacier retreat held at the 2003 annual meeting of the American Anthropological Association in Chicago. We thank the participants of this session, especially Julie Cruikshank and Sarah Strauss, for raising questions that we felt would benefit from further exploration from an interdisciplinary perspective. The Wengen Workshops on Global Change Research provided an ideal setting to discuss these questions and others closely related to them. We are especially thankful to Martin Beniston, who has ably organized these meetings since 1995, for accepting the topic. Members of the steering committee of the 2004 workshop, particularly Harald Bugmann, Paolo Burlando, and Wilfried Haeberli, contributed to the success of the meeting, as did the proprietors and staff of the Hotel Regina, who provided us with an idyllic locale for our discussions. The Swiss National Science Foundation provided financial and institutional support to the conference, as did the Department of Geosciences of the University of Fribourg. The Association Montagne 2002 in the Valais, Switzerland, generously helped defray participant travel expenses. Chapter authors also deserve our appreciation for their willingness to revise the conference presentations, which were earlier versions of most of the chapters in this volume.

We are grateful to the numerous reviewers who provided comments on individual papers in the first phase, in which the Wengen conference presentations were evaluated and a selected number of them transformed into book chapters. Blake Edgar and Matthew Winfield at the University of California Press provided many helpful suggestions throughout the second phase of revising and editing the papers after they had been submitted to the press; we thank them and the three anonymous reviewers of the volume manuscript for their detailed, thoughtful comments and critiques. Myra Kim and Julio Postigo took great care in tracking manuscripts, reviews, and revisions from contributors on five continents. Barbara Metzger's copyediting brought not only a consistency in vocabulary and usage to the volume but also greater coherence of argument and vision. As the editor for the images, Barbara Wolf ensured the high quality of the figures and tables that form an essential part of this volume.

OVERVIEW

Glaciers in Science and Society

1

The Place of Glaciers in Natural and Cultural Landscapes

Ben Orlove, Ellen Wiegandt, and Brian H. Luckman

In the first half of the nineteenth century, glaciers taught us a great lesson about the earth. In the 1820s and 1830s, Swiss naturalists established the existence of the Ice Age. Their key insight was the fact that the small glaciers found at high elevations in mountainous regions were remnants of vast sheets of ice that once had covered large portions of the earth's surface. They combined many sources—Alpine villagers' intimate knowledge of mountain landscapes, earlier research by other geologists, and their own extensive explorations—to document the remote periods of the past, when areas that now are towns, fields, and forests had lain under miles of ice. Once they understood that this now-vanished ice had transformed the earth's surface, they were able to explain features such as the parallel scratches found on rock faces that were engraved by glaciers and the long walls of rocks, stretching across valleys, that were carried by glaciers. Later researchers traced multiple Ice Ages and linked them to the cyclical fluctuations in the earth's orbit. This first lesson, then, was of the dynamic quality of the earth. Geologists found common elements in the study of the ice sheets and other discoveries that were made around the same time, such as the formation of rocks from sediments deposited on the floor of the sea. They came to understand that the earth was immensely old and always changing. This first lesson faced many challenges, particularly from those who held to a literal interpretation of the Bible, but finally received broad acceptance.

Now, at the beginning of the twenty-first century, glaciers are teaching us a second lesson, one that may be even larger than the first and that certainly is more somber and more urgent. That lesson is of the susceptibility of the earth to human impacts. The nineteenth-century picture of an earth whose surface is continually being modified by very slow natural processes is being replaced by the image of a planet that is being altered by rapid processes caused by humans. Once again, glaciers are playing a major role in shaping our understanding of our planet, since the retreat of glaciers around the world is a clear and dramatic example of this vulnerability. This retreat has come to public attention only recently. Through the 1970s, glaciologists—the intellectual heirs of the early Swiss geologists—observed that some glaciers expanded and others shrank and spoke of glacial "fluctuations" rather than a coherent, unidirectional process. By the 1980s, though, they had noticed the consistency with which many glaciers in different regions were becoming smaller. Glacier retreat became headline news in 1991. In that year two hikers discovered Oetzi, a Bronze Age man who had died high in the Alps and whose body, soon covered with snow that turned into ice, had rested frozen for 5,000 years until melting ice exposed it to view. Another such ice man appeared at the edge of a receding glacier in Canada in 1999 and received the name Kwaday Dan Ts'inchi (Long Ago Person Found) from the indigenous people of the region. Other lines of evidence pointed in the same direction: the series of measurements, released in 1998, that documented the shrinkage of the glaciers for which Glacier National Park in Montana is famed; the announcement in 2001 that scientific models predicted the complete disappearance of the famous snows of Kilimanjaro by 2020; the dramatic melting of Alpine glaciers in the unusually warm European summer of 2003. It became clear that many of the glaciers that had seemed permanent features of the landscape for millennia will not survive for many decades. Geoscientists are finding common elements in the study of glaciers and other recent discoveries, such as the warming of Arctic regions and the rise of sea level. We are coming to understand that human impacts can be more significant than we had realized. This second lesson, like the first, is also facing many challenges, particularly from those with an unwavering trust in the power of technology to solve problems.

There will not be a third lesson, or, at least, it will not come for a very long time. Current research suggests that the addition of large amounts of greenhouse gases to the atmosphere will postpone any future glaciation by tens or hundreds of thousands of years (Berger and Loutre 2002; Archer and Ganopolski 2005; Cochelin, Mysak, and Wang 2006). Looking up at mountains on the horizons, future generations will see bare rock where we still see gleaming snow and ice; traveling to the high country itself, they will find the dry beds of streams that once were filled with meltwater. One crucial and cherished part of our world will be gone.

The nature, history, and consequences of these changes form the subject of this volume. Reduced to its essence, each of the major aspects of glacier retreat can be summarized by a single term. The first term, *perception*, evokes the ways in which people know glaciers and form mental images of them, whether by seeing them or by hearing about them from others. The second, *observation*, might seem to be virtually a synonym of the first, but we use it in the specific sense of scientific observation—the systematic collection of measurements of glaciers. These two terms provide the basis for the third, *trends*, which includes both the reconstruction of glacial dynamics in the past and the projection of these dynamics into the future, indicating the probable state of glaciers in coming decades and centuries. As the fourth term, *impacts*, suggests, people are affected by glacier retreat in a variety of ways. We identify the major areas where new dangers—and perhaps new opportunities—have emerged. We highlight the vulnerability and resilience of societies in the face of significant changes to their landscapes and livelihoods. The final term, *responses*, indicates the action that people take on the basis of their perceptions of these trends and of their impacts. These responses may include different forms of accommodation and adaptation, as well as the adoption of policies by government agencies.

To guide our broad overview of glacier issues, we present some concepts and basic terms in the field of glaciology. Glaciers occur in places where, over a period of time, winter snowfall amounts exceed summer melting, so that snow accumulates on the surface and is transformed to ice. Once it reaches a critical thickness (about 30 m) and density (about 0.85 g/cm^3), this ice can deform and move downslope under the influence of gravity, forming glaciers. The state of health of a glacier can be defined by its mass balance. When annual snow and ice accumulation exceeds the loss by melting and other processes such as the calving of icebergs, the glacier has a positive mass balance and increases in mass. When the snow and ice loss exceeds the mass gain, the glacier has a negative mass balance. Generally the headward part of the glacier (the accumulation area) has a net gain, and the lower part of the glacier (the ablation area) has a net loss. The line separating these two zones is the equilibrium line, and the equilibrium line altitude (ELA) is the elevation at which the net accumulation in a given year is zero. Glacier flow transfers ice from the accumulation area to replace ice lost from the ablation area. If loss at the toe exceeds replacement by downvalley flow, the position of the toe recedes upvalley (the glacier "retreats"); if the ice delivered to the toe exceeds melt on an annual basis, the glacier front advances. Generally, glaciers with a negative mass balance exhibit frontal recession, although they may also lose considerable mass by thinning without frontal recession. Direct measurement of mass balance is expensive and time-consuming, but crude estimates of mass balance may be obtained from inspection of the equilibrium line elevation at the end of the melt season: Glaciers with a positive mass balance have, on average, more than two-thirds of their total area above the equilibrium line; this fraction is known as the accumulation area ratio (AAR).

A significant portion of the earth's land surface, roughly one-tenth of the total area, is permanently covered with ice, but it is heavily concentrated in distant and uninhabited regions (Barry 2006). The Antarctic ice sheets hold most of this ice, about 85% of the total area. About two-thirds of the remaining 15% is located in the Greenland ice sheet. In other words, all the other glaciers make up only about 5% of the world's ice-covered area.

Nonetheless, these glaciers account for a significant area, 680,000 km^2 by a recent thorough estimate (Dyurgerov 2005). These, too, are concentrated in remote areas at high latitudes and elevations (Table 1.1). Over half of this area consists of glaciers located on islands near Antarctica and in the Arctic Ocean or on Antarctica and Greenland but not contiguous with the major ice sheets. Within Europe, about two-thirds of the glacierized area is in Iceland. Patagonia holds a similar proportion of South America's glaciers, and the bulk of them are located in the more remote South Patagonian icefield rather than in the North Patagonian icefield, closer to towns and roads. New Zealand's glaciers are concentrated in the distant southwestern corner of the less populated South Island. Despite their concentration in cold places where few or no people live, however, glaciers play an important role in human societies.

PERCEPTION OF GLACIERS AND GLACIAL PROCESSES

Two attributes of glaciers shape the ways in which human perceive them: They are visible, and they are subject to cultural framing. The first attribute is a simple one. As large, slow-moving objects, glaciers can be directly seen. Though this point may seem so obvious that it does not merit being mentioned, it is quite significant. There are many other environmental concerns that involve entities that cannot be seen by the naked human eye. One cannot gaze up into the sky and tell whether ozone thinning has taken place, nor can one feel whether one is exposed to harmful levels of radioactivity. Genetically modified crops cannot be distinguished from other crops simply by looking at them. However, a person who returns to a glacier after an absence of several decades or who compares photographs of it taken at different times can easily note glacier retreat.

The direct accessibility of glaciers to human vision has helped to make them a topic of personal and public concern. Moreover, many glaciers can be seen for long distances, as can be attested by the tourist-brochure photographs of gleaming peaks rising beyond fields and forests in the southern Andes or in the Alps or of Mt. Kilimanjaro standing high above the dry plains of East Africa. People have adopted glacierized peaks as icons of particular regions: Residents of the city of Seattle have a special

TABLE 1.1

Distribution of Glaciers in the World, ca. 2000

REGION	TOTALS AND SUBREGIONS	GLACIERIZED AREA (KM^2)
World	Total	680,000
Europe	Total	17,800
	Alps	2,900
	Scandinavia	2,900
	Iceland	11,300
	West Caucasus	700
	Other	10
North Asia and Siberia	Total	3,500
Central Asia	Total	114,800
	Tien Shan	15,400
	Pamir	12,300
	Karakoram	16,600
	Himalaya	33,000
	Hindu Kush	3,200
	Other	34,300
Middle East	Total	830
	East Caucasus	780
	Other	50
Arctic islands	Total	315,000
	East Arctic islands	56,100
	West Arctic islands	36,700
	Canadian islands	151,800
	Greenland small glaciers	70,000
North America	Total	124,200
	Alaska	74,600
	Canada	49,000
	Other U.S.	530
	Mexico	10
South America	Total	25,000
	Patagonia	17,200
	Peru	1,800
	Other	6,000
New Zealand	Total	1,200
Sub-Antarctica	Total	77,000
	Sub-Antarctic islands	7,000
	Sub-Antarctic glaciers	70,000
Other	Total	9
	Africa	6
	New Guinea	3

NOTE: Figures include some rounding. Source: Dyurgerov (2005).

relationship with Mt. Rainier and will stop, when walking on a busy street, if the ordinarily cloudy skies clear and offer a view. The great height and pure whiteness of this mountain, featured on the city's Web site, evoke the natural beauty and abundant resources of the Pacific Northwest. At the center of the coat of arms of the Republic of Armenia is an image of Mt. Ararat, which now lies within Turkey but whose massive glacierized peak, visible from much of Armenia including the capital, Yerevan, serves as a symbol of the Armenian people and nation, of their greatness and of their indomitable will to survive.

However, not all glaciers capture the imagination. One may contrast the great fame of Kilimanjaro in Tanzania with the utter obscurity of the glaciers on Mts. Baker, Speke, and Stanley in the Ruwenzori Mountains of Uganda. First explored in 1906 by a team led by an Italian nobleman, they are less well known than Kilimanjaro, not only because they are somewhat smaller but also because they are much harder to view. In contrast to Kilimanjaro, a free-standing volcanic cone in an area of dry climate, these peaks lie behind other ridges in a moist region, almost constantly obscured by clouds. In other regions as well, some glaciers are seen only infrequently. Local residents in the Val Bavona in the Italian-speaking region of Switzerland only rarely view the nearby Basodino Glacier, even though it is the largest in the region (Madden 2003). It is difficult even to discern the glacier from the narrow and curved valley; over one-fifth of the valley's adult inhabitants have never seen it. No major ski facilities have been established in the high country above the valley, and therefore only the small number of men, less than one-tenth of the population, who work at a high-altitude hydroelectric power plant have any direct relation to the glacier or even see it on a regular basis.

Perception of glaciers does not rest on physical visibility alone. Cultural framing can also shape the ways in which glaciers are perceived, both by influencing the patterns of movement that can bring people close to them and by shaping their understandings (Knight 2004). Local populations in Uganda have generally considered the Ruwenzori Mountains unappealingly cold and damp. Even the Bakonjo people, who live closest to the peaks, identify themselves and are identified by others not with the peaks but with the forests of the lower and middle slopes, where they grow crops in clearings, gather wild plants, and maintain a series of shrines (Oryemoriga et al. 1995). Though the Batoro, one of the precolonial kingdoms that continued under British rule and after independence, have sought to impose their rule over this area and to tax its gardens, its rough terrain and its image as a wild, uncivilized place have supported the autonomy of the Bakonjo (Alnaes 1967, 1969; Cooke and Doornbos 1982; Horowitz 1977, 1881). Bakonjo men have occasionally obtained employment as porters and mountain guides, but this work is considered physically demanding and dangerous because of the possibility of angering the spirits that inhabit the high mountain zone (Busk 1954). If tourism brought foreign travelers more regularly to the glaciers, it might lead the local people to become more involved with them, but tourists coming to this region tend to visit the lakes and observe the wildlife in the lower, forested zones. In this case, cultural framing reinforces the limited physical visibility to keep the glaciers out of mind as well as out of sight.

In the Val Bavona, local residents may not see the nearby glacier, but they are quite conscious of glacier retreat and express concern about it. Over four-fifths of the people, including some who have never seen the Basodino Glacier, are aware that it is shrinking, since it is a topic of at least occasional conversation in the valley and those who see it often enough to notice the change have commented about it to others. Moreover, the glacier retreat has significance for them because they are generally concerned about climate change. The residents recognize other aspects of local climate change, such as drier summers, as well. They express their concern about its potential impacts elsewhere in Switzerland and throughout the world, but this concern stems from exposure to the national and international media. They are relatively untroubled by possible local repercussions such as the genuine risk that the reservoir of the power plant would receive insufficient inflow as glacier retreat continues. Nor do they attribute the increased number of rockfalls to glacier retreat, since the rockfalls that most affect them do not come from the area near the glacier. In the case of Val Bavona, then, cultural framing is influenced by global rather than local perceptions and leads to a greater level of awareness of the glacier than would result from its physical visibility alone.

Recent research thus confirms the variety of forms of cultural framing of glaciers. Cruikshank's work (2005) with the indigenous communities of the Yukon in Canada documents their frequent travel on glaciers, which they use as paths to cross the high ranges of the region in which they live. Cruikshank notes as well that they consider glaciers to be sentient beings who can observe and respond to human behavior. Native people express concern that white researchers will not demonstrate proper respect for the glaciers (by avoiding certain foods and ways of speaking), thus raising the risk of glacial surges. In this case, the local perceptions of glaciers come not only from the proximity that travel brings and the attentiveness required to avoid crevasses but also from the rich stock of stories that provide information about glacier movement in the past. Research with indigenous groups in the southern Peruvian highlands documents a long tradition of pilgrimage to a glacierized peak. In this region, native communities recognize spirits that live in the high mountains and make offerings and recite prayers to them. A glacierized peak, Sinakara, is the site of a pilgrimage that dates back to the eighteenth century (Sallnow 1987; Bolin 2001).The pilgrimage lasts several days and includes visits to a number of churches and chapels; one crucial element is the climb by costumed pilgrims to the glacier itself, where, until recently, they carved off blocks of ice, believed to have medicinal properties. The glacier itself was divided into different sections, each allocated to the pilgrims of a specific region. The retreat of the glacier has been a matter of great concern to the local people; their regular pattern of visiting ensures that they see the location and shifts of the glacial front, and the ritual significance of the travel leads them to pay close attention to it. Troubled by the thought that the mountain spirits are withdrawing the ice from them, they have resolved to stop cutting blocks from the glacier.

Cultural framing by these various indigenous groups thus influences both the ways people physically see the glacier and the ways they understand it, but it is by no means restricted to indigenous cultures and to earlier times. As Wolf and Orlove (this volume) show in their study of Mt. Shasta in California, many contemporary Americans attribute a kind of consciousness or awareness to the mountain itself. They note in particular the comments made by many locals that the mountain protects itself from encroachment by sending avalanches through current or proposed ski resorts, as it did in 1978 and 1995. In her ethnographic study of several German-speaking villages in the northern Italian province of Alto Adige, Jurt (2007) describes the economic importance of glaciers in this region heavily dependent on tourism. She notes that the villagers attribute the recession of the glaciers to many material factors, such as physical removal of surface ice and snow by skis and snowmobiles and the deleterious effect of disposing of rubbish by throwing it in crevasses. They also speak of the glaciers as being sentient; they comment that the numerous visitors "stören" (trouble, distress) the glaciers and that to avoid further retreat "man sollte sie in Ruhe lassen" (one should leave them [the glaciers] in peace).

Researchers have shown that these cultural framings of glaciers can change over time. Strauss (2003) documents the terrifying Swiss folktales that portrayed glaciers as physical sites of residence of souls of dead people, trapped in purgatory. There were stories as well of the villagers' ability to alter the glaciers through magical means. One such story told of a stranger who traveled to a valley and advised the residents to send a young maiden to the mountains at dawn;

if she removed a piece of ice from each of seven glaciers and assembled these pieces above the valley, a new glacier would form, increasing the flow in the local river.

These stories were ancient and conveyed fears of real hazards. During the nineteenth century, advancing glaciers increased the level of danger by blocking entire valleys, damming rivers to create lakes that burst out in devastating floods (Wiegandt and Lugon, this volume). Close attention to these developments led to more systematic observation of glaciers and ultimately contributed to the beginnings of the science of glaciology. Once glaciers began to be monitored regularly, the growing data sets allowed researchers to trace the movement of glaciers and to examine hypotheses to account for their shifts. The scientific observation of glaciers, deriving in part from the concerns of the populations living closest to them and in part from the interest of scholarly individuals and associations, is a new form of perception of glaciers.

OBSERVATION

The regular gathering of data about glaciers involves repeated and coordinated visits to particular glaciers, usually on a fixed schedule, and the systematic recording of their attributes. This observation is nearly always conducted in the context of scientific research, though it has significant premodern and prescientific antecedents. Cruikshank (2005) has recounted how Native peoples near the Yukon-Alaska border preserve stories describing the movement of the Lowell Glacier, including the advances that blocked the Alsek River to create lakes that accumulated behind the ice dam before bursting out in floods. In a similar vein, Quechua-speakers in highland Peru have visited a glacier in yearly pilgrimages, noting the shifting location of several tongues in relation to fixed features in the surrounding mountain landscape. Nonetheless, the stories remain largely oral and local and, as Cruikshank points out, require significant translation to be incorporated into the body of scientific knowledge.

Though measurements of glaciers had been made since the first half of the nineteenth century (Haeberli, this volume), it was not until near the close of that century that the modern systematic collection of data about glaciers began in Switzerland and Norway. The International Glacier Commission was established in 1894 and sought to coordinate the recording of the length and areas of glaciers (Radok 1997; Braithwaite 2002). The particular form of glacier monitoring reflected the nature of the groups involved: People living close to glaciers were often concerned about glacial hazards, while the members of national scientific communities and international scientific organizations were interested in understanding the motion and characteristics of glaciers and their history. In countries such as Switzerland and Norway, extensive monitoring is a measure of the importance of glaciers to the national identity, economy, and society. In Switzerland, densely populated valleys are exposed to glacier hazards, and major hydropower resources depend on the amount and timing of glacier meltwater. Cantonal forest services, natural-hazard managers, and university scientists work together to observe glacier dynamics. The mix of factors is similar in Norway, where the research has been conducted by national institutes, particularly those involved with polar exploration and with the extensively ice-covered archipelago of Svalbard, an international zone of whaling and coal mining over which Norway came to exercise a predominant role. The country's great pride in its accomplishments in ice-cap exploration (including the first crossing of the interior of Greenland and the first successful expedition to the South Pole) initially contributed to support for glacier research, though the national hydropower authority has played an increasingly important role (Andreassen et al., this volume). Glacier monitoring in Peru, by contrast, began several decades later. Disastrous floods, in which thousands of people died, directed attention to the monitoring of glacial lakes in the regions in which the risks of floods are greatest (Carey, this volume). Government services have sought to gather data on glaciers to assess stream flow in areas where glacial meltwater has high economic value as a source of hydropower and the supply of irrigation water to arid regions (Carey 2005).

Though glacier monitoring began over 100 years ago, its history has not been smooth (Macdougall 2004). Governments sometimes fail to provide steady funding, and the contribution of effort by local volunteers has similarly proved to be unreliable in a number of cases. For example, the Canadian government began an ambitious program of glacier inventory in the 1960s but abandoned it in the 1970s; it has yet to be resuscitated. International organizations were disrupted during the two world wars, and funding was often scanty in the intervening decades. Some expansion of glacier monitoring began after world war II, particularly with advances in mass balance studies. In contrast to earlier work, which concentrated on measuring the frontal positions of glaciers and their surface areas and lengths, this research examined the changes in volume of glaciers by measuring accumulation and ablation (Schytt 1962). Many such studies were set up during the International Hydrological Decade of 1965–74.

Glacier observation has expanded greatly in the past two decades (Knubel, Greenwood, and Wiegandt, this volume; Bowen 2005). The concern about global warming in the 1980s brought attention to glaciers as a valuable indicator of environmental change. The creation of the World Glacier Monitoring Service (WGMS) in 1986, centered at the University of Zurich and the Swiss Federal Institute of Technology, combined earlier organizations, systematized data collection, and extended it to new areas. The WGMS coordinates closely with the United Nations Environment Program, the United Nations Educational, Scientific, and Cultural Organization (UNESCO), and international bodies in the geosciences. The International Geosphere-Biosphere Program (IGBP), also founded in 1986, has vigorously promoted research linking glacial data with climate change. The data sets assembled by the WGMS have documented the dramatic glacier retreat around the world and have received specific mention in such major reports as the Intergovernmental Panel on Climate Change (IPCC) Fourth Assessment Report of 2007.

As glacier observation has become an increasing priority and attracted more funding, new techniques have allowed more efficient and detailed data gathering and analysis. These methods provide much better information than the aerial photographs, some dating back to the 1930s, which offered poor resolution and uneven coverage. New remote sensing methods have improved the collection of data on glacier area. Aircraft-based laser altimetry and satellite-based kinematic global positioning system (GPS) measurement allow mass balance to be studied by observing shifts in the height of glacial surfaces. The rapid advances in computational power permit analysis of the new data sets. New modeling techniques have led to better projections of glacial mass balance by integrating data on glacier area and elevation with meteorological data on temperature and precipitation and energy-input data from solar radiation at a high degree of spatial resolution (Paul et al., this volume). This modeling work has also led to assessments of the effects of different scenarios for climate change on hydrology in glacierized watersheds (Corripio, Purves, and Rivera, this volume; Schneeberger et al. 2003) and on changing hazards such as debris flows (Huggel et al. 2004).

TRENDS

Much of the current interest in glaciers rests on the linkages between trends in climate and trends in glacier extent (length, area, or volume). The consensus reported in the most recent IPCC report is that glaciers around the world are shrinking primarily because of global warming (Lemke et al. 2007). The linkage between glacier extent and temperature is not quite as direct and immediate as this account suggests. The effects of warming can be influenced by other variables such as topography and cloud cover, which both affect exposure to sunlight, and the nature of the bed of the glacier, which can favor or delay the downslope flow of ice. Changes in precipitation, also associated with climate change, can influence accumulation in the upper portions of a glacier and ablation in the lower portions. Moreover, many glaciers are quite large, so it can take a long time for them to respond to shifts in temperature; though warming takes place on their surfaces, their interiors can remain cold for some time. Nonetheless, the simplest account—glaciers are melting because temperatures are rising—is a fair summary.

The recent improvements in glacial observation and the growth of longer-term data sets have supported more detailed studies of glacier history, recent recession, and modeling of future changes. Recent glacier history can be reconstructed from a variety of geologic and geomorphic techniques that rely on the morphologic and stratigraphic evidence. Steiner, Zumbühl, and Bauder (this volume) present a classic study of the reconstruction of glacier history from a wide variety of documentary sources and show how, using modern cartographic techniques, some of the early, high-quality topographic mapping can be used to develop accurate three-dimensional images of the glacier surfaces over 100 years ago for comparison with detailed modern records of changes. Tree-ring data have been used in the past to date glacier fluctuations and to provide reconstructions of temperature and precipitation from mountain areas. For example, Watson et al. (this volume) use tree-ring data from several tree species to develop proxy records for summer and winter mass balance that can be used to examine the controls and past history of glacier fluctuations over the last 300 years in the Rocky Mountains in North America. These reconstructed mass balance records can be verified against the discontinuous record of glacier fluctuations based on more traditional geomorphic techniques and dating of moraines. This work allows the authors to reconstruct glacial fluctuations at a finer time scale for the eighteenth and nineteenth centuries than is possible from the morphological record. Their results demonstrate that warming in the twentieth century has been an important cause of glacial retreat, though fluctuations in precipitation have also had an effect.

Similar efforts to link climate variables with mass balance models have supported connections between global warming and glacier retreat in other regions. In particular, models indicate that shifts in temperature—especially in temperate regions but in many high-latitude regions as well—have a stronger effect on glacier extent than shifts in precipitation, which tend to be more localized. As Oerlemans (2005) documents, the small increase in temperature that has taken place in many parts of the world, on the order of 1°C, is sufficient to create significant melting. It would require a large increase in precipitation, about 25%, to offset it; no model of climate change suggests long-term increase on such a scale. In the Alps, the area in the world for which the most detailed records are available, the basic pattern—general retreat for the past 150 years, with an accelerated pace in the past two decades—is consonant with atmospheric warming, though individual glaciers show different rates of response to this process, depending on orientation, topography, and several other factors (Zemp et al., this volume). These studies have also shown that small glacial advances in the 1990s in a few regions of the world, such as the South Island of New Zealand, far northern California, and portions of Norway, occurred in response to augmented winter precipitation due to changes in storm trends in areas of heavy precipitation. Glaciers in these areas are nevertheless sensitive to temperature increases and have also seen retreat since 2000 or so (Andreassen et al. 2005; Chinn et al. 2005) or are projected to retreat in coming decades (Wolf and Orlove, this volume).

Temperature increases are also associated with glacier retreat in the tropics, particularly in the Andes, where the great majority of the tropical glaciers in the world are found (Wagnon et al. 1999; Francou et al. 2003). However, the situation is more complex for Kilimanjaro. As Mölg et al. (this volume) show, the well-known shrinking of the glaciers there has more to do with changing precipitation patterns in East Africa linked to shifts in Indian Ocean circulation than with increasing temperature.

Recent research points to the significant impact of warming on glacier size and mass balance in the future. From estimates in the 2007 IPCC Fourth Assessment Report, glacier retreat will cause a rise in sea level of 0.076–0.152 m between 1990 and 2090 (Meehl et al. 2007). A recent study points to a figure just below the lower end of that estimate (Raper and Braithwaite 2006), but even this level would be accompanied by further reduction in glacier extent far beyond what has been experienced to the present. Projections for specific regions of the world also indicate such shrinkage (Hall and Fagre 2003). On the basis of different scenarios for greenhouse-gas emissions, one article forecasts that 73–94% of mountain glacier volume will be gone by 2400 (Wigley and Raper 2005). Any projection that far into the future must necessarily rest on assumptions about changing levels of greenhouse-gas emissions, a difficult matter to anticipate. Here we will not adopt such a distant time horizon but will instead look at the impacts of the glacier retreat that has already taken place and note the responses to them.

IMPACTS

In general, the study of physical processes associated with glacier retreat is more advanced than the study of the impacts of these processes on human societies; indeed, the question of impacts constitutes an important gap in our knowledge. Nonetheless, current research has progressed to the point of being able to recognize four broad categories of impacts: global environmental change, economic resources, hazards, and cultural landscapes.

The most frequently discussed impact of glacier melting on global environmental change is its contribution to sea-level rise, a factor with the potential for major effects in coastal regions. There is strong consensus that sea level rose about 15 cm in the twentieth century and that the rate increased in the last decades of that century. This rise can be attributed partly to the melting of glaciers, though other factors, particularly the small but positive tendency of water to expand as it warms, have been important. The IPCC report suggests, with some uncertainty, that glaciers contributed about 27% of the rise in the last decades of the twentieth century (Bindoff et al. 2007). Sea levels are projected to continue to rise in the twenty-first century, with an estimated increase of three times the twentieth-century rate. Of this rise perhaps 30% will be attributable to glacier melt. Though glaciers will shrink in later centuries, projected sea-level rises are higher at that time, since the ice sheets in Greenland and Antarctica, which hold considerably more water than glaciers, are likely to melt more rapidly by then. Recent research has pointed to faster rates of current melting of the Greenland ice sheet (Rignot and Kanagaratnam 2006; Ekström, Nettles, and Tsai 2006), and it is likely that new findings will continue to alter our understandings of sea-level rise (Parkinson 2006).

Glacier retreat may also lead to other forms of global environmental change, particularly through its impact on biodiversity, including plants and invertebrates in high mountain regions. Two major research efforts are monitoring these changes in biodiversity. The first of these is Global Change in Mountain Regions (GLOCHAMORE), with many international partners under the auspices of several UN programs, the University of Vienna, and the Mountain Research Institute. The other, the Global Observation Research Initiative in Alpine Environments (GLORIA), is linked to the International Geosphere-Biosphere Program (IGBP) and the University of Vienna (Pauli et al. 2005).

The economic sectors directly affected by glacial retreat are primarily related to water and, to some degree, tourism. Changes in water resources occur both in time (since an early phase of glacier retreat increases water flows as glacial ice is converted to water but then reduces these flows when the glacial volume is much reduced) and in space (since the water is used in different ways in different parts of watersheds). All users may receive considerable water in the short run but face likely shortages in the future. The specific economic impacts and possible adjustments will be different for hydroelectric power generation, often in the upper portions of watersheds, and for irrigation and urban consumption, usually in the middle and lower sections. Tourism may also be affected by glacier retreat, since in some glacierized regions, especially in the Alps and Scandinavia, in western North America, and in New Zealand, tourists visit glaciers for hiking and sightseeing. The capacity of glaciers to extend the ski season also attracts many recreational visitors (Smiraglia et al., this volume). However, the direct relation between tourism and particular landscapes must also be put in the context of price factors and changing tastes.

Hazards in the form of floods and landslides constitute another important impact of glacier retreat. Though these floods have several causes, including expansion of rivers, swollen beyond their banks by glacial melt, one particular form is important. The lakes that form below glaciers are often dammed by relatively weak glacial moraines that can collapse or be breached and create devastating floods. These are known either by their Icelandic name, *jökulhlaups*, or as glacial lake outburst floods. A single outburst flood that occurred in Peru in 1970 killed more than 10,000 people (Carey, this volume). Other disasters caused by glacier retreat include landslides, rockfalls, and debris falls. As the sides of valleys formerly filled with ice become exposed, the material in their slopes is prone to slip and fall. Moreover, temperature changes may result in the degradation of mountain permafrost that can also lead to instability in accumulations of loose surface debris, for example, moraines or talus, formerly bonded by ice. These also provide loose debris accumulations that can be mobilized during intense summer rainfall events and produce major debris flows that destroy communities in the valleys below (Wiegandt and Lugon, this volume).

These three impacts of glacier retreat have economic as well as physical consequences, since they reduce or destroy productive resources and interfere with economic activities (Agrawala, this volume). In addition, there are cultural impacts of glacier retreat. Many human societies have strong attachments to glaciers, as they do to other features of the natural environment. These features have strong symbolic significance, and people identify with them. Brenning (this volume) discusses a case in Chile in which many people opposed the destruction of a glacier by a mining company that proposed to exploit ore deposits not only because of the impacts of such actions on water resources but because they cared about the glacier itself. If a rather extreme parallel may be drawn, many people experienced deep distress over the attacks on the World Trade Center in New York not merely because thousands of people were killed, not merely because valuable property was destroyed, but also because of the symbolic importance of the buildings themselves and because of the key place of the buildings in the built urban landscape. In a similar vein, the white summits of lofty peaks play a special role in natural landscapes—landscapes that, because of the meanings connected with them, are also cultural landscapes. In some instances, this role is clearly religious and ritual, as is shown by the cases of pilgrimages to glacierized regions. In other instances, glaciers become representations of regional or national identity. The glacierized peaks of Illimani lie within clear view of La Paz, the capital of Bolivia; they also are featured prominently on the city's official shield, on the seal of the major university in the city, on the labels for the beer brewed there, and in countless other forms. The residents identify not only with the city and with the nation but also with the highland region of the country, embodied in the summits of Illimani. As do the residents of Seattle and Yerevan mentioned earlier, the Paceños identify not only with a mountain but with the glaciers that make it distinctive. Though glacier retreat in Illimani will reduce the hydropower that La Paz uses and greatly limit the skiing season for local enthusiasts, the cultural impacts of a dark Illimani rather than a gleaming white mass that watches over the city seem as serious as the economic ones (Vergara et al. 2007).

These different areas of impact are closely related. Tourism, for example, links the economic aspects of purchasing goods and services with the cultural aspects of establishing and maintaining social identities. Case studies in this volume indicate that outburst floods and other glacial lake hazards have destroyed hydroelectric plants in Nepal and Peru and periodically injured or killed villagers in Switzerland and tourists in Italy and New Zealand. It is therefore difficult to disentangle linkages between hazards and resources, since economic activities can be disrupted by natural catastrophes.

RESPONSES

Of the different themes discussed here, the question of human responses to glacier retreat is the one for which there is the smallest amount of empirical research, but we can make some general observations and present some examples.

Of the observations, three are particularly important. First, the global scale of climate change means that the people most directly affected by glacier retreat make a very small direct contribution to the worldwide emissions of greenhouse gases that are its root cause. As a consequence, their behavior will have little impact on the future course of this shrinkage. In contrast to the situation with many other environmental issues, local or regional responses cannot directly affect the causes of the problem. This fact can contribute to what might be termed a narrative of victimhood, in which the people affected by glacier retreat passively suffer the consequences of actions of others (Adger et al. 2001). Such accounts rightly capture certain aspects of climate change and its effects, but they also tend to reduce complex historical changes to an overly simple story. Carey's account of glacial hazards in the Peruvian highlands (2005 and this volume) offers a particularly good example of moving beyond simple narratives of victimhood by considering the wide range of actors, interests, and beliefs involved in shaping responses to changing glacierized landscapes.

A second reality about responses is that people who live in glacierized watersheds respond not only to glacier retreat but also to other environmental, economic, and political factors. Luterbacher et al. (this volume) present a particularly clear instance in which Central Asian republics in general and Kyrgyzstan in particular make decisions about water use based not only on projections of glacier retreat but also on their own and other nations' economic goals and geopolitical interests in the current unstable post-Soviet period. These elements all become part of complex negotiations with neighboring countries about future glacier-generated water use.

Third, responses to glacier retreat are complicated by the fact that they involve balancing multiple concerns (Agrawala 2004 and this volume). The long time horizon of glacier retreat and the uncertainty about its trajectory make it difficult to calculate the economic value of glaciers. Moreover, the extraeconomic value represented by the cultural significance of glaciers means that decisions are not made with economic criteria alone. Institutional aspects are yet another factor that intervene to shape responses to changing landscapes and resources. Property rights are particularly ambiguous and ill defined for glaciers. In remote areas, they often constitute a no-man's-land. Incidents such as the example cited by Agrawala (this volume) of a Chinese military truck's being washed into Nepal by a glacial lake outburst flood highlight dilemmas of transboundary adjudication. The problem of glaciers' water resources is even more general and compelling because of their considerable value. The attribution of rights is therefore particularly important. However, the domain of water rights is fraught with complexities, ambiguities, and hence potential conflict. Although agreements exist for specific river systems or lakes, there is no general international water regime. The Kyrgyz case (Luterbacher et al., this volume) illustrates the general problem of resolving asymmetric power relations inherent in upstream-downstream relations, where upstream users can simply prevent access by downstream users. There have been many efforts to have glaciers themselves designated as protected areas managed by agencies concerned with conservation and tourism, but the question then arises of what happens to a protected entity that is disappearing because of global climate change. This problem parallels the global commons issue that climate change represents.

Despite these complicating factors, some responses can be noted in each of the impact areas discussed earlier. The most widely noted aspect of global environmental change is sea-level rise, though this is due to a number of factors and not only to glacier retreat. Planning for this rise has been active in many countries, such as the United Kingdom and the Netherlands. Recent unusual and extreme events such as the tsunami of December 2004 and the deaths and destruction associated with Hurricane Katrina in August 2005 brought attention to the vulnerability of coastal regions and led to efforts to develop early warning systems for tsunamis and to improve defensive works against storm surges.

Responding to the shifts in water resources that follow glacier retreat poses significant challenges. Andreassen et al. and Rhoades, Zapata Ríos, and Aragundy Ochoa (this volume) offer interesting counterpoints. Andreassen et al. show that Norway carefully manages the reservoirs that produce hydroelectric power, the country's principal source of electricity, and this monitoring forms one piece of complex planning that includes consideration of incorporating other energy sources such as wind power (Kristiansen 2006; Hagstrom, Norheim, and Uhlen 2005). The country seeks to balance economic efficiency and environmental sustainability in its energy policy (Johnsen 2001). The Ecuadorian case described by Rhoades, Zapata Ríos, and Aragundy Ochoa is a sobering instance of glacier retreat that has proceeded to the point of glacier disappearance. The indigenous farmers in Ecuador who relied on glacial meltwater for irrigation have invested considerable amounts of labor and capital in linking with other irrigation systems. Their expenses have increased, but their access to water is less secure, since they join these systems, themselves under stress, as latecomers with less effective claims to water. These examples, taken together, point to the great human capacity for adaptability but also to its limits and to the substantial differences among countries. Ecuador is much poorer than Norway. Its possibilities for water storage are limited by the configuration of the Andean volcanic cones, with fewer reservoir sites than the deep canyons of Scandinavia, and its ability to develop strategies for substitution is considerably limited by economic factors. However, in an increasingly globalized world, robust national economies may not be sufficient to withstand the parallel pressures of climate change and market forces, as the resistance to electricity market liberalization in Switzerland demonstrates (Wiegandt and Lugon, this volume).

Tourism constitutes another sector responsive to natural conditions and wider market forces. The effects of glacier retreat on skiing can be dramatic. In the summer of 2005, the Swiss ski resort of Andermatt covered a small portion of a glacier, less than a hectare, with a reflective blanket to slow its melting; this unusual effort attracted considerable attention in the press. Other ski resorts make different efforts. As Smiraglia et al. (this volume) document, one Italian resort on the Vedretta Piana glacier moves firn (snow that has been partially transformed into ice) from the higher portion of the glacier to the lower sections in which skiing takes place, a policy that is in effect accelerating the shrinkage of the glacier. In the United States, a number of resorts, including those on the glaciers of Mt. Hood in Oregon, participate in "Keep Winter Cool," a program run jointly by the U.S. National Ski Areas Association and a large environmental nongovernmental organization (NGO), the Natural Resources Defense Council, encourages skiers to drive in carpools to the resorts and to purchase tags that subsidize the efforts of ski resorts to run lifts with power produced from renewable sources. Though this program, at least in its early stages, appears to set weak standards and to lack enforcement, it does point to the growing awareness of climate change and glacier retreat (Rivera, de Leon, and Koerber 2006). Other ski resorts, however, seek other adaptations, such as emphasizing the use of mountain bikes on snowless trails. All these efforts must also be viewed in the context of comparative costs of travel and lodging that will influence tourists' choices of venues in ways completely unrelated to landscape and activity.

Hazards associated with glacier retreat have very local effects and must be addressed to protect residents as well as visitors. In some instances, the hazards can be directly reduced. Glacial lakes that present high risks of jökulhlaups can be drained, as has been done in Nepal (Agrawala, this volume) and Peru (Carey, this volume). These efforts are extremely expensive, however, and cannot keep up with all the dangerous glacial lakes in the world. The uncertainties associated with these outburst floods also make it difficult to prioritize lakes for draining. Warning systems are another possibility, though these, too, are expensive and difficult to implement in regions in which populations are scattered. Moreover, they are prone to the problem of false warnings. For example, U.S. National Aeronautics and Space Administration (NASA) scientists issued a warning in 2003 after reviewing satellite imagery that appeared to indicate a fissure in a glacier above Lake Palcacocha in Peru, the site of an earlier outburst flood. Many people fled the zone after this warning, and tourism declined. This panic was followed by frustration when it was discovered that the interpretation of the imagery was erroneous (Georges 2005; Vilimek et al. 2005). This case raises the problem of the differences between expert and lay assessments of risk, a topic reviewed by Wiegandt and Lugon (this volume).

A response to hazards that appears to be simpler than direct reduction or warning systems is the use of hazard mapping to indicate zones where travel, construction, or other activities should be limited. Though such mapping has proved an effective means of addressing other geophysical hazards, such as flooding, the evidence for its use to address glacial hazards is limited. Personal interests, as well as objective knowledge, can be involved in making these judgments. Often the desire to create value for sale to tourists or to attract industry leads to the minimization of risks in key zones. Agrawala (this volume) offers the problem of siting hydroelectric power plants because few sites in Nepal present both sufficiently high rates of return of energy production to investment and satisfactorily low vulnerability to outburst floods. In cases where restrictions are imposed, there are enforcement problems; it can be difficult to curb travel in such zones, as is apparent from the accidents that have occurred in areas clearly marked with warnings in New Zealand (Hay and Elliot, this volume) and Italy (Smiraglia et al., this volume). Limits on construction are also very difficult to enforce, as shown by studies in Georgia (Huggel, Haeberli, and Kääb, this volume), Peru (Carey, this volume), and Switzerland (Wiegandt and Lugon, this volume).

The cultural resonance of glaciers creates additional challenges in formulating responses to glacier retreat. Rhoades, Zapata Ríos, and Aragundy Ochoa (this volume) note the great sadness with which highland villagers describe the darkness of the formerly glacierized peak of Cotacachi and their efforts to ensure that younger generations will know of the beauty of the peak when it was still white. Other sources describe a strong concern for retreat in other, less traditional societies, such as Switzerland (Haeberli, this volume) and the United States (Wolf and Orlove, this volume). As Brenning (this volume) notes, many people are troubled when they learn that mining operations are destroying glaciers in remote regions of the world. Though some of their regret comes from the waste of water that could otherwise be used for irrigation, drinking, or other purposes, they can be distressed even when this destruction occurs in remote areas with few inhabitants. As Ehrlich (2004) points out, glaciers often seem to have an intrinsic value beyond any immediate usefulness that they might offer; their loss, whether through unregulated mining or global warming, constitutes a wanton destruction of nature that borders on or crosses into the realm of immorality.

Indeed, human responses to glacier retreat may be shaped as much by this cultural attachment as they are by the economic issues of resources and hazards. The economic impacts should certainly not be minimized, but the cultural impacts, felt outside mountain regions, merit attention as well. It is not only the inhabitants of La Paz or Yerevan or Seattle for whom glacierized peaks are a sign of their connection to their home. Many people around the world care deeply about seeing these mountains on the horizon or in photographs or even in their mind's eye. Eternally white, rising high above the places where people live, they are one of the treasures of our world, a sign for all of a connection to the planet that is our common home.

We may try to imagine the world as it will be a few hundred years from now. It may well be that waves will lap the lower stories of skyscrapers in former coastal cities and that people will understand that the ocean's rise, due principally to the great shrinkage of the ice caps in Greenland and Antarctica, began with the reduction of mountain glaciers to small fragments. But it may also be that new technologies and new patterns of consumption, elements of which can be discerned at present, will avert this change and others just as extreme. If the world does address the great challenge of global warming, it will be in part because of the way that glaciers serve as icons to make this challenge visible.

ACKNOWLEDGMENTS

We thank the participants in the conference at which earlier versions of most of the chapters in this volume were presented, the Wengen-2004 International and Interdisciplinary Workshop on Mountain Glaciers and Society, held in Wengen in the Berner Oberland of Switzerland in October 2004. We particularly appreciate the efforts of Martin Beniston, Paolo Burlando, and Wilfried Haeberli to organize the discussion at that conference.

We also appreciate the comments of the three anonymous reviewers and the recommendations of the University of California Press Editorial Committee and of Blake Edgar, our editor at the press. Blake Stimson's useful insights into the role of visuality in human social life encouraged us to consider the visual aspects of glacier retreat that form an important element in this chapter. Julie Cruikshank provided helpful comments on a late draft on very short notice. In addition, we received useful advice from a group of graduate students at the University of California, Davis (Alexis Jones, Christine Jurt, Myra Kim, Patricia Pinho, Julio Postigo, and Kevin Welch), and a group of writers in Davis (Alexander Cameron, Alan Elms, Karen Joy Fowler, Don Kochis, Clint Lawrence, Debbie Smith, and Sara Streich). Above all, we have been fortunate that Barbara Metzger carried out the copyediting. Her care in correcting errors and her numerous helpful proposed rewordings have significantly improved this chapter.

REFERENCES CITED

Adger, W. N., T. A. Benjaminsen, K. Brown, and H. Svarstad. 2001. Advancing a political ecology of global environmental discourses. *Development and Change* 32:681–715.

Agrawala, S. 2004. Adaptation, development assistance, and planning: Challenges and opportunities. *IDS Bulletin—Institute of Development Studies* 35(3):50–54.

Alnaes, K. 1967. Nyamayingi's song: An analysis of a Konzo circumcision song. *Africa* 37:453–65.

———. 1969. Songs of the Rwenzururu rebellion: The Konzo revolt against the Toro in western Uganda. In *Tradition and transition in East Africa*, ed. P. H. Gulliver, 243–72. Berkeley: University of California Press.

Andreassen, L. M., H. Elvehøy, B. Kjøllmoen, R. V. Engeset, and N. Haakensen. 2005. Glacier mass balance and length variation in Norway. *Annals of Glaciology* 42:317–325.

Archer, D., and A. Ganopolski. 2005. A movable trigger: Fossil fuel CO_2 and the onset of the next glaciation. *Geochemistry, Geophysics, Geosystems* 6, Q05003, doi:10.1029/2004GC000891.

Barry, R. C. 2006. The status of research on glaciers and global glacier recession. *Progress in Physical Geography* 30:288–306.

Berger, A., and M. F. Loutre. 2002. Climate: An exceptionally long interglacial ahead? *Science* 297:1287–88.

Bindoff, N. L., J. Willebrand, V. Artale, A, Cazenave, J. Gregory, S. Gulev, K. Hanawa, C. Le Quéré, S. Levitus, Y. Nojiri, C. K. Shum, L. D. Talley, and A. Unnikrishnan. 2007. Observations: Oceanic climate change and sea level. In *Climate change 2007: The physical science basis. Contribution of Working Group I to the Fourth Assessment Report of the Intergovernmental Panel on Climate Change*, ed. S. Solomon, D. Qin, M. Manning, Z. Chen, M. Marquis, K. B. Averyt, M. Tignor, and H. L. Miller, 385–432. Cambridge, UK, and New York: Cambridge University Press.

Bolin, I. 2001. When *apus* are losing their white ponchos: Environmental dilemmas and restoration efforts in Peru. *Development and Cooperation* 6:25–26.

Bowen, M. 2005. *Thin ice: Unlocking the secrets of climate in the world's highest mountains.* New York: Henry Holt.

Braithwaite, R. J. 2002. Glacier mass balance: The first 50 years of international monitoring. *Progress in Physical Geography* 26:76–95.

Busk, D. 1954. The southern glaciers of the Stanley group of the Ruwenzoris. *Geographical Journal* 120:137–45.

Carey, M. 2005. Living and dying with glaciers: People's historical vulnerability to avalanches and outburst floods in Peru. *Global and Planetary Change* 47:122–34.

Chinn, T., S. Winkler, M. J. Salinger, and N. Haakensen. 2005. Recent glacier advances in Norway and New Zealand: A comparison of their glaciological and meteorological causes. *Geografiska Annaler Series A, Physical Geography* 87A:141–57.

Cochelin, A. S. B., Mysak, L. A., and Z. M. Wang. 2006. Simulation of long-term future climate changes with the green McGill paleoclimate model: The next glacial inception. *Climatic Change* 79:381–401.

Cooke, P., and M. Doornbos. 1982. Rwenzururu protest songs. *Africa* 52:37–60.

Cruikshank, Julie. 2005. *Do glaciers listen? Local knowledge, colonial encounters, and social imagination.* Vancouver: University of British Columbia Press.

Dyurgerov, M. 2005. *Glacier mass balance and regime measurements and analysis, 1945–2003*, ed. M. Meier and R. Armstrong. Boulder, CO: Institute of Arctic and Alpine Research, University of Colorado, distributed by National Snow and Ice Data Center.

Ehrlich, G. 2004. *The future of ice: A journey into cold.* New York: Pantheon.

Ekström, G., M. Nettles, and V. C. Tsai. 2006. Seasonality and increasing frequency of Greenland glacial earthquakes. *Science* 311:1756–58.

Francou, B., M. Vuille, P. Wagno, J. Mendoza, and J.-E. Sicart. 2003. Tropical climate change recorded by a glacier in the central Andes during the last decades of the twentieth century: Chacaltaya, Bolivia, 16° S. *Journal of Geophysical Research*, 108, D5, 4154, doi:10.1029/2002JD002959.

Georges, C. 2005. Recent glacier fluctuations in the tropical Cordillera Blanca and aspects of the climate forcing. Ph.D. diss., University of Innsbruck.

Hagstrom, E., I. Norheim, and K. Uhlen. 2005. Large-scale wind power integration in Norway and impact on damping in the Nordic grid. *Wind Energy* 8:375–84.

Hall, M. H. P., and D. B. Fagre. 2003. Modeled climate-induced glacier change in Glacier National Park, 1850–2100. *Bioscience* 53:131–40.

Horowitz, D. L. 1977. Cultural movements and ethnic change. *Annals of the American Academy of Political and Social Science* 433:6–18.

———. 1981. Patterns of ethnic separation. *Comparative Studies in Society and History* 23:165–95.

Huggel, C., W. Haeberli, A. Kääb, D. Bieri, and S. Richardson. 2004. An assessment procedure for glacial hazards in the Swiss Alps. *Canadian Geotechnical Journal* 41:1068–83.

Johnsen, T. A. 2001. Demand, generation, and price in the Norwegian market for electric power. *Energy Economics* 23:227–51.

Jurt, C. 2007. Risk perception in a mountain zone: The case of Stilfs (South Tyrol) in northern Italy. Ph.D. diss, University of Bern.

Knight, P. G. 2004. Glacier: Art and history, science and uncertainty. *Interdisciplinary Science Reviews* 29:385–93.

Kristiansen, T. 2006. Hydropower scheduling and financial risk management. *Optimal Control Applications and Methods* 27(1):1–18.

Lemke, P., J. Ren, R. B. Alley, I. Allison, J. Carrasco, G. Flato, Y. Fujii, G. Kaser, P. Mote, R. H. Thomas, and T. Zhang. 2007. Observations: Changes in snow, ice, and frozen ground. In *Climate change 2007: The physical science basis. Contribution of Working Group I to the Fourth Assessment Report of the Intergovernmental Panel on Climate Change*, ed. S. Solomon, D. Qin, M. Manning, Z. Chen,

M. Marquis, K. B. Averyt, M. Tignor, and H. L. Miller, 337–83. Cambridge, UK, and New York: Cambridge University Press.

Macdougall, D. 2005. *Frozen earth: The once and future story of ice ages.* Berkeley and London: University of California Press.

Madden, E. 2003. Perceptions of environmental change and climate change in the Val Bavona, Switzerland: Social inquiry in a scientific arena. M.S. thesis, University of California, Davis.

Meehl, G. A., T. F. Stocker, W. D. Collins, P. Friedlingstein, A. T. Gaye, J. M. Gregory, A. Kitoh, R. Knutti, J. M. Murphy, A. Noda, S. C. B. Raper, I. G. Watterson, A. J. Weaver, and Z.-C. Zhao. 2007. Global climate projections. In *Climate change 2007: The physical science basis. Contribution of Working Group I to the Fourth Assessment Report of the Intergovernmental Panel on Climate Change.* ed. S. Solomon, D. Qin, M. Manning, Z. Chen, M. Marquis, K. B. Averyt, M. Tignor, and H. L. Miller, 747–845. Cambridge, UK, and New York: Cambridge University Press.

Oerlemans, J. 2005. Extracting a climate signal from 169 glacier records. *Science* 308:675–77.

Oryemoriga, H., E. K. Kakudidi, A. B. Katende, and Z. R. Bukenya. 1995. Preliminary ethnobotanical studies of the Rwenzori mountain forest area in Bundibugyo District, Uganda. *Bothalia* 25(1):111–19.

Parkinson, C. L. 2006. Earth's cryosphere: Current state and recent changes. *Annual Review of Environment and Resources* 31:33–60.

Pauli, H., M. Gottfried, D. Hohenwallner, K. Reiter, and G. Grabherr. 2005. Ecological climate impact research in high mountain environments: GLORIA (Global Observation Research Initiative in Alpine Environments)—its roots, its purpose, and the long-term perspectives. In *Global change and mountain regions: An overview of current knowledge,* ed. U. M. Huber, H. K. M. Bugdmann, and M. A. Reasoner, 383–92. Dordrecht: Springer.

Radok, U. 1997. The International Commission on Snow and Ice (ICSI) and its precursors, 1894–1994. *Hydrological Sciences Journal* 42:131–40.

Raper, S. C. B., and R. J. Braithwaite. 2006. Low sea level rise projections from mountain glaciers and icecaps under global warming. *Nature* 439:311–13.

Rignot, E., and P. Kanagaratnam. 2006. Changes in the velocity structure of the Greenland ice sheet. *Science* 311:986–90.

Rivera, J., P., de Leon, and C. Koerber. 2006. Is greener whiter yet? The sustainable slopes program after five years. *Policy Studies Journal* 34:95–221.

Sallnow, Michael J. 1987. *Pilgrims of the Andes: Regional cults in Cusco.* Washington, DC, and London: Smithsonian Institution Press.

Schneeberger, C., H. Blatter, A. Abe-Ouchi, and M. Wild. 2003. Modelling changes in the mass balance of glaciers of the northern hemisphere for a transient $2 \times CO_2$ scenario. *Journal of Hydrology* 282:145–63.

Schytt, V. 1962. Mass-balance studies in Kebnekasje. *Journal of Glaciology* 33:281–89.

Strauss, S. 2003. Weather wise: Speaking folklore to science in Leukerbad. In *Weather, climate, culture,* ed. S. Strauss and B. Orlove, 39–59. Oxford: Berg.

Vergara, W., A. M. Deeb, A. Valencia, R. S. Bradley, B. Francou, A. Zarzar, A. Grünwaldt, and S. M. Haeussling. 2007. Economic impacts of rapid glacier retreat in the Andes. *Eos* 88:261, 264.

Vilímek , V., M. Luyo Zapata, J. Klime, Z. Patzel, and N. Santillán. 2005. Influence of glacial retreat on natural hazards of the Palcacocha Lake area, Peru. *Landslides* 2:107–15.

Wagnon, P. W., P. Ribstein, B. Francou, and B. Pouyaud. 1999. Annual cycle of energy balance of Zongo Glacier, Cordillera Real, Bolivia. *Journal of Geophysical Research* 104, D4:3907–23.

Wigley, T. M. L., and S. C. B. Raper. 2005. Extended scenarios for glacier melt due to anthropogenic forcing. *Geophysical Research Letters* 32, L05704, doi:10.1029/2004GL021238.

PART ONE

Societal Perceptions

Cultures and Institutions

2

Changing Views of Changing Glaciers

Wilfried Haeberli

Glacier changes have been observed for centuries. Throughout historical times, the perception of these often striking changes in high-mountain environments has shifted from legends about holy peaks and fear of punishment for worldly misbehavior via natural catastrophes to curiosity about movements from the "icy sea" of mountains and romantic enthusiasm for and realistic documentation of local phenomena to the discovery of past Ice Ages and the corresponding ideological disputes about the origin and evolution of the earth. Modern scientific investigation and worldwide coordination of glacier observations began toward the end of the nineteenth century, and the scientific observation of glacier fluctuations went through deep crises before evolving toward modern, integrated, multilevel concepts and advanced technologies. Today, glacier changes are increasingly recognized as a key phenomenon of global change in climate and living conditions on earth. Working Group 1 of the Third Assessment Report of the Intergovernmental Panel on Climate Change (IPCC) lists the shrinking of mountain glaciers, together with instrumentally measured sea-surface and land air temperatures, as the highest-confidence temperature indicators in the climate system (Figure 2.39a in Houghton et al. 2001).

This chapter concentrates on the increasing interest in the glacier during roughly the past three to four centuries in the European Alps, where rich material exists, where systematic monitoring programs were first initiated, and where discussions have sometimes been intense. The primary emphasis of this chapter is on the major directions of change rather than the details of a development that, in reality, has a great many facets. Unavoidably, therefore, it must remain a rather sketchy and somewhat subjective outline. A more detailed overview of early developments has been presented by Haeberli and Zumbühl (2003). Much more differentiated analyses would be not only possible but also worthwhile; the power of images and views profoundly influences our perception of environmental change in a critical period for human life on earth.

CURIOSITY AND ROMANTIC ENTHUSIASM FOLLOWING ROUSSEAU AND VON HALLER

The early period of glacier perception is characterized by an antagonism between fear and fascination. The widespread advance of Alpine glacier tongues to their greatest Little Ice Age extent around AD 1600 was in places perceived as a threat to mountain dwellers (Zumbühl 1980). In fact, legends about glacier advances over flowering meadows as a heavenly punishment for the worldly misbehavior of humans exist in various high-mountain valleys and reflect the early religious or mythological attribution of causes and effects. The first scientific descriptions, those of J. Scheuchzer, J. Altmann, and G. Gruner, appeared in the eighteenth century (Haeberli and Zumbühl 2003). In the context of the contemporaneous call of J.-J. Rousseau ("back to nature") for a new approach to life and the poetry of A. von Haller (which created the historical "myth of the Alps"), the growing scientific interest was accompanied by romantic views about the "pure," "eternal," and "untouched" firn fields and the water- and life-giving glaciers. The unique appeal of high-mountain scenery in the European Alps was the extreme contrast between glaciers, as an expression of "wild, indestructible" nature, and the carefully cultivated Alpine landscape, with its assumed "hard, simple, good, and healthy" living.

The painting by Pierre-Louis de la Rive depicting Mont Blanc (Figure 2.1) illustrates this notion of sustainable living in an intact high-mountain environment (cf. the extensive discussion of historical developments by Bätzing 2003). The view is upward through a green, garden-like landscape with open forest to the intact and "clean" white of firn and ice on a beautiful mountain peak. With the glacier-clad peak of the Eiger in the background and a cow in the foreground (Figure 2.2), Maximilien de Meuron portrays a combination of symbols that represents this paradise-like image and indeed became part of the reputation of Switzerland as "a beautiful mountain/glacier country in the heart of Europe" for centuries. The detail of the tongue of the Unterer Grindelwald Glacier (Figure 2.3) from a painting by Caspar Wolf perfectly reflects the romantic enthusiasm and curiosity about the spectacular phenomenon of moving ice that led to much deeper-reaching questions about glaciers and life on earth.

FIGURE 2.1. Mont Blanc, French Alps, from Sallenches in 1802. Painting by Pierre-Louis de la Rive (from Rasmo et al. 1981).

THE ONSET OF SCIENCE: FROM A NATURAL CATASTROPHE TO THE ICE AGE DEBATE

The initial impulse for thinking in a completely new dimension about humankind and nature originated at least partially outside scientific circles (Knight 2004). On the occasion of the 1818 Mauvoisin catastrophe, caused by the outburst of a lake dammed by ice avalanche debris from the advancing Giétro Glacier (Figure 2.4; Kasser and Haeberli 1979; Röthlisberger 1981), the farmer and mountain-goat hunter Jean-Pierre Perraudin met with the leading scientists of the time (Agassiz, de Charpentier, Escher von der Linth) and told them that the glaciers must have been much larger in the past to have been able to deposit large rocks far down the valley.

Such ideas may have been quite widespread among the "primitive" Alpine population living close to and experiencing glaciers, but they were new and astonishing in the academic circles

FIGURE 2.2. The Eiger near Grindelwald, Swiss Alps, painted from the alpine meadows of Wengernalp by Maximilien de Meuron, probably in 1821 (from Rasmo et al. 1981).

FIGURE 2.3. The advancing front of the Unterer Grindelwald Glacier, 1774–77. Detail of a painting by Caspar Wolf, the most important eighteenth-century painter of the Swiss Alps. (Photo H. J. Zumbühl)

FIGURE 2.4. Broken and collapsed debris cone of ice-avalanche deposits from the advancing Giétro Glacier after the sudden lake outburst and devastating flood wave at Mauvoisin/Bagnes, Swiss Alps, painted probably in 1818 by Hans Conrad Escher von der Linth (from Rasmo et al. 1981; cf. details in Kasser and Haeberli 1979).

FIGURE 2.5. The Agassiz team in the hut of Hugi, which later became the "Hôtel des Neuchâtelois" (the large boulder serving as shelter) on the medial moraine of Unteraar Glacier. Lithograph after a drawing by Joseph Bettanier published by Agassiz (1840–41) (from Rasmo et al. 1981).

developed in cultural centers in the lowlands. Louis Agassiz and his colleagues, detecting traces of glacier erosion and erratic boulders far from the Alps or other mountain chains, formulated the Ice Age theory, and this immediately excited a heated debate. Because the Bible contains no description of any smaller or larger glaciers, the idea of "Ice Ages" was widely considered offensive. The debate went on for decades and increasingly encouraged accurate field studies. The first paintings and illustrations showing scientists setting foot on and living or working on glaciers date from this time (Figure 2.5). The thorough investigations of the Unteraar Glacier by an interdisciplinary group of scientists led by Agassiz in the 1840s can be considered to constitute the beginning of modern experimental glacier research (Agassiz, Guyot, and Desor 1847). The results of such research on the geometry, material characteristics, and flow of a valley glacier and the first description of a large ice sheet in Greenland by E. Kane in the 1850s (cf. Bolles 1999) established the recognition of dramatic changes in climatic and environmental conditions for life in the most recent earth history. This revolution in our scientific understanding unavoidably led to questions of the physical causes and possible future repetitions of such events.

THE BEGINNING OF SYSTEMATIC OBSERVATIONS: CLIMATIC PERIODICITIES AND CATASTROPHES

In 1893, François Alphonse Forel established the first systematic glacier observation network in Switzerland. Using this network as a model, the Sixth International Geological Congress in Zurich in 1894 initiated coordinated worldwide observation of glacier changes with the International Glacier Commission under Forel as the leading board of the newborn network. The goal was to learn more about the factors—whether internal or external to the earth system—that govern climate changes and cause Ice Ages to begin and end (Forel 1895).

The monitoring strategy consisted of regular and exact surveys at selected glacier tongues but also included indigenous knowledge about earlier glacier stages collected by scientists through communication with the mountain people. It was clearly oriented toward a better understanding of large-scale and long-term processes and therefore thought to require patience in order to bear fruit for future generations. It was this generous and "transdisciplinary" concept that helped develop one of the longest existing series of environmental observations, a real treasure of climate-related geoscience. During the twentieth century, the evolution of the international glacier-monitoring program and the corresponding views of glacier changes is marked by four distinct phases (see Haeberli, Hoelzle, and Suter 1998). The first phase of international glacier observation, around the turn of the century, was characterized by the search for regular oscillations in the climate/glacier-system, as is illustrated by the titles of the corresponding reports ("Les variations *périodiques* des glaciers") issued by various members of the multilingual commission. The short glacier readvances in the Alps around 1890 and 1920 seemed to confirm the impression that climate and glaciers fluctuated in a periodic or at least quasi-periodic way. At the same time, a number of important glacier catastrophes in the Alps—a water outburst and devastating debris flow from the Tête Rousse

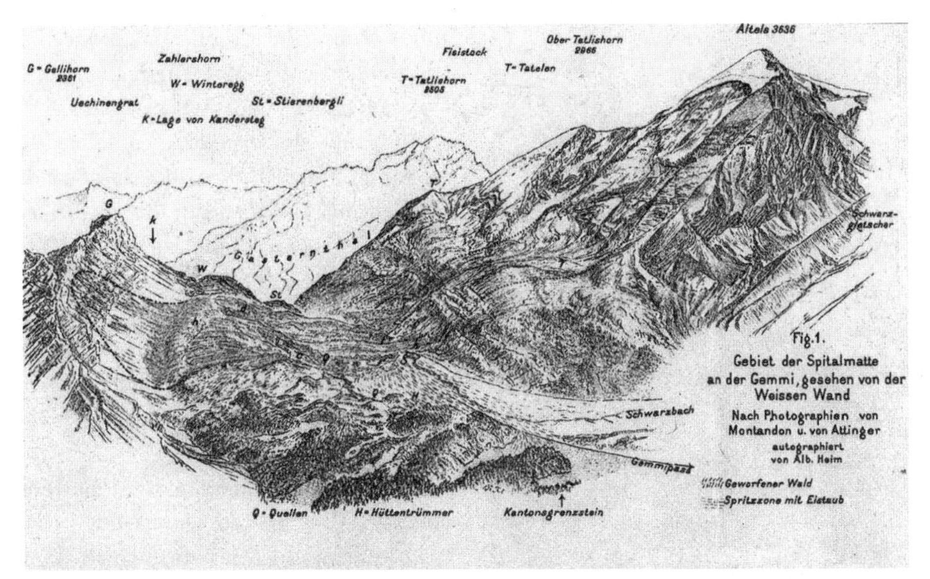

FIGURE 2.6. Drawing by the geologist Albert Heim of the traces of the large ice avalanche in 1895 at the Altels, Kandersteg, Swiss Alps (from Heim 1896).

Glacier on Mont Blanc in 1892, the large ice avalanche from the Altels in the Bernese Alps in 1895 (Figure 2.6), the large rock/ice avalanche from the Fletschhorn at Simplon Pass in 1901, the rapid advances of Vernagt Ferner in the Austrian Alps, and the massive disappearance of ice in Glacier Bay, Alaska—captured the interest of specialists as extraordinary events to be documented and analyzed. This first "golden" phase of worldwide glacier observation was unfortunately soon to be confronted with the dark shadows of global politics, economics, and even science.

HIGH TECHNOLOGY, WARS, AND A SCIENTIFIC CRISIS

One important product of the "golden" phase was the compilation of high-precision topographic maps, some specifically prepared for glaciers (Rhône, Vernagt, Guslar, Schnee, Hintereis; cf. Mercanton 1916) in the Swiss and Austrian Alps. These first maps are comparable in accuracy to modern topographic maps and can therefore be used to determine long-term volume and mass changes of the Alpine glaciers (Steiner et al., this volume). Thus high technology was used from the very beginning, and even today these maps constitute a unique basis for scientific comparison. The mean century-scale mass balance estimates (–0.2 to –0.6 m water equivalent per year) derived from them and confirmed today by modern model calculations (Haeberli and Hoelzle 1995; Haeberli and Holzhauser 2003; Hoelzle et al. 2003) represent a standard against which glacier changes in other mountain ranges can be compared.

The second phase of international glacier observation spans the two world wars and the period of economic crisis between them, when glacier observations were reduced to a minimum. The number of glaciers observed in the Alps, for example, decreased by about 50% during each of the wars (see Zemp et al., this volume), with the Italian and Austrian glaciers being primarily affected. As a consequence, the intense global warming and glacier shrinkage of the 1930s and 1940s passed virtually unnoticed in the scientific literature. It was through the meritorious efforts of P. Mercanton in Switzerland that multiyear reports, though somewhat thin, continued, keeping the core of the worldwide network alive, albeit at a low level of intensity and scientific analysis. In addition to presenting numerical data on length changes of glaciers in the Alps and Scandinavia, these reports made

FIGURE 2.7. Synthetic oblique view of the Morteratsch Glacier, Bernina, Swiss Alps. The satellite image is a fusion of Landsat TM (1999, resolution 25 m) with the panchromatic channel of IRS 1C (1997, resolution 10 m). The retreat of the glacier tongue is 2 km from 1850 to 1973 and an additional 300 m to 1997. Since then, the glacier tongue has further retreated by more than 100 m. (Data and image processing by F. Paul; DHM25: © 2005. Reproduced by permission of swisstopo [BA057490]; satellite imagery © Eurimage/NPOC. 9.)

reference to various interesting national reports. Matthes (1934), for instance, reported a series of length variation measurements of the Nisqually Glacier since 1857. Signs of shrinking and glacier retreat clearly predominated, with the exception of a short but marked advance of glaciers in the Alps around 1920.

After this period of neglect, the third phase saw the reorganization of the international network under the umbrella of UNESCO by P. Kasser in Switzerland. In 1967, the Permanent Service on the Fluctuations of Glaciers (PSFG) was established as one of the services of the Federation of Astronomical and Geophysical Services (FAGS) of the International Council of Scientific Unions (ICSU). This resulted in the publication at five-year intervals of reports on the fluctuations of glaciers. Mass balance data from various countries, including the Soviet Union, the United States, and Canada, were included in these reports for the first time, forming the essential link between climate fluctuations and glacier length changes. Length variation data from the United States, the Soviet Union, and other countries completed the corresponding records from the Alps, Scandinavia, and Iceland. Glacier readvances were reported from various parts of the world, especially from the Alps, where mass balances were predominantly positive in the late 1960s and 1970s. For the first time, therefore, empirical information about glacier responses to well-documented and strong signals in mass balance history started to become available.

This promising development, however, was soon faced with another crisis. Problematic theories about glacier mechanics (kinematic wave theory with unrealistic century-long reaction times [Nye 1960]) and the reporting of observed glacier changes as percentages of annually advancing/retreating glaciers (suppressing the essential cumulative effects so clearly visible in nature) reduced the credibility of these data in scientific, governmental, and public circles. This led to a brief interruption of the worldwide monitoring program when, after the sudden death of Fritz Müller (who had taken over the leadership of the program), the International Commission on Snow and Ice (ICSI), the responsible ICSU body for the network, did not immediately see the need to continue gathering worldwide glacier observations. This situation changed, however, with the growing interest in glacier fluctuations as a measure of ongoing climate change.

The fourth phase of international glacier monitoring took place during the past two decades, when improved theories and numerical models of changes in climate, energy and mass balance, ice thickness and flow, and glacier fluctuations finally became available (e.g., Jóhannesson, Raymond, and Waddington 1989; Haeberli and Hoelzle 1995; Oerlemans 2001; Hoelzle et al. 2003). Excellent international collaboration and advanced observational technologies (remote sensing, geoinformatics [see Bishop et al. 2004]) were now involved, but, more significantly, growing awareness of the continued and accelerating loss of mountain glaciers in most parts of the world had finally led to the recognition that glacier changes were key indicators of global climate change and were therefore important to international assessments (the IPCC) and global observing systems (the Global Climate Observing System [GCOS]/ Global Terrestrial Observing System [GTOS]) of the changing earth system.

Given these developments, attempts could now be made to build up a modern observational service. In the 1970s a World Glacier Inventory (WGI) had been planned under the guidance of Müller to be a snapshot of ice conditions on earth during the second half of the twentieth century. Within the framework of the Global Environment Monitoring System (GEMS) of the United Nations Environment Program (UNEP), a temporary technical secretariat began operations in 1976 as another service of ICSI. Detailed and preliminary regional inventories were compiled all over the world to update earlier compilations (see especially Field 1975; Mercer 1967) and to form a modern statistical basis for the geography of glaciers. The year 1986 finally saw the start of the World Glacier Monitoring Service (WGMS), combining and integrating the two ICSI services (PSFG and TTS/WGI). The *Glacier Mass Balance Bulletin* was issued at two-year intervals to speed up and facilitate access to information concerning mass balances of selected reference glaciers. International efforts were also made to collect and publish short abstracts on special events such as glacier surges, ice avalanches, glacier floods or debris flows, drastic retreats of tidewater glaciers, rock slides onto glaciers, and glacier-volcano interactions.

Modern glacier views (Figure 2.7) combine satellite imagery with digital terrain information; they show glaciers at high resolution from (virtual) elevated viewpoints and clearly document the striking ice losses and the bare ground left by the retreating and decaying ice. Recent and ongoing glacier changes are being documented through integrated multilevel strategies combining (1) information on mass balance, length change, and glacier inventories and (2) in-situ measurements with remote sensing for large/ representative samples and GIS-based numerical modeling of distributed mass balance and flow for interpolation and extrapolation in space

FIGURE 2.8. The retreating Stein Glacier, Sustenpass, Swiss Alps. The top two photographs are from 1947 and 2003; the bottom is a visualization of an assumed landscape in Roman times based on the hypothesis—in the strongest contradiction to existing scientific knowledge about Holocene climate, glacier, and forest fluctuations (see, e.g., Haeberli and Holzhauser 2003; Haeberli et al. 1999)—that glaciers had been almost completely replaced at that time by meadows and forests. The inherent message embedded in this third image (and understood by large parts of the public in central Europe) is that extremely warm temperatures (comparable to extreme IPCC scenarios for the twenty-first century) existed during historical times, remained unnoticed by contemporaneous people as well as by modern science, and did not cause any serious problems (from *Weltwoche*, August 12, 2004, based on Schlüchter and Jörin 2004).

and time (Haeberli 2004). Today wide public attention can be drawn to the accelerated disappearance of glaciers as dramatically documented in repeated glacier inventories (Paul et al. 2004). The spectacular find of the roughly 5,000-year-old and perfectly preserved body of the Oetztal ice man emerging from a small, probably cold (and now disappearing) miniature ice cap of the Austrian/Italian Alps confirmed that the "warm" or "high-energy" limit of Holocene, preindustrial glacier and climate variability may have been reached if not surpassed. The possibility can no longer be excluded that anthropogenic influences on the atmosphere could now, and for the first time, represent a major contributing factor to the observed glacier shrinkage (Haeberli et al. 1999; see also Houghton et al. 2001 for detailed discussion of anthropogenic influences on climate change). If these predictions are correct, many mountain ranges could lose their glacier covers within decades.

FIGURE 2.9. Two lakes (the upper one almost completely in shadow) filling collapse holes in the tongue of Palü Glacier, Grisons, Swiss Alps, summer of 2005.

PROSPECTS

Projections into the future can be based on simple extrapolation (with or without acceleration) of observed trends, on numerical model calculations of realistic scenarios, or on pure imagination. Extrapolation of developments documented by repeated glacier inventories for the Alps (Kääb et al. 2002; Paul et al. 2004) and provided by numerical models that combine glacier mass balance and flow (Oerlemans et al. 1998) both confirm that the disappearance of many mountain glaciers is quite likely to be a matter of a few decades. Pure imagination, of course, opens the possibility of optimism concerning such prospects (Figure 2.8). Such visualizations constitute a surprising revival of the romantic views originally developed during the eighteenth century: the Alps are again seen as a paradise-like white/green landscape with an open (probably cultivated) forest/meadow pattern surrounding some pretty remains of surface snow without any visible debris or moraines. Dreams of this type are obviously based on the hope that soils and vegetation will be able to immediately follow the disappearing ice, that high-mountain ecosystems will remain in equilibrium even with rapid climate forcing, and that changes in atmospheric temperatures of several degrees (causing timberline changes of many hundreds of meters) will occur without adverse effects on society. In reality, the developments of the coming decades may include the complete disappearance of small mountain glaciers and down-wasting rather than retreat for long valley glaciers. This already discernible trend may be accompanied by the development of extreme and long-lasting disequilibria in the abiotic as well as the biotic aspects of ecosystems and habitats, not only in high-mountain areas but elsewhere (Watson and Haeberli 2004).

The extreme summer of 2003 in the Alps removed an estimated 5–10% of the remaining glacier volume (Zemp et al. 2005). As a consequence, many glacier tongues have started to collapse rather than to retreat (Figure 2.9). The resulting dark and dusty mountain chains, free of snow and ice, provide an impression of the (summer) landscape that is likely to develop as a consequence of such a scenario of climate

FIGURE 2.10. View from Piz Corvatsch to the Albula Mountains, Grisons, Swiss Alps, in the extremely hot, dry summer of 2003.

and glacier change (Figure 2.10). It is definitely no longer a question of "variations *périodiques* des glaciers." Formerly glacierized mountains now becoming "black," "dry," and geomorphologically unstable may increasingly be perceived as primary testimony to general and very serious human impacts on the global environment (Watson and Haeberli 2004). In this sense, glacier observation is becoming an activity of ever-increasing socioeconomic and political importance. Accurate views of changing glaciers may therefore help to change our attitude toward accelerating changes in the global human environment, a challenge of historic dimensions.

REFERENCES CITED

Agassiz, L. 1840–41. *Untersuchungen über die Gletscher*. Solothurn and Neuchâtel: Deutsche Bearbeitung C. Vogt.

Agassiz, L., A. Guyot, and E. Desor. 1847. *Système glaciaire ou recherches sur les glaciers, leur mécanisme, leur ancienne extension et le rôle qu'ils ont joué dans l'histoire da la terre*. Pt. 1. *Nouvelles études et expériences sur les glaciers actuels, leur structure, leur progression et leur action physique sur le sol*. Paris/Leipzig: Victor Masson.

Bätzing, W. 2003. *Die Alpen: Geschichte und Zukunft einer europäischen Kulturlandschaft*. München: C. H. Beck.

Bishop, M. P., J. A. Olsenholler, J. F. Shroder, R. G. Barry, B. H. Raup, A. B. G. Bush, L. Copland, J. L. Dwyer, A. G. Fountain, W. Haeberli, A. Kääb, F. Paul, D. K. Hall, J. S. Kargel, B. F. Molnia, D. C. Trabant, and R. Wessels. 2004. Global land ice measurements from space (GLIMS): Remote sensing and GIS investigations of the Earth's cryosphere. *Geocarto International* 19(2):57–84.

Bolles, E. B. 1999. *The ice finders: How a poet, a professor, and a politician discovered the Ice Age*. Washington, DC: Counterpoint.

Field, W. O. 1975. *Mountain glaciers of the Northern Hemisphere*. 2 vols. Hanover: CRREL.

Forel, F-A. 1895. Les variations périodiques des glaciers: Discours préliminaire. *Archives des Sciences Physiques et Naturelles* 34:209–29.

Haeberli, W. 2004. Glaciers and ice caps: Historical background and strategies of world-wide monitoring. In *Mass balance of the cryosphere*, ed. J. L. Bamber and A. J. Payne, 559–78. Cambridge, UK: Cambridge University Press.

Haeberli, W., R. Frauenfelder, M. Hoelzle, and M. Maisch. 1999. On rates and acceleration trends of global glacier mass changes. *Geografiska Annaler* 81A:585–91.

Haeberli, W., and M. Hoelzle. 1995. Application of inventory data for estimating characteristics of and regional climate-change effects on mountain glaciers: A pilot study with the European Alps. *Annals of Glaciology* 21:206–12.

Haeberli, W., M. Hoelzle, and S. Suter, eds. 1998. *Into the second century of worldwide glacier monitoring: Prospects and strategies*. UNESCO Studies and Reports in Hydrology 56.

Haeberli, W., and H. Holzhauser. 2003. Alpine glacier mass changes during the past two millennia. *PAGES News* 1(11):13–15.

Haeberli, W., and H. J. Zumbühl. 2003. Schwankungen der Alpengletscher im Wandel von Klima und Perzeption. In *Welt der Alpen: Gebirge der Welt*, ed. F. Jeanneret et al., 77–92. Bern: Haupt.

Heim, A. 1896. Die Gletscherlawine an der Altels am 11. September 1895. *Neujahrsblatt der Naturforschenden Gesellschaft in Zürich*, 98.

Hoelzle, M., W. Haeberli, M. Dischl, and W. Peschke. 2003. Secular glacier mass balances derived from cumulative glacier length changes. *Global and Planetary Change* 36:295–306.

Houghton, J. T., et al., eds. 2001. *Climate change 2001: The scientific basis. Contribution of Working Group 1 to the third assessment report of the Intergovernmental Panel on Climate Change*. Cambridge, UK: Cambridge University Press.

Jóhannesson, T., C. F. Raymond, and E. D. Waddington. 1989. Time-scale for adjustment of glaciers to changes in mass balance. *Journal of Glaciology* 35:355–69.

Kääb, A., F. Paul, M. Maisch, M. Hoelzle, and W. Haeberli. 2002. The new remote-sensing-derived Swiss glacier inventory. 2. First results. *Annals of Glaciology* 34:362–66.

Kasser, P., and W. Haeberli, eds. 1979. *Die Schweiz und ihre Gletscher.* Zürich: Schweizerische Verkehrszentrale/Bern: K+F Geographischer Verlag.

Knight, P. 2004. Glaciers: Art history, science, and uncertainty. *Interdisciplinary Science Reviews* 29:85–393.

Matthes, M. F. E. 1934. Committee on glaciers of the American Geophysical Union, report for 1931/32. *IAHS Bulletin* 20:251–64.

Mercanton, P-L. 1916. *Vermessungen am Rhonegletscher.* Neue Denkschriften der Schweizerischen Naturforschenden Gesellschaften 52.

Mercer, J. H. 1967. *Southern Hemisphere glacier atlas.* U.S. Army Natick Laboratories Technical Report 67–76-ES.

Nye, J. F. 1960. The response of glaciers and ice-sheets to seasonal and climatic changes. *Proceedings of the Royal Society of London Series A* 256:559–84.

Oerlemans, J. 2001. *Glaciers and climatic change.* Rotterdam: Balkema.

Oerlemans, J., B. Anderson, A. Hubbard, P. Huybrechts, T. Johannesson, W. H. Knap, M. Schmeits, A. P. Stroeven, R. S. W. van de Wa, J. Wallinga, and Z. Zuo. 1998. Modelling the response of glaciers to climate warming. *Climate Dynamics* 14:267–74.

Paul, F., A. Kääb, M. Maisch, T. W. Kellenberger, and W. Haeberli. 2004. Rapid disintegration of Alpine glaciers observed with satellite data. *Geophysical Research Letters* 31:L21402.

Rasmo, N., M. Röthlisberger, E. Ruhmer, B. Weber, and A. Wied. 1981. *Die Alpen in der Malerei.* Rosenheim: Rosenheimer.

Röthlisberger, H. 1981. Eislawinen und Ausbrüche von Gletscherseen. *Jahrbuch der Schweizerischen Naturforschenden Gesellschaft* 1978:170–212.

Schlüchter, C. and U. Jörin. 2004. Holz- und Torffunde als Klimaindikatoren: Alpen ohne Gletscher? *Die Alpen* 6:34–46.

Watson, R. T., and W. Haeberli. 2004. Environmental threats, mitigation strategies, and high-mountain areas. In *Royal Colloquium: Mountain areas, a global resource,* 2–10. Ambio Special Report 13.

Zemp, M., R. Frauenfelder, W. Haeberli, and M. Hoelzle. 2005. Worldwide glacier mass balance measurements: General trends and first results of the extraordinary year 2003 in Central Europe. *Materialy Glyatsiologicheskii Issledovanii* 99:3–12.

Zumbühl, H. J. 1980. *Die Schwankungen der Grindelwaldgletscher in den historischen Bild- und Schriftquellen des 12. bis 19. Jahrhunderts: Ein Beitrag zur Gletschergeschichte und Erforschung des Alpenraumes.* Denkschriften der Schweizerischen Naturforschenden Gesellschaft 92.

3

Challenges of Living with Glaciers in the Swiss Alps, Past and Present

Ellen Wiegandt and Ralph Lugon

Glaciers dominate mountain landscapes and the lives of the people who live among them. Sources of precious resources and causes of natural catastrophes, they inspire both fear and reverence. In periods of climate cooling as well as those of climate warming, glacier-related catastrophes have destroyed lives and property. Yet people in many parts of the world have lived continuously near glaciers, accepting risks that might seem intolerable in other environments in order to benefit from exploiting glacier waters. Their perceptions of risk are shaped by this effort to balance the negative and positive aspects of glaciers. For mountain populations living among glaciers in the past, these trade-offs seemed rational. They suffered the periodic flood or ice avalanche but were able to harness glacier waters to grow crops that would otherwise not survive in the harsh mountain environment. Today, glacier retreat and the related permafrost melting, due in part to global climate warming, may intensify these hazards.[1] But glacier waters remain central to mountain economies even as they evolve. Agriculture has been joined by hydroelectricity generation and tourism as new sources of livelihood. Multiple risks also confront these domains. On the one hand, the frequency, intensity, or localization of natural hazards is likely to change as a result of global warming. On the other, new and different risks are being introduced by broad socio-economic changes. Trade in agricultural goods and energy draws mountain populations into wider economic networks; the influx of foreign tourists makes them dependent on economies beyond their own. Political horizons have also expanded because the details of these relationships are often negotiated at level higher than that of the community, as are many rules and regulations governing land use and hazard management.

Our puzzle is to understand how mountain populations achieved a balance between the negative and potentially positive effects of glaciers in the past and how they are likely to adjust to the similarly mixed results of climate and social change in the future. We first describe the institutional arrangements for distributing resources. Of particular importance in this regard are the property rules that have governed water resources. Historically rights have been rooted in local communities, but decision making has become progressively more centralized at higher levels. How people view profound modifications to glaciers and how they respond to new hazards and new opportunities are filtered through these cultural constructs. They are also related to perceptions of risk, and therefore understanding how people accommodate high levels of danger requires an analysis of factors influencing perception of risk and the way these shape behavior. Whether traditional strategies for monitoring glacier dynamics and harnessing water from glacier runoff will be appropriate in a transformed world is a fundamental question we raise here.

Accepting uncertainty and risking danger to seize opportunities reflect the dual nature of glaciers, qualities that are starkly apparent in the canton of the Valais, Switzerland (Figure 3.1). The region is emblematic because individual and collective strategies to confront risks and exploit opportunities have a long history here. So too does the science of glaciology, which has permitted people to incorporate new knowledge about glacial phenomena, thereby altering their perceptions of risk and improving their abilities to respond effectively. In our case study of the Valais we focus primarily on institutional strategies to manage risk and the role of scientific knowledge in shaping risk perception and behavior. We begin in the past, focusing on some institutional adaptations that allowed local communities to both capture critical water resources from glaciers and respond appropriately to the catastrophes that inevitably befell them. In our subsequent discussion of the present, we describe how changing understandings of local glacier and permafrost dynamics provide the potential to adapt to new hazards due to climate warming but also introduce new uncertainties about the intensity, frequency, and location of glacier hazards. We surmise that it is likely that the Valaisan population will try to meet these challenges because there are still important economic and social benefits to living among glaciers. Energy production and tourism have largely replaced agriculture as the major uses of glaciers and their water, but these activities demand institutional modifications that will be as difficult to achieve as the adaptations to evolving physical dangers. The historical perspective provides a context for speculating on whether the particular social and institutional context in the Valais is adequate to face the uncertain future.

RISKS AND RESPONSES IN PREINDUSTRIAL VALAIS SOCIETY

Mountain environments are fraught with dangers and unknowns and have always been so. Indeed, the earliest inhabitants of the Valais settled in the uplands in order to escape the unruly Rhône River, which regularly flooded the lowlands. They merely exchanged one form of hazard for another, however, because at higher altitudes they confronted challenges imposed by steep slopes and rugged climate as well as by random catastrophic events. Glaciers were the cause of some of these disasters, figuring prominently in a rich body of legend and stories (Schüle 1987). Glacier-related events appear here as punishments for collective sins (Luyet 1924, 1929). From the available evidence it is difficult to judge how deeply these religious analogies affected daily lives. On the one hand, glaciers were mostly in areas where few people ventured. However, their presence was acutely felt, especially when catastrophes produced occasional but severe and widespread devastation. Moreover, at times, such as during the Little Ice Age (Payot 1950), glaciers became a closer presence, expanding to encroach on inhabited and cultivated lands (Le Roy Ladurie 1983).

For all their negative impacts, however, glaciers in the Valais were the source of important benefits, such as the seasonal glacial runoff that provided water for domestic and agricultural use. The bottom of the Rhône Valley in the central and upper Valais is a rain-shadow area and one of the driest in Switzerland. In Sion, for example, the average annual rainfall for the period 1966–94 was 620 mm (Rebetez, Lugon,

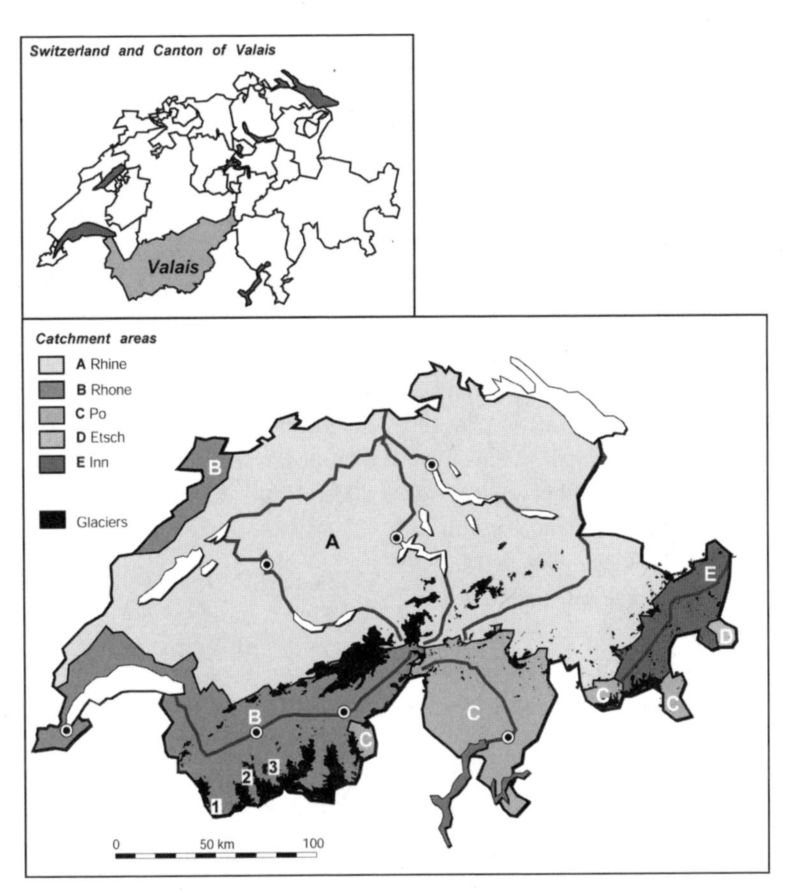

FIGURE 3.1. Catchment areas and glaciers of Switzerland. 1. Dolent Glacier, Ferret Valley. 2. Giétro Glacier, Bagnes Valley. 3. Tsarmine Glacier and Upper Aiguilles Rouges Glacier, Arolla Valley. (Adapted from Maisch et al. 1999).

and Baeriswyl 1997), insufficient to sustain the characteristic agro-pastoral production system. To remedy this annual and seasonal deficit, Valaisans relied on extensive irrigation networks. The story of the origin and maintenance of these canal systems unfolds as a negotiation between individual preferences and collective interests in which the stakes were efficient resource management in the face of considerable uncertainty with regard to the physical properties of water in mountain regions.

Irrigation systems (*bisses*) became widespread in the Valais in the fifteenth century, when there was a shift from grain cultivation to a mixed production system based on herding in addition to cereal grains (Dubuis 1999). The new production system demanded greater amounts of water to sustain meadows and hay production and required that sufficient quantities be available during critical periods of the growing season as well as reliable from year to year. The task of harnessing and managing water faces several kinds of problems that are difficult to overcome. Flowing water creates asymmetries between upstream and downstream users because those upstream can divert flow and deprive downstream users of what they need. The solution adopted in the Valais was to declare the water a common property resource and to resolve allocation issues on the basis of private property principles, a solution consistent with the general organization of the Alpine agro-pastoral system based on autonomous households producing for their own needs. Each household required access to a complete range of resources, including irrigation water. The standard way of distributing goods and resources is through property rights. In the Valais we find both private and common property rules. Water use drew on both sets of principles. The collective resource was distributed to households according to the length of time it flowed through each small tributary canal. Landowners had individual, private rights to this water time, rights that they could buy, sell, and pass on to their heirs (Netting 1981; Wiegandt 2007).

The system of water use favored collective needs over individual preferences. A mix of private and common property regimes spread risks and ensured widespread access to resources. Private rights to some parts of the system created incentives for efficient use and management.[2] Collective management over other parts distributed costs as well as benefits throughout the community. This was important to minimize risks from limited and incomplete knowledge about climate trends and natural disasters. To compensate for negative impacts of uncertainty, such as unusually long winters, low precipitation, or catastrophes, various mechanisms created and enforced solidarity and led to burden sharing. Private plots, for example, were small and distributed throughout the communal territory, providing insurance against losses from microclimatic conditions and localized catastrophes. A buffer was created against temporary shortfalls by the maintenance of private and collective stocks of food and reliance on stored rather than fresh food.

Recent work on risk suggests that acquiring knowledge through experience leads to a particular form of risk perception that underweighs risks with low probabilities, making people relatively risk-preferring when they are familiar with particular kinds of dangers (Hertwig et al. 2004).[3] If we adopt this perspective, we can better understand why and how Alpine populations, with their particular resource management system, persisted in regions exposed to frequent natural hazards and a variable and forbidding climate. We can indeed document that for several centuries population size even increased with no apparent degradation of resources; measures of household wealth remained constant or increased without any obvious negative impact from climate fluctuations or extreme events (Wiegandt 1977, 1980). This leads us to hypothesize that the set of interacting buffering mechanisms successfully distributed risks and smoothed potentially disruptive variations in production between the fifteenth and the mid-twentieth century.

THE GIÉTRO DISASTER OF 1818: AN INTERSECTION OF NATURAL PROCESSES AND SOCIAL NORMS

The patterns and trends of ordinary lives were nevertheless occasionally disrupted by dramatic and deadly natural events. These tell us how the system was tested and how people responded. The story of Giétro is exemplary. It was a landmark moment when science took a leap forward, leading to a shift in social attitudes toward risk and changes in response behavior. In 1818, glacier science was just beginning. With it came the realization that understanding a process could allow more effective intervention. Scientific knowledge fostered the development of preventive measures and set in motion the establishment of regional and national disaster response policies in place of the intensely local response patterns of earlier periods.

On June 16, 1818, a huge ice dam broke at the Giétro Glacier in the Val de Bagnes (see Figure 3.1). The resulting flood into the valley below leveled buildings and killed dozens. Giétro had a history of devastation. A probable cataclysm had occurred in 589 and another in 1549, and in 1595 a rupture of an ice dam holding back the Giétro Glacier lake had released an estimated 20 million m^3 of water that destroyed 500 buildings and killed 140 people (Musée de Bagnes 1988). All these disasters were linked to advances and retreats of the glacier as climate cooled and warmed as well as to Giétro's particular configuration. The glacier face extended over the edge of a high cliff, and occasionally gigantic pieces of ice broke off, falling several hundred meters into the river valley below and creating a secondary glacier.

In the years just before 1818, in the midst of glacier advance due to the Little Ice Age (approximately 1650–1850, though some scholars date its beginnings to as far back as 1350), local inhabitants and glacier scholars alike became increasingly alarmed. Several especially cold winters had created a huge secondary ice dam that prevented spring meltwater from flowing down the valley, trapping it instead in a massive lake. This development was monitored and measured by the Valais cantonal engineer, Ignace Venetz, using knowledge and technology gained from the new scientific attention to glaciers. As he evaluated the volumes of ice and water, he was increasingly convinced that a momentous event was pending. In the spirit of the times, he also believed that he had the means to intervene. He directed a channel to be built through the glacier to evacuate the accumulated water toward the river. The channel was successfully completed, but the water flowing through it melted away the remaining ice barrier. When the ice dam broke, it sent 18 million m^3 of water tearing through the valley, picking up huge boulders, ripping out bridges, flooding fields, destroying harvests, sweeping away houses, barns, and factories, and killing scores of animals and 40 people.

Response was immediate. Other villages in the Valais sent food or took in the homeless. Funds were sent from elsewhere in Switzerland and from abroad. These expressions of solidarity marked a change in response to disasters that is observed elsewhere in Switzerland during this period (Pfister 2002). It is consistent with the fundamental shift from fate-based to probability-based perceptions of nature that is described by Bernstein in his comprehensive study of the evolution of risk perception (1996). Bernstein shows that, beginning with the Renaissance, the search for explanations of phenomena progressively replaced attribution of catastrophe to divine punishment. With this change, feelings of impotence gave way to a growing sense that people could exert at least some control over the forces of nature. Unpredictable and therefore unavoidable though they were, disasters were now perceived not as evidence of the wrath of God but as outcomes of natural processes. Humans could therefore systematically study and confront them, just as Ignace Venetz did. This meant that everyone, and not only sinners, would experience consequences from natural disasters. Part of the charitable response reflected the realization that the next time disaster struck, one might oneself be a victim, dependent on the help of fellow citizens, and that therefore a kind of preemptive reciprocity would ensure future protection.

Reinforcement of state institutions paralleled progress in science and the growing feelings of social responsibility. Cantonal authorities in the Valais coordinated much of the response to the Giétro hazard, first engaging Venetz to evaluate the situation and then authorizing him to implement preventive strategies. Once the catastrophe occurred, they formed a committee to oversee aid and reconstruction. The committee's mandate was to establish rules defining eligibility for compensation and to receive and distribute relief funds (Comité de Bienfaisance 1820). The procedures and institutions became part of an official response strategy.

The Giétro catastrophe tells us several important things about society and glaciers as well as about risk and solidarity. It underscores that scientific knowledge is crucial in reducing uncertainty. With knowledge comes the possibility of devising appropriate actions. However, reducing uncertainty does not mean eliminating it. Appropriate levels of risk acceptance, along with behavior and policies for sharing risk, remain essential for social resiliency. Local institutions seem to have functioned adequately, given a certain level of environmental variability and change in the Valais, but were overwhelmed by huge fluctuations. In those moments, the response of networks at the cantonal and national levels proved indispensable. The response to the 1818 Giétro tragedy marked an early step in the development of disaster relief and prevention based on both increasing scientific knowledge and the growing role of Swiss government institutions.

Since 1818, natural dynamics at Giétro have continued to pose challenges to the people living in the glacier's shadow and to the authorities responsible for their protection. As the Little Ice Age drew to a close in the mid-nineteenth century, the climate warmed and the Giétro Glacier began to retreat. Rather than eliminating dangers, however, this created new types of hazards. Toward the end of the century, large lakes formed at the confluence of two retreating glaciers. Outburst floods in 1894 and 1898 destroyed all the bridges in the upper valleys down to the valley floor at Martigny. They were the most serious floods the Val de Bagnes had experienced since the 1818 disaster. The site of the lake continued to be monitored for possible flood events until the 1930s (Huggel, Kääb, and Haeberli 2002; Raymond, Wegmann, and Funk 2003). During a period of cooling in the 1960s and 1970s the Giétro Glacier advanced once again, leading to fears of ice avalanches. Now, however, it threatened to fall into the retention lake behind the Mauvoisin Dam, which had been completed in 1957. The phenomenon was closely monitored, but the danger disappeared with a new phase of retreat that began in the 1980s. Current potential risks concern ice avalanches, but these are linked to the Tournelon Blanc Glacier rather than to the Giétro.

The Giétro disaster and those that preceded and followed it graphically illustrate the perils of living near glaciers. The physical dangers waxed and waned partly because of shifting climates but also because of people's ability to adapt and respond to environmental challenges. Local populations through history accepted significant risks of occasionally deadly hazards because there were parallel benefits to living near the large stocks of water held in the glaciers. These also changed. During the period we have detailed so far, agriculture was the main livelihood. Insufficient rainfall led to early reliance on glacier meltwater, whose contribution was maximized by the construction of extensive irrigation networks beginning in the Middle Ages. Needs and opportunities evolved, however, and these also required adaptation of water resource management systems. By the end of the nineteenth century, water took on a different and increasingly important role as the primary source of energy for nascent industries and electric power. Energy was produced first from the natural flow of streams and rivers and then, particularly after the late 1940s, increasingly from water retained behind high Alpine dams. Dams play a dual role, in parallel with the glaciers that have fostered their construction. They allow accumulation of water that, when discharged, generates electricity, and they also hold back waters that would be released by an outburst flood and generally moderate stream floods. Still, they have not eliminated danger. With a new advance of the Giétro Glacier in the 1970s, the risk of an ice avalanche's falling into the dam's retention lake increased. This would have overwhelmed even the considerable safety net the lake provided. This risk has decreased only with the glacier's recession since the 1980s.

The history of Giétro is a window not only to the past but to the future. Global warming, already visible in the significant retreat of Alpine glaciers, may well substitute new natural hazards for the ones that have diminished. It may also provide new opportunities. The "culture of risk" shaped by history and institutional developments and the evolution of scientific knowledge will determine how effectively populations can adjust to the natural and social environments of tomorrow in the context of climate change and increasingly integrated markets and societies.

LIMITS AND POSSIBILITIES OF MOUNTAIN CRYOSPHERE HAZARD MANAGEMENT

Even as scientific understanding advances, new uncertainties threaten our capacity to deal with environmental changes. Current climate warming may alter types of hazards, and we need to understand unfolding processes, just as the scientists observing the Giétro Glacier had to alter their way of thinking about nature in order to respond to the new dangers introduced by the changes of the Little Ice Age. Today we base our definition of hazardous glaciers in Switzerland primarily on an inventory assembling accounts of past events by local inhabitants, glacier reports, newspapers, and scientific publications. This inventory includes 82 glaciers, 55 of which are located in the Canton of Valais (Raymond, Wegmann, and Funk 2003). Produced in the context of the Glaciorisk Project of 2001–03 (see Richard and Gay 2006), it is the basis for a participatory program of hazard-risk management that incorporates residents' assessments of the levels of danger of glaciers (Charly Wuilloud, personal communication, 2005). However, such historical sources may or may not be able to identify potentially hazardous glaciers under future conditions.

This initiative uses a network of specially trained local observers, taking advantage of their knowledge and experience.[4] Additional technical measures and tests, such as fixed surveillance cameras and continuous monitoring of ice movements with the help of global positioning systems, to anticipate ice avalanches have been installed at several sites. It is an indication of the official perception of these kinds of dangers, however, that budget allocations for glacier hazard monitoring pale in comparison with the resources devoted to other hazards affecting the Valais (Charly Wuilloud, personal communication, 2005).

The need to know about various natural hazards derives from the legal requirement that the canton develop hazard maps. Hazard maps indicate different levels of hazard in terms of the intensity and probability of occurrence of given events. There are, however, no maps of glacial hazards, because the scenarios for creating them involve too much uncertainty. For example, hazard maps indicate potential disasters that have return times of from zero (snow avalanches) to several hundred years (floods). In the case of glacier-related events such as ice avalanches or outburst floods, however, return times are unknown.

Thus, while knowledge and experience gained from previous and historical events is useful and important, it is insufficient to predict events in highly dynamic glacier environments in the context of accelerated socioeconomic and environmental changes for several reasons: (1) Glacier hazards do not necessarily recur in the same place, especially as basic climate conditions change. Historical experience thus increasingly loses relevance for predicting the location and impact of hazardous processes, in particular in the context of climate warming (Haeberli and Beniston 1998). (2) Numerous events that seem unpredictable are in fact the result of high-magnitude/low-frequency geomorphic processes. We do not have enough cases because of the rarity of their occurrence, and categorization as a hazard on the basis of a single event at a given location is therefore unjustifiable (Huggel, Kääb, and Haeberli 2002). (3) Some types of hazards, for example, those associated with the degradation of permafrost (Harris, Davies, and Etzelmüller 2001) and interactions between glacial processes and related permafrost hazards (Kääb et al. 2005; Kääb, Reynolds, and Haeberli 2005), are unknown to local people. Scientists have only very recently identified these new types of hazards and know little about how they function.

In sum, the mountain cryosphere response to climate change is difficult to predict on the local scale because it is highly variable. The geomorphic response, which is delayed with respect to a climatic perturbation, thus becomes doubly difficult to predict. Extrapolation to the future therefore remains hypothetical and vague.

Ideally, frequency-magnitude relationships should be applied to events with repeat cycles, but such relationships have not yet been established (Huggel 2004). Frequency-magnitude relationships fail to account for events that occur only once or for the first time. Better predictions will depend both on deeper understanding of processes and on improved detection methods. Current research on the whole alpine cryosphere (including glacier ice, permafrost ice, and snow) is making important advances by identifying interactions among sets of processes that lead to instabilities in high mountain environments. Identification of potential hazard zones and the risk management strategies that derive from this information will depend on a comprehensive approach that takes these relationships into account (Haeberli, Wegmann, and Vonder Mühll 1997). Geoscientists are also developing new methods of hazard assessment, including remote sensing, numerical modeling, and monitoring programs to assess and mitigate glacier and permafrost hazards in a context of accelerated change (see Hewitt 2004; Huggel 2004; Huggel et al. 2004; Kääb et al. 2005; Kääb, Reynolds, and Haeberli 2005).

These new analyses will be effective for hazard management, however, only if the phenomena being studied are recognized as potential dangers under these new conditions. As in the case of glacier hazards, this is often not the case for mountain permafrost because potential environment- and climate-linked catastrophic events involving permafrost changes result from processes that are either not well understood or different from those that occurred in the past. Moreover, there are significant uncertainties about their likelihood, intensity, and frequency. We have noted that these features can influence behavior, and therefore, learning about the nature of the phenomena is central to better hazard management.

Transmission of findings to decision makers and populations at risk is also an important part of the process. Kääb et al. (2005; Kääb, Reynolds, and Haeberli 2005) underline the need to improve communication between the scientific and political communities. This will not always lead to an optimal solution, however, because attitudes toward risk are dependent on the source of knowledge and the nature of the event. Experience leads people to be more risk-preferring, according to Hertwig et al. (2004), but if events are extremely rare, they will have to rely on descriptive knowledge and will be risk-averse (as predicted by Kahnemann and Tversky 2000). Familiarity with some level of danger can lead to risk acceptance against losses, and this may explain why people in disaster-prone areas often refuse to evacuate their homes when threatened by impending catastrophes. The contradictory responses that we observe among different groups may come about because some are relying on experience and others on indirect knowledge, a situation that is most likely to obtain under rapidly changing conditions. Popular perceptions of phenomena such as glaciers are based on knowledge of dangers or absence of dangers in the very recent past. People are therefore less likely to respond to warnings about events that have not yet occurred, are different from what they know, or whose probabilities are not fully known. The design of appropriate mitigation policies must take into account the complexities of risk perception and acceptance as well the nature of the phenomena themselves.

Two examples illustrate the potential hazards from permafrost and glaciers that could result from climate warming and underscore the need to go beyond the projection of uniquely historical data and to apply new theoretical understandings to the identification of sites of potential danger.

On July 10, 1990, part of the historical terminal moraine of the Dolent Glacier collapsed (Figure 3.2). This triggered a debris flow of approximately 40,000 cubic meters that took the inhabitants of L'A Neuve and its camp grounds completely by surprise. One person was hurt, a scout camp destroyed, and a chalet in the hamlet damaged (Lugon, Gardaz, and Vonder Mühll 2000). Inhabitants of the Ferret Valley are aware of the risk associated with debris flows. In fact, debris flows can be expected to occur yearly on the western slopes of the valley, where the more dominant schistose rocks furnish an abundant mass of rocky debris and the fine silty sediments are easily set in motion during a storm. However, the triggering factor is usually an intense rainfall event. The partial collapse of a moraine in good weather in 1990 was unknown to valley residents until it happened.

FIGURE 3.2. Dolent Glacier, Ferret Valley, Mont Blanc Range. (A) Historical moraine dam and the flow path of the July 10, 1990, debris path. (B) Scar of the debris flow with a maximum depth of 15 meters. (C) Deposit on the L'A Neuve alluvial fan. Note the two deflection dams built to protect the hamlet of L'A Neuve against snow avalanches. (Photos J.-D. Rouiller.) (D) Historical moraine dam of the upper Aiguilles Rouge Glacier in the Arolla Valley. (Photos R. Delaloye.)

Lugon, Gardaz, and Vonder Mühll (2000) have studied in detail the conditions that produced this moraine failure. The instability of the moraine was not induced by an intense rainfall event. Among the possible explanations of the debris flow is a combination of a slow outburst of a water pocket in the glacier, snowmelt that soaked the moraine with water, and a change in the direction of the subglacial flow. Geophysical soundings showed that the moraine dam was well compacted, without extremely loose layers of high porosity and without groundwater-saturated layers. Currently, there is no permafrost near the surface in the sediment. Permafrost in the deeper layers (15–20 m deep) cannot be excluded, but the moraine was probably never perennially frozen, and therefore permafrost was not a factor in the chain of processes leading to the moraine failure.

Since the 1990 event, the Dolent Glacier has been considered a short-term hazardous glacier (Raymond, Wegmann, and Funk 2003). The local authorities have developed an efficient system of management of debris flows based on simple structural measures and a verbal/telephone warning system. Nevertheless, a partial collapse of the moraine could occur again at any time during the summer months. This sudden and unexpected event shows that future surprises cannot be excluded. The upper Aiguilles Rouge Glacier in the Arolla Valley (Figure 3.2, D) is not considered hazardous in the official inventory because there is no historical evidence of its having caused damage to persons or property. However, the hamlet of Satarma could be threatened if a Dolent-type debris flow were to occur there. Hazard assessment is conditioned not only by likely consequences but also by social, economic, and political considerations, and this raises the question whether to allocate resources to deal with high-magnitude (potentially catastrophic) but low-frequency hazards that may never happen.

The Tsarmine Glacier margin and the Tsarmine rock glacier are not considered hazardous in the official inventory because there are no historical records of events and, moreover, rock glaciers are not included in the inventory. The geomorphology of the Tsarmine glacial margin is dominated by a debris-covered glacier, a moraine dam, push moraines, and a rock glacier (Figure 3.3, A and C). Geophysical prospecting has demonstrated the presence of buried glacier ice and frozen sediments (permafrost ice) in the sediments (Lambiel et al. 2004; Reynard et al. 2005). Permafrost conditions support the long-term preservation of dead ice bodies that, upon melting, leave behind underground cavities. The frontal moraine, where the proglacial lake is located, is free of ground ice. This could have implications for potential geomorphological hazards. The presence of the lake in the unfrozen morainic bastion may be a factor in triggering debris flows.

The Tsarmine rock glacier is characterized by an unstable topographical position dominating a steep slope (Figure 3.3, A and C). Permafrost creep measurements on its surface point to the probable recent reactivation of a former inactive rock glacier or, at least, to a severe acceleration of this landform (Delaloye, Lambiel, and Lugon 2005; Lugon et al. 2005). It furnishes an important amount of debris that accumulates in a gully. There are no historical records of recent debris flow from the rock glacier, but, given the quantity of loose sediment and the slope angle, a debris flow of unpredictable magnitude is likely to occur in the near future (years to decades). According to recent observations in the European Alps (Kääb, Frauenfelder, and Roer 2007) and in the Valais Alps in particular (Delaloye, Lambiel, and Lugon 2005; Lambiel and Delaloye 2004), a surprisingly large number of alpine rock glaciers have shown an increase in speed during recent years. Kääb, Frauenfelder, and Roer (in press) suggest that "this large number points to other than solely local influences," in particular "some regional-scale impact such as the observed increase in air temperatures." The number of rocks transported over the rock-glacier front and the total horizontal mass flux increase with increasing (surface) speed. As a consequence, local rockfall hazards and the buildup of potential debris-flow-triggering volumes might increase under certain topographic conditions, such as those of the Tsarmine rock glacier. In this case, accelerated permafrost creep in coarse-grained sediments is a potential hazard and one with which local people have no direct experience.

During the 2003 summer heat wave that affected most of Europe, numerous superficial rockfalls occurred along the northwestern face of the Blanche de Perroc, which dominates the small Tsarmine Glacier (Figure 3.3, B).[5] Melting of superficial layers of permafrost on a mountain face in which cracks are still filled with ice favors surface instability in the rock wall.

The Tsarmine area thus illustrates, within the same spatial environment, four types of potential but generally unrecognized glacier and related permafrost hazards: (1) increase of superficial rockfall activity originating from deglaciated parts of a rock face (the northwestern face

FIGURE 3.3. Blanche de Perroc, Tsarmine Glacier, with its proglacial margin, and Tsarmine Rock Glacier, Arolla Valley. (A) View of the Tsarmine slope. (B) Rock-falls from the northwestern face of the Blanche de Perroc. (C) Aerial view of the proglacial margin of the Tsarmine Glacier. (Photos C. Lambiel.)

of Blanche de Perroc), (2) increasing creep rates leading to frequent rockfalls along the steep front of a rock glacier and consequent increased risk of debris-flow formation (the Tsarmine rock glacier), (3) frozen or unfrozen debris slopes as another source of debris flows (the Tsarmine glacial margin, with a morainic dam and a rock glacier, the Tsarmine rock glacier and scree slope), and (4) potential outbursts of glacier lakes and consequent debris flows (the Tsarmine moraine dam). Hazards 1 and 2 exist but are not well known to the people or to the authorities. The potential for hazards 3 and 4 is not recognized at all because they have never occurred. The establishment of a surveillance program for potential hazards will depend on whether such risks are acknowledged by the local inhabitants and the political authorities.

SOCIAL CHANGE, RISK PERCEPTION, AND RESPONSE

The difficulties of projecting natural risks into the future as conditions change have their parallel in responses to social developments. Glacier retreat is expected to affect the role of glaciers as sources of crucial water resources, but these natural processes are not the only factors that will influence water use. Economic and political transformations will have equal impact and are also sources of significant uncertainty. From the twelfth to the twentieth century, glacier meltwater played an important role in the agricultural system, but since the mid-1950s the importance of agriculture to the Valais has declined. This sector now employs only 3.3% of the population full-time and an additional 8.6% as part-time workers (Lehmann et al. 2000). However, the many activities that are replacing agriculture are also dependent on glaciers and their water. Tourism, for example, which now dominates the Valaisan economy, relies on glaciers as iconic features in the landscape and on their snow cover and water for the ski industry. Glaciers also furnish water for hydroelectric power generation, which provides 60% of the country's electricity supply. These new activities have not only changed the landscape but introduced competition for water. Tourism increases domestic water consumption throughout the year but particularly during winter snow manufacturing. The huge dams for power generation divert water from natural streams and irrigation canals. Although irrigation has lost its importance with the shrinking of the agricultural sector, the consequences of the changes are nevertheless far-reaching. Local management of the irrigation systems underpinned the local authority structure, and with changing water use these prerogatives have eroded.

In addition, perceptions of glaciers and landscapes and of risks and risk sharing have evolved. On the one hand, physical space is used more extensively, putting people and buildings in areas that had previously been viewed as hazardous. In addition, new employment sectors have brought people with different cultural backgrounds, incomes, and education levels into contact with each other. This geographic expansion and social diversity have modified people's evaluations of probabilities, outcomes, and consequences concerning major social and natural changes.

The development of hydroelectric power encapsulates many of these new developments and is linked, as a contemporary example, to our earlier discussion of the old story of adaptation to shifting climate conditions and social environments. As in the Middle Ages, the past few decades have seen a new use of glacier meltwater competing with traditional uses, in this case with irrigation (itself the new pattern in the fifteenth century). Predicting its future involves recurrent problems of decision making under uncertainty. Climate change will alter water dynamics, and evolving regulatory and institutional frameworks will alter the basic conditions of its use, creating a situation not unlike that during the Middle Ages, when changing property rights influenced resource use and shaped communities' institutional frameworks.

Notable among these current changes is the progressive loss of local control over critical resources. This process began long ago.

The handling of disasters such as the Giétro signaled the beginning of significant state intervention in resource management. Throughout the nineteenth century, the canton of the Valais undertook many infrastructure projects, among them large dams for electricity production. Rules were needed to regulate this new use of water, which more than ever transcended community boundaries. The 1916 federal law on the use of hydroelectric power gave the canton ultimate authority over the power of the waterways. Upland communities maintained their property rights to the water, and they negotiated with the power companies to permit use of specified quantities of water within limits imposed by the canton. In the 1916 law and successive legislation, communes ceded important powers to the canton or the Confederation, which acquired authority to set maximum prices, define time limits for the concessions granted by communes to electricity companies, and regulate minimum water flows to rivers and streams.

Eroding communal prerogatives are only one of many changes in Valais mountain communities. An integrated system organized around household agriculture has given way to a more diversified economy in which interest groups representing industry or tourism have different views about risks and choices. Attitudes about the future of the energy sector and its evolving legal and regulatory framework reflect these conflicting positions and highlight the uncertainties and risks not only for mountain populations but for Switzerland as a whole. Swiss electricity production is under pressure to become integrated into the European market. In 2002, however, consistent with their reluctance to join the European Union, the Swiss turned down a new energy law that sought to foster efficiency by introducing competition among suppliers, including those from outside Switzerland (with safeguards ensuring nondiscriminatory prices and equitable distribution). At about the same time they also rejected a tax on nonrenewable energy sources intended to compensate for stranded investments (the costs of wrong decisions made under uncertainty). The revenues would have been returned to locals to allow them to keep the price of hydroelectric power competitive in the liberalized market. The rejection of this tax left unsettled the proposed energy law's impacts on the hydropower sector in mountain regions, cantonal revenues, and regional autonomy. Though the question remained moot because of the defeat of both measures, the issues will undoubtedly resurface, since Switzerland must ultimately come to an agreement with its neighbors about the conditions of electricity transmission.

There were many reasons for the Valais's double rejection. The proposed policies would have entailed major shifts in power relations. Community ownership of resources and local control over their revenue-generating possibilities would have been sharply reduced. Dependence on higher authorities would have increased because of the need for subsidies. Traditional patterns of local resource management tend to be incompatible with the increasingly centralized energy sector across Europe. New laws and regulations thus threaten fundamental values and decision-making behavior that evolved to preserve resources in the face of climate uncertainties and potential threats to universal access. Under these conditions, the Swiss in general and Valaisans in particular voted for the status quo over innovation, postponing decisions about how to manage water, energy, and risk under climate and social uncertainty.

The future holds new challenges. Climate change is expected to alter the nature of physical hazards, but important uncertainties remain about exactly how it will do so. In this context, we have suggested the importance of improving scientific understanding about how warming will modify the location, magnitude, and frequency of debris flows, rockfalls, and floods and affect the availability of water resources, whose uses are additionally influenced by socioeconomic changes that have similar levels of uncertainty. Our observations of Valais history and its people's perceptions of and responses to risk suggest ways in which they might respond to these environmental and societal transformations.

It will be important to seek to harmonize various forms of knowledge about physical dangers and economic and political uncertainties so that assessments of probabilities of various outcomes and acceptable levels of risk are more homogeneous across the population, fostering cooperative strategies and common goals. Risk sharing will be a critical adaptive mechanism. Strikingly important in the historical case was the existence of collective safety nets. Innovation and the acceptance of higher levels of risk associated with the construction of irrigation systems were compensated for by mechanisms for sharing these risks across the community. Today there need to be equivalent ways to prevent small minorities from either bearing the brunt of natural catastrophes or monopolizing the windfall benefits of new water uses.

Given the changes in political organization that have led to increased centralization, these policies will be organized not locally but at the cantonal level. The monitoring systems described previously are one way of anticipating dangers; rescue and rebuilding aid are policies for responding to disasters that are not anticipated. Technological solutions are also part of the policy mix. The recently launched joint cantonal-federal plan to reengineer the Rhône River channel over the coming decades is a salient example of infrastructure changes intended to reduce flood risks. Fundamental changes in economic patterns also destabilize traditions and behaviors. Liberalization of the energy market has been decided at the European level, but given the high degree of decentralization in Switzerland, many political decisions are subjected to popular referendum, often translating into resistance to change. The rejection of the electricity law and the energy tax proposal shows that there were not sufficiently convincing incentives to compensate for new risks of competition.

Glaciers are unique physical features of the landscape that play a central role in shaping perceptions and influencing behavior. They have provided essential resources even as they have periodically caused death and destruction. In the Valais, people have devised ways to minimize negative impacts in order to benefit from their bounty. Today, glaciers are melting, potentially altering the rhythm and amount of available water and causing new natural hazards. This is occurring amid profound social changes. History shows that relations with glaciers and the social context in which they unfolded were constantly changing and that people more or less successfully adjusted to new conditions. In this regard, the future will resemble the past. New hazards and new uses for water resources are emerging. The institutional context for confronting these mutations is also evolving. Long accustomed to decentralized ownership and management, inhabitants of the Valais and Switzerland may find their institutions ill-suited to these changes.

Our case study has shown how Valaisan society moved from a homogeneous population to one with greater economic and social diversity. Competition over how to allocate resources has increased, and evaluations of social and natural risks have become more varied. Very local response is no longer a viable strategy. In fact, it had already begun to show its limits in the Giétro disaster, which overwhelmed the community-level response mechanisms. Before this event, risks and resources were regulated mostly at the village level. Because of changes in perceptions, knowledge, and social institutions, new ways of dealing with risk emerged after that period. Systematic observation of natural phenomena and a growing understanding of the physical processes producing catastrophes were accompanied by efforts to anticipate them. This necessitated wider coordination and more resources, which went hand in hand with a more top-down and centralized risk management structure that accompanied the state consolidation occurring at the same time. Administration of resources followed a similar evolution from community control to cantonal and federal oversight. These new institutional arrangements affected the nature of responses to disasters and forms of adaptation to social change.

A particularly telling present-day example is the way in which risk perception, level of knowledge, and definition of responsibilities have come together to pose new dilemmas for those responsible for security. A recent court case condemned the authorities of a commune for having failed to protect buildings and people in an avalanche that caused several deaths in the Valais in 1999. Such decisions assume that uncertainty about some physical process has been reduced, obviating the need for traditional risk sharing and introducing a new level of institutional responsibility for damages. It is an open question whether these policies will be an effective response to disaster management, but they illustrate how society has redefined its perception of risk and its mechanisms for confronting it. In a similar way, the risks engendered by societal transformations are eliciting new arrangements, notably the centralization of administrative structures. These are often perceived as reducing individual or local autonomy and increasing social and economic disparities. Under such conditions, people may become risk-averse with regard to the very innovations that could help lessen impacts of inevitable environmental and social changes, such as regulation of the electricity market. If the rejection of this regulation proves to have negative consequences (a matter of considerable political debate), it will be an argument for the need for better information that incorporates people's attitudes toward risk. It also highlights the disagreements that surround the choice of policy to mitigate or adapt to change. In the case of the Valais, our historical perspective suggests that adaptation to glacier risks and opportunities has been an ongoing process. The adoption of new social arrangements and institutional constructs has depended on individuals' and groups' perceptions that their way of life or status could be preserved or enhanced and that economic disparities resulting from social and environmental change would be adequately compensated for. The details are local, but the dilemmas confront all populations struggling to deal with new environmental conditions in changing social contexts.

ACKNOWLEDGMENTS

We thank Raymond Dacey and Emmanuel Reynard and three reviewers for helpful comments on an earlier version. We are grateful to Charly Wuilloud for providing details on the citizen glacier monitoring program being established in the Valais.

REFERENCES CITED

Bernstein, P. 1996. *Against the gods: The remarkable story of risk*. New York: Wiley.

Comité de Bienfaisance. 1820. *Compte rendu par le comité de bienfaisance établi à Martigny par le gouvernement valaisan sur la répartition des secours provenans des dons et collectes qui ont eu lieu en faveur des individues des communes riveraines de la Drance, victimes de l'inondation du 16 juin 1818*. Sion: Antoine Advocat.

Dubuis, P. 1999. Le bisse, témoin d'une civilisation alpine en mutation. In *Le rôle de l'eau dans le développement socio-économique des Alpes*, ed. K. Bösch, 83–89. Sion: Institut Universitaire.

Delaloye, R., C. Lambiel, and R. Lugon. 2005. *Service for Landslides Monitoring SLAM InSAR Bas-Valais, validation of InSAR data in permafrost zone, final report, prepared at the Department of Geosciences, Geography, University of Fribourg, Switzerland*. Bern: Federal Office of Water and Geology.

Haeberli, W., and M. Beniston. 1998. Climate change and its impacts on glaciers and permafrost in the Alps. *Ambio* 27:258–65.

Haeberli, W., M. Wegmann, and D. Vonder Mühll. 1997. Instability problems related to glacier shrinkage and permafrost degradation in the Alps. *Eclogae Geologicae Helvetiae* 90:407–14.

Harris, C., M. Davies, and B. Etzelmüller. 2001. The assessment of potential geotechnical hazards associated with mountain permafrost in a warming global climate. *Permafrost and Periglacial Processes* 12:145–46.

Hertwig, R., G. Barron, E. Weber, and I. Erev. 2004. Decisions from experience and the effect of rare events in risky choice. *Psychological Science* 15:534–39.

Hewitt, K. 2004. Geomorphic hazards in mountain environments. In *Mountain geomorphology*, ed. P. Owens, N. Wake, and O. Slaymaker, 187–218. London: Hodder Arnold.

Huggel, C. 2004. Assessment of glacial hazards based on remote sensing and GIS modeling. Ph.D. diss., University of Zurich.

Huggel, C., W. Haeberli, A. Kääb, D. Bieri, and S. Richardson. 2004. Assessment procedures for glacial hazards in the Swiss Alps. *Canadian Geotechnical Journal* 41:1068–83.

Huggel, C., A. Kääb, and W. Haeberli. 2002. Glacier hazards. http://www.glacierhazards.ch/ (accessed November 2006).

Kääb, A., R. Frauenfelder, and I. Roer. 2007. On the response of rockglacier creep to surface temperature increase. *Global and Planetary Change* 56:172–87.

Kääb, A., C. Huggel, L. Fischer, S. Guex, F. Paul, I. Roer, N. Salzmann, S. Schlaefli, K. Schmutz, D. Schneider, T. Strozzi, and Y. Weidmann. 2005. Remote sensing of glacier- and permafrost-related hazards in high mountains: An overview. *Natural Hazards and Earth System Sciences* 5:527–54.

Kääb, A., J. M. Reynolds, and W. Haeberli. 2005. Glacier and permafrost hazards in high moutains. In *Global change and mountain regions: An overview of current knowledge*, ed. U. Huber, H. Bugmann, and M. Reasoner, 225–34. Dordrecht: Springer.

Kahnemann, D., and A. Tversky. 2000. *Choices, values, and frames*. Cambridge, UK: Cambridge University Press.

Lambiel, C., and R. Delaloye. 2004. Contribution of real-time kinematic GPS in the study of creeping mountain permafrost: Examples from the western Swiss Alps. *Permafrost and Periglacial Processes* 15:229–41.

Lambiel, C., E. Reynard, G. Cheseaux, and R. Lugon. 2004. Distribution du pergélisol dans un versant instable: Le cas de Tsarmine (Arolla, VS). *Bulletin de La Murithienne* 122:89–102.

Le Roy Ladurie, E. 1983. *Histoire du climat depuis l'an mil*. Paris: Flammarion.

Lehmann, B., E. Stucki, N. Claeyman, V. Miéville-Ott, S. Réviron, and P. Rognon. 2000. *Vers une agriculture valaisanne durable: Étude réalisée à la demande du Valais par l'Antenne romande de l'Institut d'économie rurale de l'École polytechnique fédérale de Zurich*. Zurich: Eidgenössische Technische Hochschule Zürich.

Lugon, R., J.-M. Gardaz, and D. Vonder Mühll. 2000. The partial collapse of the Dolent Glacier moraine (Swiss Alps, Mont Blanc Range). *Zeitschrift für Geomorphologie* 222:191–208.

Lugon, R., C. Lambiel, G. Cheseaux, E. Reynard, and R. Delaloye. 2005. Recent reactivation of an inactive rock glacier (Arolla, Swiss Alps) *Terra Nostra, 2nd European Conference on Permafrost, June 12–June 16, 2005, Potsdam, Germany, Programme and Abstracts*, 71–72. Terra Nostra 2005/2. Berlin: GeoUnion Alfred-Wegener-Stiftung.

Luyet, B. 1924. *Légendes de Savièse*. Archives Suisses des Traditions Populaires 25.

———. 1929. *Contes de Savièse*. Cahiers Valaisans de Folklore 7.

Maisch, M., A. Wipf, B. Denneler, J. Battaglia, and C. Benz. 1999. *Die Gletscher der Schweizer Alpen: Gletscherhochstand 1850, aktuelle Vergletscherung, Gletscherschwund-Szenarien*. Final report of the Swiss National Research Program 31. Zurich: Eidgenössische Technische Hochschule Zürich.

Musée de Bagnes. 1988. 16 juin 1818: Débacle de Giétro. In *Collection du Musée de Bagnes*, vol. 1. Le Châble, Switzerland.

Netting, R. McC. 1981. *Balancing on an Alp*. Cambridge, UK: Cambridge University Press.

Payot, P. 1950. *Au royaume de Mont-Blanc*. Bonneville, France: Plancher.

Pfister, C. 2002. *Le jour d'après: Surmonter les catastrophes naturelles, le cas de la Suisse entre 1500–2000*. Bern: Paul Haupt.

Raymond, M., M. Wegmann, and Martin Funk. 2003. *Inventar gefährlicher Gletscher in der Schweiz*. Mitteilungen der Versuchsanstalt für Wasserbau, Hydrologie und Glaziologie der ETH Zürich 182.

Rebetez, M., R. Lugon, and P.-A. Baeriswyl. 1997. Climatic change and debris flows in high mountain regions: The case study of the Ritigraben Torrent (Swiss Alps). In *Climatic change at high-elevation sites*, ed. M. F. Diaz, M. Beniston, and R. Bradley, 139–57. Dordrecht: Kluwer Academic Press.

Reynard, E., C. Lambiel, G. Cheseaux, and R. Lugon. 2005. Permafrost-glacier relationships in an unstable sediment complex, Tsarmine, Arolla Valley, Western Swiss Alps. In *Terra Nostra, 2nd European Conference on Permafrost, June 12–June 16, 2005, Potsdam, Germany, Programme and Abstracts*, 101. Terra Nostra 2005/2. Berlin: GeoUnion Alfred-Wegener-Stiftung.

Richard, D., and M. Gay. 2006. Glaciorisk, final report, 01.01.2001-31.12.2003, Section 6: Survey and prevention of extreme glaciological hazards in European mountainous regions. http://glaciorisk.grenoble.cemagref.fr (accessed July 2007).

Schüle, R.-C. 1987. Il vaut mieux souffrir du froid maintenant: Imaginaires de la haute montagne. *Documents d'Ethnologie Régionale, Centre Alpin et Rhodanien d'Ethnologie* 9:31–40.

Société d'Histoire du Valais Romand. 1995. *Actes d'un colloque international sur les bisses, 15-18 septembre 1994*. Annales Valaisannes 70, series 2. Sion: Société d'Histoire de Valais Romand.

Wiegandt, E. 1977. Inheritance and demography in the Swiss Alps. *Ethnohistory* 24:133–48.

———. 1980. Un village en transition. *Ethnologica Helvetica* 4:63–93.

———. 2007. From principles to action: Incentives to enforce common property water management. In *Mountains: Sources of water, sources of knowledge*, ed. E. Wiegandt. Dordrecht: Springer.

4

Environment, History, and Culture as Influences on Perceptions of Glacier Dynamics

THE CASE OF MT. SHASTA

Barbara Wolf and Ben Orlove

Mt. Shasta offers an interesting case for considering the factors that shape human perceptions of glacial processes. The mountain is unusual in many ways. It is distinctive as a natural feature. A tall, free-standing volcano in the far northern part of California, it contains the largest glaciers in the state. These glaciers, moreover, have demonstrated dynamics different from those in many other parts of the world, most notably a pattern of advance rather than retreat in the second half of the twentieth century.

The mountain is also distinctive for the resources it affords and the risks that it presents to the human populations that live near it. It contains abundant forests and contributes significantly to surface streams and to groundwater. It also is the site of avalanches, mudslides, debris flows, floods, and volcanic eruptions. In addition, its location close to the largest north-south valley in the mountainous western United States favored its early integration into regional transportation networks. These resources, risks, and location have shaped its involvement in resource extraction economies; its forests, snows, and rugged formations have also favored recreation-oriented tourism and settlement.

Moreover, the mountain is distinctive for the ways in which it is perceived by human populations. The economic activities associated with its physical nature influence perceptions but do not wholly determine them. The people in the region give the mountain cultural meanings, often of an aesthetic, religious, and spiritual nature, that also affect their perceptions of it and their views of the different forms of human activity it supports. Glaciers contribute to the mountain's resources and risks, and human perceptions of the mountain involve glaciers, though they include other elements as well. Indeed, the difficulty of disentangling the specific contribution of glaciers to the resources, risks, and perceptions is the final and perhaps the most striking of the ways in which Mt. Shasta is distinctive.

GEOLOGY, RESOURCES, AND HAZARDS

Mt. Shasta, one of the southernmost peaks of the Cascades chain, is a 4,316-m compound stratovolcano located in northern California just

FIGURE 4.1. View of Mt. Shasta with Shastina and glaciers, from the northwest, near Weed. July 2004. (Photo Barbara F. Wolf.)

61 km south of the Oregon border (Blodgett et al. 1996). The glacier-frosted mountain consists of four major volcanic cones: Hotlum, Shastina, Misery Hill, and Sargents Ridge (Miesse 2005) (Figures 4.1, 4.2, and 4.3). It rises above the snow line, the only mountain in California to retain permanent snow year-round even on exposed slopes (Guyton 1998). Considered a dormant volcano, Shasta is thought to have last erupted about 200 years ago (Miesse 2005). Geologic records indicate that eruptions have occurred about every 600 years for the past four or five millennia (Guyton 1998). The current glaciation began a relatively recent 3,000 years ago. Earlier ice would have been melted by lava flows.

The mountain has a very striking appearance. The naturalist John Muir described "the colossal cone of Shasta, clad in ice and snow, the one grand unmistakable landmark—the pole star of the landscape" (Muir 1918). A local businessman noted "the drama of the lone mountain on the horizon." Visible from more than 150 km away, its 350-km^3 mass rises more than 3 km from the surrounding forested Shasta Valley (Callaghan 2000), a contrast that gives it greater visual impact at a distance than that of taller mountains located among peaks of similar size, such as Mt. Whitney. Its location close to the head of the Sacramento Valley has also placed it near major railroads and roads that connect California with Oregon and Washington in the Pacific Northwest.

Clarence King of the U.S. Geological Survey first described Shasta's glaciers in 1870, although others are said to have seen them before then (Leviton and Aldrich 1997). King named the Whitney Glacier, and the explorer John Wesley Powell, on an 1879 expedition, named the others, calling one Wintun for a local indigenous group and using words from their language for others: Hotlum (steep rock), Bolam (great), and Konwakiton (muddy one) (Powell 1885). These five are generally considered to be the major glaciers, but over the years arguments have been made that there are in fact seven,[1] nine, or ten glaciers (Rhodes 1987; Freeman 1997; Guyton 1998). Whitney is California's only valley glacier and is also the longest glacier in the state, approximately 3 km in length (Biles 1989). Hotlum has the greatest ice volume of the five (Guyton 1998).

The total volume of ice on Mt. Shasta is said to be 0.13 km^3, with an area of 6.9 km^2 (Dreidger and Kennard 1986). These figures are widely cited but have been criticized because they involve significant extrapolation from limited data. More recent research concentrating

FIGURE 4.2. Southeast face of Mt. Shasta, from McCloud, showing Konwakiton Glacier. July 2004. (Photo Barbara F. Wolf.)

on Whitney and Hotlum Glaciers has drawn on photographic as well as climate records to trace their fluctuations. They shrank significantly in the 1920s and 1930s but advanced in the second half of the twentieth century (Howat et al. 2006). The western lobe of Whitney Glacier, the largest on Mt. Shasta, advanced downslope 955 m between 1944 and 2003, reaching the elevation of 2,780 m. Much of the variation can be attributed to El Niño and the Pacific Decadal Oscillation, both known to affect winter snowfall in this region. An examination of snow water equivalent trends at different elevations between 1948 and 2003 shows an increase of precipitation (occurring principally as snow) at 2,200 m and above and a reduction at lower elevations. This pattern has permitted increased winter accumulation of snow that, to date, more than compensates for any effects of warming on ablation. Numerical models combined with scenarios derived from regional climate models suggest, however, that the effects of warming will outpace increases in precipitation in this century, leading to considerable shrinkage of Whitney Glacier by 2080 and the complete disappearance of Hotlum Glacier by 2065 (Howat et al. 2006). Mt. Shasta is thus at present an exception to the general pattern of retreat, found nearly everywhere else in the world, though it will not remain so indefinitely. It resembles the cases of the advancing glaciers of the late twentieth century in Norway (Andreassen et al., this volume) and New Zealand (Hay and Elliott, this volume), where increases in winter snowfall, linked partly to interdecadal variability, also compensated, at least for some years, for ablation due to warming.

The mountain's volcanic nature has a significant influence on its glaciers. While the precise hydrology is not well understood, it is known that water from melted snow and glacial ice percolates into the mountain's unconsolidated volcanic soils, which act as a filter. These waters arise as springs farther down the mountain or seep down to the groundwater table and supply nearby towns with pure drinking water. Shasta's meltwaters feed the Klamath and Sacramento River watersheds, expanding its influence into Oregon and central California. The springs have also attracted international water bottling companies to the small mountain towns, stimulating both the local economy and local controversy (Sanchez 2001; Benda 2003, 2004; Kennedy 2003; Breitler

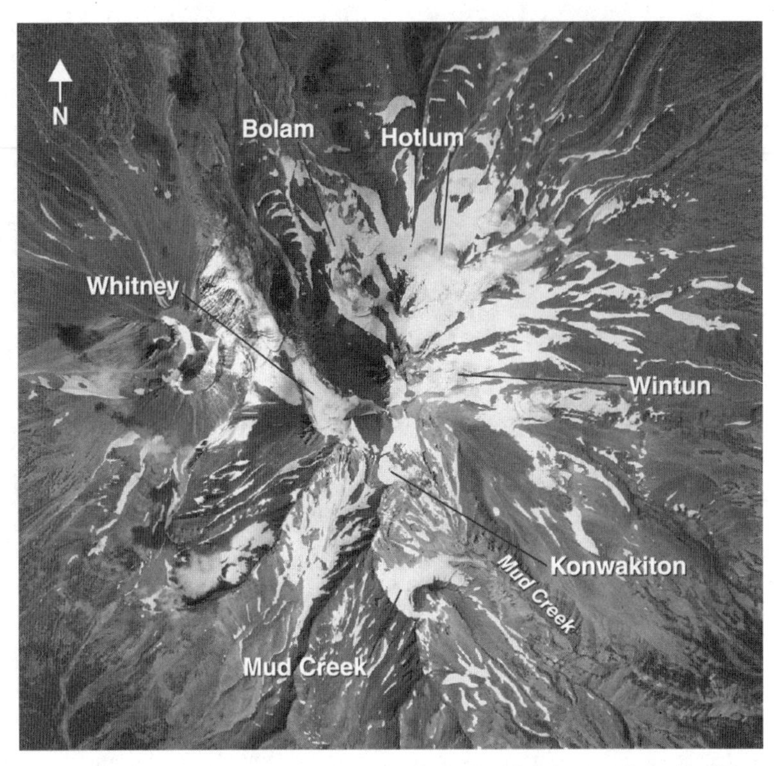

FIGURE 4.3. Glaciers of Mt. Shasta. Imagery: USDA-FSA Aerial Photography Field Office, Salt Lake City, UT. August 2005.

2004; Le Guellec 2004; Save McCloud's Water Group 2004; Vance 2004).

The mountain's height influences its value for human society. The cool, moist climate and rich volcanic soils support extensive forests, which have provided timber. Its dramatic character and rugged topography draw tourists, especially hikers and climbers; there have also been efforts to develop ski resorts on the mountain. Its height, symmetry, and isolation give it iconic qualities recognized, in different ways, by Native Americans from the precontact period to the present and by settler and contemporary populations.

Though many people appreciate Mt. Shasta's value as a natural wonder, a tourist attraction, and a source of valuable resources, only a few have voiced concern about the risks that it poses. A geologist based at a nearby college noted that its history of eruptions suggests a 1:3 or 1:4 chance that the mountain will erupt in a person's lifetime (Hirt n.d.), yet real estate agents do not inform newcomers of the potential risks of building in certain areas. A vacation rental agent mentioned the fact that constructing government buildings is prohibited in certain high-risk locations but did not know whether the prohibition applied to other development.

More frequent and more certain, though perhaps smaller in scale, than volcanic eruptions are mud (or debris) flows or glacial outburst floods, also known as *jökulhlaups* (see Table 4.1). In the historical period, these have occurred about once every 10 years. Such events were recorded in 1881, 1920, 1926, and 1931, and prehistoric flows are documented as well (Blodgett et al. 1996). The most destructive of these flows, a jökulhlaup, took place in August 1924, damaging the McCloud River Railroad's tracks and McCloud's city water system. The Konwakiton Glacier was the culprit in this event; during a

TABLE 4.1

Historic Debris Flows and Economic Impacts

YEAR	LOCATION	EVENT TYPE[a]	ECONOMIC IMPACT
1881	Mud Creek	Debris flow[b]	Unknown
1920	Mud Creek	Debris flow[b]	Unknown
1924	Mud Creek	Jökulhlaup	Damage to McCloud water system and railroad tracks, environmental change. Forest closed to public. Damaged McCloud River fishery, silted Sacramento River.
1926	Mud Creek	Debris flow[b]	Broke through a diversion channel and blocked a road. Occupied workers to divert the flow.
1931	Mud Creek	Debris flow[b]	Unknown
1935	Whitney Creek	Debris flow[b]	Buried U.S. Highway 97 by 1.2 m for distance of 370 m.
1977	Mud Creek	Debris flow[b]	Silt and ash dumped into McCloud River and Reservoir. Hampered fishing and threatened endangered trout species.
1977	Ash Creek (Wintun Glacier)	Mud flow[c]	Stream downcutting; no threat to human life or structures.
1985	Whitney Creek	Debris flow[b]	Unknown
1997	Whitney Creek	Debris flow[d]	Blocked U.S. Highway 97.

[a]Some events described as debris flows may have been jökulhlaups. Blodgett, Poeschel, and Osterkamp. (1996) and others note that snow bridges or collapsed snow tunnels catch earth and silt that wash down with meltwater, forming temporary dams that eventually break with the force of water and material behind them, "creating surges of water and debris that may cause debris flows downstream."
[b]Blodgett, Poeschel, and Osterkamp. (1996).
[c]Frank (1997*b*).
[d]U.S. Geological Survey (1997).

record drought season, a chunk of overhanging ice broke off, moved down Mud Creek Canyon, and dammed it. Meltwater and debris built up behind the ice dam until it burst and sent mud, trees, and boulders the size of automobiles washing down the mountainside. It plastered an area about 20 km long by 1.5 km wide, according to newspaper accounts (Frank 1990; *San Francisco Chronicle*, August 29, 1924). Because Mud Creek flows on the mountain's loose volcanic surface, the event sent silt into the McCloud River and from there into the Sacramento River, turning the latter "milky" all the way to San Francisco Bay (Stuhl 1924, 94–103).

A month later, in September 1924, on the west side of the mountain, a chunk of the Whitney Glacier above Weed broke off and slid down to timberline in a cloud of dust kicked up by the boulders it displaced, raising the concern that melting might turn it into "another river of mud" (*Siskiyou News*, September 11, 1924). Nine major debris flows occurred in this drainage during the twentieth century (Callaghan 2000). A mudflow in 1977 was probably the largest of the century, but its location in the unpopulated Ash Creek drainage meant that it attracted little notice. A Whitney Creek mudflow in 1997 reached the flats below Whitney Glacier, closing a regional highway and affecting the railroad (U.S. Geological Survey 1997). Conditions common to the debris flow or jökulhlaup events were light winter snowfall followed by dry conditions and summer heat, which accelerated

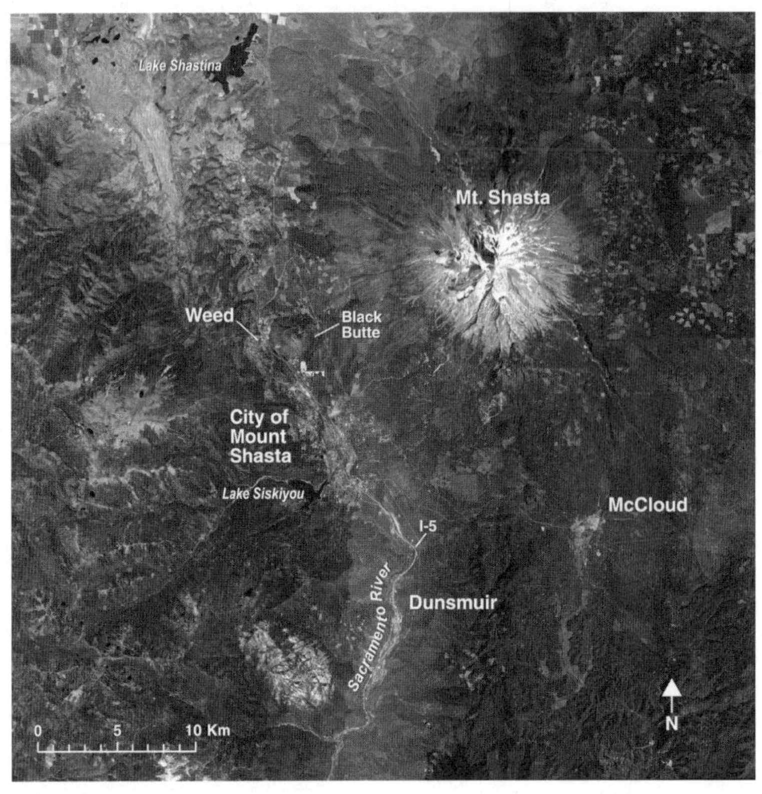

FIGURE 4.4. Mt. Shasta and surrounding towns. Imagery: USDA-FSA Aerial Photography Field Office, Salt Lake City, UT, August 2005.

melting and created "rotten ice." Subsequent flows "correlated on both major warm rain events and with very warm temperatures" (Callaghan 2000).

HISTORY OF HUMAN SETTLEMENT

Our review of the human history of the Mt. Shasta region (Figure 4.4) covers four periods: a long early time of indigenous settlement, a brief period in the mid-nineteenth century of Euro-American conquest, a later period, stretching into the second half of the twentieth century, of lumber extraction and interregional transportation development, and a recent period of water bottling plants and tourism. The mountain's great height and abundant water have shaped human presence and activity in each period; the glaciers have also had a specific influence at various times. We review the last two periods by covering each of the towns in the region individually to underscore the great importance of particular corporations and government agencies in this region dependent on natural resources and dominated for many decades by company towns. The particular characteristics of the towns, established by the early twentieth century, have endured into the present.

The mountain's impressive beauty was not lost on the aboriginal inhabitants of the region. Shasta forms the hub of the ancestral territories of five native groups—Shasta, Wintu, Modoc, Klamath, Okwanuchu, and Karuk. It is a crucial site in the origin stories of some and, along with the volcanic landscape of the nearby Medicine Lake Highlands, harbors vision quest and ceremonial sites for other nations within its view, among them the Achomawi/Atsugewi (Pit River) tribes (Theodoratus and Evans 1991; Eargle 2000). The discovery of gold in 1851 brought an

FIGURE 4.5. A mural in Mt. Shasta honoring Muir and mountain man/writer Joaquin Miller. (Photo Barbara F. Wolf.)

influx of non-Indians to the area, leading to conflicts deadly mainly for the tribes. Today, conflict persists over the appropriate uses of the mountain and surrounding landscape as sacred area and public domain (Huntsinger and Fernández-Giménez 2000; Little 2001; Sacred Land Film Project 2002; Mount Shasta Bioregional Ecology Center 2004).

In the Shasta region, a greater resource than gold was the region's dense forests. The present towns of Weed, City of Mt. Shasta, McCloud, and Dunsmuir all owe their existence to the timber industry in one way or another. Today the mountain's higher slopes and summit are within the 8,500-km^2 Shasta-Trinity National Forests (USDA Forest Service 1997). The Mt. Shasta Wilderness was established in 1984 and now comprises 160 km^2 from the 10,000-ft. (about 3,000-m) contour to the top (USDA Forest Service 2001). Little logging takes place on the mountain, but surrounding U.S. Forest Service lands are checkerboarded with private holdings, where cutting continues.

Before we trace the history of the lumber towns, it bears noting that John Muir, who, like many other late nineteenth-century nature writers, found an enduring, transcendent value in wilderness, climbed Mt. Shasta with a companion in April 1875 to survey the summit, following a route that led them up Whitney Glacier. Trapped overnight close to the summit by an unexpected snowstorm, they took refuge among volcanic hot springs, where the sulfuric "hissing, sputtering fumaroles" helped them avoid freezing to death. The clouds suddenly passed, revealing starry skies of great beauty. In a stupor with cold and exhaustion, Muir and his companion had "blissful visions" of extraordinary imaginary lands, survived the night, and hiked down safely the next morning (Muir 1877). The mountain provided profound experiences to other seekers as well in subsequent decades, and a culture of "spiritual tourism" has evolved on Mt. Shasta (Figure 4.5).

WEED

In the 1860s Maine lumberman Abner Weed noticed the strong drying winds at a location on the west side of Mt. Shasta that later became the town bearing his name. With countless board-feet of timber nearby, it was ideal for drying lumber, and in 1897 Weed acquired land and a mill there. The Weed Lumber Company was established in 1903, and within two years Weed was a thriving mill town, albeit with a mostly male population (*Weed Press*, Summer 1993). The company changed hands a number of times through the twentieth century but operated until 1982, when International Paper (IP), which had acquired the mill in 1956, shut it down. Weed's population hit its twentieth-century zenith of 4,686 in 1968; in 2000 it was 2,978.

Weed struggles to reach adequate levels of employment (Sellers 1988). The College of the Siskiyous, a two-year state college that opened in 1958, provides both direct employment opportunities and job training (Sellers 1988). Roseburg Lumber, a timber products company with a veneer mill in Weed, took over the IP mill site and offers some of the last forest-products industry jobs. Attempts have been made at tourism promotion, but Weed does not compete well with the City of Mt. Shasta five miles south. There is no appealing central street with lodgings, shops, and restaurants and no easy road access up the mountain. Lake Shastina, a "community service district" of Weed about 30 miles distant, is growing with retirement and vacation homes, contributing to the figure of 5,000 people living "just outside the city limits" (Weed Chamber of Commerce 2004). Mt. Shasta's role as a resource for Weed consists of providing residents with pure water and a beautiful view, which from some locations includes part of the Whitney Glacier. The mountain graces the label of locally bottled Crystal Geyser Spring Water, which opened a plant in Weed in 1998 and draws water from the same source spring from which Weed gets its water supply. A water company executive said that the water and the majestic beauty of the mountain are what brought this national business to little Weed. With water bottling plants already in Mt. Shasta and Dunsmuir and one planned for McCloud, the mountain's water appears to be the focus of the next extractive resource boom for the region. The environmental impacts and economic benefits of water bottling are of concern to some in each of the communities, but most people interviewed in Weed had a positive view of the company's presence.

CITY OF MT. SHASTA

The City of Mt. Shasta became the mountain's namesake only in 1924, after a popular vote in 1922. When the first post office opened in 1870, it was called Berryvale. Since 1886 it had been known as Sisson after Justin Hinckley Sisson, a businessman who arrived in the 1850s (Arnold 2004; Mount Shasta Visitors Bureau n.d.) and settled in the area then known as Strawberry Valley. He eventually expanded his home into a popular inn and tavern that served stagecoach travelers, initiating the region's tourist trade. It was Sisson's efforts and his donation of land to the Central Pacific Railroad for the station and town site that brought the railroad in 1886. The new city was, not surprisingly, called Sisson, and as part of the deal streets were named after Sisson's family members (Mount Shasta Visitors Bureau n.d.).

From the start, the city's economy was more diversified than that of neighboring towns. The railroad brought in tourists and other business and enabled the transport of natural resources such as lumber and hatchery-raised fish, as well as the locally brewed beer. The fish hatchery, built on Sisson's land in 1888, is still operated by the State of California Department of Fish and Game. Hotels, bars, and restaurants catered to woodsmen with a weekend off and pay in their pockets. Local lumber also supplied mills and manufacturers.

Mt. Shasta Boulevard and downtown streets today sport cafés and restaurants, outdoor outfitters, galleries, real estate offices, bed-and-breakfasts, and gift shops catering to both conventional and New Age visitors. The mountain attracts spiritual seekers who may arrive individually, in workshop groups, or en masse for events such as the Harmonic Convergence of August 1987 and the Rainbow Gathering of August 1995. Several businesses cater to them with activities described as sacred-site treks, guided vision quests, and healing earth journeys. Signs point the way to Mt. Shasta City Park, the site of Big Spring (see Figure 4.6). Though the locals' claim that the spring is the headwaters of the Sacramento River has no basis in hydrology (Gabrukiewicz 1999), the spring is a tributary to the great river and a significant local attraction as well. Surrounded by woods and lush shrubbery, it flows from rock and earth, and both locals and tourists fill jugs with confidence in the water's purity.

The mountain dominates the view from downtown, rising 3,300 m above the town, which, like Weed and McCloud, sits on old lava flows at an elevation of about 1,100 m. There are few places from which its ever-changing aspect cannot be seen. The City of Mt. Shasta

FIGURE 4.6. Big Spring in Mt. Shasta City Park, thought by some to be the headwaters of the Sacramento River. (Photo Barbara F. Wolf.)

is the main access point to the mountain, with the convenience of an interstate-highway exit and the longest paved road up the mountain. The road winds through forest past Forest Service campgrounds, trailheads, and tree plantations to a parking lot above treeline at the base of an old ski area. The "Ski Bowl" operated from 1959 to 1978, when it was destroyed by an avalanche that some Native Americans and Euro-Americans viewed as a sign from the mountain. A proposal in 1988 to build an elaborate new ski resort development generated a heated controversy that ended in 1998 when the Forest Service, which administers the mountaintop as a wilderness area, denied the permit. The City of Mt. Shasta's population has been growing steadily since the 1930 census, to the 2000 figure of 3,621 (Figure 4.7).

DUNSMUIR

Dunsmuir is a narrow linear town nestled in the canyon of the upper Sacramento River, along the railroad tracks that line the riverbank and parallel U.S. Interstate 5. The town's history is tied to the railroad, first the Central Pacific and then

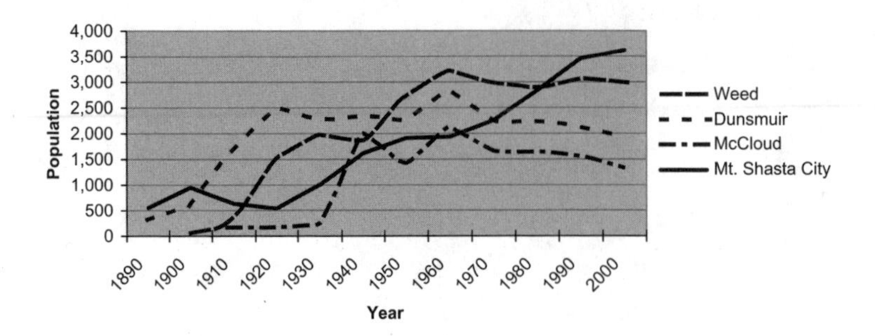

FIGURE 4.7. Population change in the Mt. Shasta region.

the Southern Pacific, which acquired it in 1885. Dunsmuir was originally called Pusher because it was the place where locomotives were attached to trains to push them up the steep northbound grade. A Canadian coal baron passing through was, however, taken by the location and the people and promised that he would donate a fountain if they named the town after him. The town's name was changed in 1887, and the fountain arrived; it now stands in the city park (Dunsmuir Chamber of Commerce n.d.). Donor Alexander Dunsmuir never returned.

In 1916 Dunsmuir became the headquarters of the Shasta Division of the Southern Pacific Railroad, which ran from Gerber, California, to Ashland, Oregon. It is now the nearest Amtrak station to Mt. Shasta. The railroad still runs regularly through town, snaking alongside the steep canyon wall. Train bells, whistles, and chuffing create a noisy backdrop. The railroad gives the town a promotional tag line: "historic railroad town." Although a railroad museum is no longer operating, the hardware store ("Dunsmuir's Oldest Retail Business—Since 1894") doubles as one, with shelves and walls lined with railroad photographs, memorabilia, tools, and assorted other antiques alongside diverse modern merchandise.

Locals proudly say that one can get down to the river from anywhere in town. Visitors pay local fishing guides to take them to the best spots along the Sacramento and its tributary, the McCloud. As important as the river is to the town's heritage and identity, it is the pure spring water that flows from Mt. Shasta that gives it its second promotional tagline, "home of the best water on earth" (Dunsmuir Chamber of Commerce 2004). The water emerges in several places as waterfalls, which are popular tourist attractions. Dunsmuir, too, has (or had) its water bottler. A local company marketed Dunsmuir water under the Castle Rock Spring Water label but was bought out by Danone/Coke, which now pipes the water to its Mt. Shasta bottling plant. The parent company, Group Danone, bottles glacier-derived water in France under the Evian brand.

The attraction of the waters is not just recent; the railroad used to stop at Shasta Springs, between Dunsmuir and Mt. Shasta City. There travelers would debark and get a drink of the naturally carbonated spring water or stay at a luxury resort and hotel. Eventually the waters from Shasta Spring were bottled and, later still, shipped in tank cars to San Francisco. With added flavoring and recarbonation, the enterprise in time became Shasta Soda, a brand that still exists but has no remaining links to its origins as mountain spring water. Dunsmuir's population grew from 1,719 in 1910 to a high of 2,873 in 1960, declining gradually to 1,923 as of the 2000 census.

McCLOUD

McCloud was born as a company town and remained so until the 1970s. It was dominated by a regional lumber and railroad company whose lines tied into the much larger Southern Pacific Railroad. The McCloud River Railroad Company, later the McCloud River Lumber Company, literally built the town, founded in 1897. Only company employees could live there, and they came to view the company as their beneficent caretaker, denoted by the nickname "Mother McCloud" (Keith and Niemann 2002). With the sale of the mill, railroad, and town to U.S. Plywood in 1963 and the latter's subsequent merger with Champion International, the parental relationship came to an end. Champion put its McCloud holdings up for sale in 1965, and residents lost the security of a company that had seen them through the Depression and employed generations of family members. Even though residents were allowed to purchase their homes at low cost, an economic development report characterized the town's transition as going "from being wards of the Company to wards of the County" (Keith and Niemann 2002). The Champion lumber mill closed for good in 1979 after logging off its remaining timber holdings in the 1960s and 1970s. Another company purchased and operated the mill on a smaller scale for a few years but has since closed. With mill jobs gone, many people left town, and in 2000 nearly 25% of the population were 62 or older (U.S. Census Bureau 2000). From a population high of about 2,140 in 1968, McCloud's residents as of 2000 numbered 1,343 (U.S. Census Bureau 2000; Keith and Niemann 2002).

McCloud has made efforts at developing tourism. Several buildings have been renovated into bed-and-breakfasts. The Heritage Junction Museum showcases items from the town's history, featuring newspaper clippings, photos, artifacts, clothing, and equipment of the lumber business among oddities such as permanent-wave machines, old typewriters, and an antique barber chair. The McCloud River Railroad Depot now hosts a gift shop in a railroad car, and tourists can ride a dinner train through the forest. The old Mercantile Building has been renovated and houses shops, galleries, and a café, with room for additional offices and businesses. The Ski & Board Park opened in the mid-1980s and provides some year-round and several hundred seasonal jobs. Lower on the mountain than the controversial proposed Ski Bowl resort development (see following), the Ski Park did not have significant opposition and draws skiers from as far as San Francisco and Medford, Oregon.

The mill property was sold in 2003 to Nestlé Waters of North America, which plans to construct a spring-water bottling plant, joining those in Weed, Mt. Shasta, and Dunsmuir (Benda 2003, 2004) and reinforcing the impression that water is the new timber in this region. As a local real estate agent commented, "The mountain will supply commerce again, indirectly—water bottlers."

OVERVIEW

In sum, the inhabitants of the Mt. Shasta region have drawn their livelihood from natural resources, particularly lumber and water, which are both abundant and, given the location near the railroads and roads that link California with the Pacific Northwest, easily shipped to areas of demand. In the lumber era, the towns experienced a high degree of dependency on particular corporations. Tourism, stimulated by the mountain's forests, streams, and visual beauty and also facilitated by transportation links, has grown in recent decades. The presence of Mt. Shasta has supported these economic activities. The glaciers contribute, in ways that are significant though difficult to measure precisely, to the regional hydrology and the distinctive beauty that support these economic activities.

LOCAL PERCEPTIONS OF MT. SHASTA

We used three principal methods to examine local residents' perceptions of Mt. Shasta. The first was a survey with both open-ended and structured questions. Though the sample of only 33 individuals was small, it was large enough to allow useful descriptive statistics and certain

TABLE 4.2

Number of Respondents Reporting Values of Mt. Shasta by Birthplace

	BEAUTY	SPIRITUALITY	WATER	SNOW	GLACIERS
Locally born ($n = 6$)	1	1	4	4	3
Migrant ($n = 23$)	12	13	7	21	19
Unidentified ($n = 4$)	0	1	3	3	1
Total ($n = 33$)	13	15	14	28	23

statistical tests. The second was an analysis of the two major controversies associated with the mountain in recent years: over the development of infrastructure on the mountain and over commercial bottling of water. The third was an analysis of print and electronic archives of materials about the mountain, including records of meetings of public agencies and community groups.

SURVEY RESULTS

Our survey population of 33 included 23 individuals born outside the region; 6 individuals born in Weed, Mt. Shasta City, McCloud, Dunsmuir, or the nearby county seat of Yreka; and 4 who did not mention their birthplaces. When asked to describe their recreational and leisure activities, the respondents indicated considerable participation, reporting about two activities per respondent. Though there were many activities mentioned by only one or two people (e.g., gathering sage, searching for Indian artifacts), the five most common activities were skiing (21), hiking (19), climbing (11), fishing (8), and meditating (4). The mountain supports all these activities well. It is the major site in the region with good snow-covered slopes for skiing and a high peak for climbing. According to participants, its peaceful beauty and spiritual power make it a preferred site for meditation. It also offers attractive trails for hiking and feeds streams for fishing.

We invited the respondents to talk about the importance of the mountain in general and allowed them to speak as long as they wished; we also asked them whether the white summit of the mountain was important or not and let them respond in an open-ended fashion. We then coded their responses for the presence or absence of specific values (Table 4.2). The values that they mentioned clustered in five attributes of the mountain: its beauty, its spiritual significance, the water that it provides, the snows that cover it, and the glaciers near its summit. Of these, the most frequently mentioned were the snow and the glaciers; this may in part be a consequence of our specific question on "whiteness" and clues about our interest in glaciers that we may have unconsciously given. (When asked, later in the interview, whether there were glaciers on Mt. Shasta, 26 of the 33 respondents gave a positive answer.) Nonetheless, we note some patterns[2] in these responses. The migrants were more likely ($P < 0.1$) than locally born individuals to report spirituality and glaciers as values and less likely ($P < 0.1$) to report water as a value; there were no significant differences in their likelihood to report snow or beauty. These results are consonant with the strong "New Age" and environmentalist orientation of many newcomers and the intimate familiarity that locally born residents have of the region's long history of reliance on resource extraction with its cyclical downturns. These two groups shared a deep appreciation of the mountain's striking beauty; they also both included many avid skiers and many individuals who earned a living by providing services to skiers or for other outdoor activities.

TABLE 4.3

Number of Respondents Reporting Levels of Concern for Different Potential Environmental Changes

	LOSS OF SNOW COVER	GLACIAL RETREAT	VOLCANIC ACTIVITY
Very concerned	20	15	5
Somewhat concerned	12	14	24
Not concerned	1	4	5
Total	33	33	33

We also asked questions about the level of concern about several environmental changes: loss of snow cover, glacial change,[3] and volcanic activity (Table 4.3). Though there is no significant difference between overall levels of concern for the first two changes, the respondents were significantly ($P < 0.05$) less concerned about volcanic activity.[4] As we noted in an earlier section, there is a genuine risk of volcanic eruption. The lack of concern about it may reflect the fact that the history of European and American settlement in the region is quite short and that Native American knowledge of eruptions of Mt. Shasta in the late eighteenth century has generally not passed into wider knowledge. Nonetheless, the possibility of volcanic danger could have been raised by the two major eruptions in the same mountain chain in the twentieth century (Mt. Lassen, only 120 km to the southeast, from 1914 through 1917, and Mt. St. Helens, 530 km to the north, in 1980). Greater concerns may stem from the high year-to-year variability in snow levels. Another source of concern is the general reduction in winter snow at lower elevations on Mt. Shasta, which has caused the white covering of the mountain to become smaller, despite the increase in length of several glaciers higher on the mountain (Howat et al. 2006). The extensive public discussions of global warming may also have increased concerns about snow and glaciers. Examining the responses of the interviewees by birthplace, we note that the people who migrated in to the region have a significantly higher level of concern about glacial retreat, as measured on a three-point scale, than the locally born individuals ($P < 0.1$). We found no other significant differences in concerns or activities.

CONTROVERSIES

Two major controversies in recent decades also demonstrate public perceptions of Mt. Shasta. They reveal the strong commitments of different groups in the region to particular features of the regional environment and to the long-term future of the region. They reflect economic interests and other cultural values as well. Some residents seek to attract jobs to the area through the development of new commercial activities, while others oppose such efforts in favor of preserving the natural integrity of the region. The distrust of outside corporations, developed during the collapse of the lumber companies that once seemed to be a permanent element of the social landscape, also contributes to a concern about new activities. Local people, who directly observe the changes in the mountain's cover of snow and who hear of global warming and other environmental problems through the media, also fear that the mountain is vulnerable to negative change.

THE USE OF PUBLIC LANDS. Since the nineteenth century, Mt. Shasta has seen a series of disputes between supporters of commercial use and conservationists. The latter have argued for preservation and protection against excessive construction and timber extraction. As early as 1888, a bit more than a decade after his cold night on the mountain, John Muir proposed that Shasta be protected as a national park on a par with Yosemite and Yellowstone. In 1889 the forester Charles Shinn, the first supervisor of

Shasta National Forest and an apprentice of the noted conservationist Gifford Pinchot, made a similar plea that the mountain be protected before it was logged off completely (Mount Shasta Companion 1888; Shinn 1889)—an unusual position for a forester of his era or those to follow. Shasta National Forest was established by proclamation in 1905 by President Theodore Roosevelt. A bill for national park status for Shasta, which would have protected the mountain against resource extraction, was pending in 1914 when the nearby volcano Mt. Lassen erupted, diverting attention from Shasta and garnering national park status for itself.

More recently, a citizens' group calling itself the Mount Shasta Resource Council formed in 1976. It published a newsletter, *Shasta*, that highlighted such concerns as the expansion of the ski area, recent glacier activity, drought, and a USDA Forest Service wilderness proposal that was in progress (Rhodes 1977*a*). Philip Rhodes, editor of *Shasta* and author of several scientific articles about Mt. Shasta's glaciers, argued for the group's alternative wilderness proposal in a Sierra Club bulletin in 1977 (Rhodes 1977*b*), making a case against the expansion of the Shasta Ski Bowl and calling for public comment on the Forest Service proposal. Wilderness designation was finally achieved in 1984, protecting 155 km^2 (USDA Forest Service 2001).

In 1988, a developer proposed a ski resort with a number of condominiums at the site of the Old Ski Bowl at 2,400 m, with plans for ski runs through Panther Meadows, a site considered sacred by the Wintu and other tribes in the region and highly valued and extensively used by other visitors. Proposed for a site on Forest Service land, the project called for major landscape modification, including significant removal of rock and soil from buttes, in addition to the construction. Michelle Berditschevsky, a local resident, founded an organization named Save Mount Shasta to challenge the project. She formed alliances with five Native American tribes for which the mountain has major cultural significance and drew in major environmental organizations as well, including the Sierra Club, the Wilderness Society, and the California Wilderness Coalition. Community sentiment was divided, with some individuals favoring the project, largely on grounds of its economic benefits, and others opposing it for its negative environmental impacts and for its assault on the aesthetic and spiritual qualities of the mountain. After 10 years of public hearings and heated debate, in 1998 the Forest Supervisor rejected the proposal, a decision later upheld by the Regional Forester and not appealed by the developer (Berditschevsky 1998). This battle appears to have been fought largely in the court of public opinion. The federal National Environmental Policy Act of 1969 promotes public comment on the development of federal lands and provided channels for the expression of such comment.

Save Mount Shasta evolved into the Mount Shasta Bioregional Ecology Center, an umbrella organization that sponsors projects related to its core goal of "protecting the environment of Mt. Shasta and its surroundings" (Mount Shasta Bioregional Ecology Center 2004). It extends this protection to include cultural and spiritual as well as ecological values. The Center collaborates with residents, Indian tribes, environmental organizations, and the Forest Service and promotes the active participation of citizens in public debate and in the management of the mountain.

In 1994, in the course of the challenge to the ski area development, Mt. Shasta was declared a National Historic District from the 4,000-ft. level to the summit, making it eligible for the National Register of Historic Places and the protections that come with that designation. Later that year, political and business interest pressure resulted in reducing the National Historic District by nearly 90% to about 75 km^2, starting above the treeline at 8,000 ft. This excluded many areas of cultural importance to the Native American tribes that had worked along with other organizations to obtain the designation, as well as important plant communities and geological features (Berditschevsky 1998).

In February 2006 the Forest Service approved a "mountain-thin" project that involves timber harvest and sale on the mountain. The project seeks to reduce the risk of forest fires, both at lower elevations near towns and a ski facility and at sites higher on the mountain. The Save Mount Shasta project opposed the "thin." Though recognizing the need for fuels reduction in both areas, the group was concerned about lack of emphasis on wildlife and cultural values in the environmental assessment for the plan. The Mount Shasta Bioregional Ecology Center gave comments on the proposed sale from its inception in 2001. Its appeal of the assessment in September 2004 was denied (Berditschevsky and McCoy 2004). As evidence of implementation appeared on the ground in fall of 2005, Save Mount Shasta maintained its vigilance and continued appeals to the district ranger to make sure that management decisions respected the spirit of statements in the environmental assessment about preserving wildlife habitat, diversity, big trees, and cultural values. It may have a new legal tool available to press its cause with a 2004 U.S. District Court ruling against the Forest Service for failing to provide adequate information to the public before approving three other timber sales in the region. The organization argues that the mountain-thin project similarly failed to provide timely information to the public.

COMMERCIAL WATER BOTTLING. Given the decline of the timber industry, the former mainstay of the regional economy, it might be expected that local residents would welcome the entry of a new resource extraction industry, especially a relatively clean and spatially compact one such as water bottling. Serious concerns have arisen, however. As early as 1977, the mayor of the City of Mt. Shasta refused a water bottler's request for access across city property to obtain water rights from Big Spring for a small bottling plant (*Herald Press News*, October 2, 2002). The applicant eventually built a plant in 1989 on the edge of the steep canyon where he taps an underground river (Breitler 2004).

Though water bottling had been going on in Dunsmuir since the late nineteenth century, city officials and community members objected to a proposal before the Siskiyou County Planning Department in 2003 for a water bottling plant to be built on private property, citing concerns about its proximity to the city's water supply (an underground source originating on Mt. Shasta) and the impacts of the septic tank storage that would be required (Bolender 2003). This request was presented again before the community in 2005 and remains controversial.

In 2001, about 15 Mt. Shasta residents commented before the California regional Water Quality Control Board on a request by Danone Waters of North America to discharge bottle rinse water onto a leaching field on its property above Big Spring. Their concerns included increased traffic and noise, impacts on wetlands, and potential loss of water rights, as well as the bright lighting at the plant, which is located adjacent to a residential area (Kennedy and Gerace 2001). The plant opened in January 2002, and at a July hearing that year residents offered more testimony opposing the discharge plan. A petition to the water board signed by 223 residents called for further environmental review, citing concerns for fish in a nearby stream should the permit allow Danone to put in more wells and withdraw more water (Sanchez 2001; Hearden 2002).

Residents have addressed broader ethical matters as well as ecological and quality-of-life concerns about water bottlers. Local control and the privatization of a resource essential for human life are issues in the City of Mt. Shasta and McCloud (Breitler 2004), as they are for communities throughout the arid regions of the western United States. Local residents express concern about their exclusion from the decision to sell water that they view as a vital public and natural resource (Lowe 2005) and about the unwillingness of bottlers to share information about their business (Breitler 2004).

Respondents to our interviews mentioned the potential for water extraction to affect residents' wells and lower the water table. The water bottlers say that they withdraw a very small fraction of the underground flow, and many believe them. A businessman in McCloud pointed out that the Nestlé plant would use far less water than the mill did to wet down its stacked lumber 24 hours a day. Nonetheless, concern remains because of the scientific uncertainty about the hydrology of the region; little is known about the movement of water from higher regions on the mountain to the water table or spring at which commercial extraction takes place. (The growth of glaciers high on the mountain may or may not compensate for the reduction in snow volume at lower elevations.) Proponents, however, see the water plants as clean, sustainable economic development and an important source of jobs in their underemployed communities. Moreover, more than one resident in each of the three towns with bottling plants commented smugly that what others have to pay for, they get free just by turning on their taps.

Property taxes from Weed's Crystal Geyser and Danone's plant in Mt. Shasta contribute about $730,000 yearly to the county (Breitler 2004); Nestlé's plant in McCloud is predicted to add about $1 million more, and in addition the company will make payments to the McCloud water district, pay an exclusivity fee, and $100,000 each year to a community enhancement fund (Kennedy 2003; Benda 2004). Nestlé officials claim that the plant could create about 240 jobs (Carlton 2005). Despite the apparent economic benefits, there is strong opposition to the Nestlé plant in McCloud. Some residents were "furious" that the McCloud Community Service District (a utility) signed the contract before an environmental impact study was made to assess what potential impact the plant would have on their springs (Carlton 2005). Concerned McCloud Citizens, another project of the Mount Shasta Bioregional Ecology Center, challenged the Nestlé contract in court. In March 2005, the Siskiyou County Superior Court set aside the contract, declaring it in violation of the California Environmental Quality Act because an environmental review had not been completed prior to the signing. Nestlé appealed the decision up through the courts and was rejected each time. In the summer of 2006, as Nestlé worked with the county and other state entities on mitigation plans, citizens were awaiting the draft environmental impact report, which will finally give them an opportunity to respond formally to the project.

CONCLUSIONS

Mt. Shasta's forests attracted transportation and lumber corporations in the late nineteenth century. The residents of the company towns that grew up alongside railroad stations and lumber mills generally favored commercial development of the mountain, though they also participated in fishing, hiking, and other forms of recreation. The mountain's size and elevation, the beauty of the region, and the pure waters have attracted visitors since early on, especially once railroad access became available. The role of glaciers in these economic activities is hard to measure precisely but is likely to be small, though they do contribute to the flow of the streams and rivers and may help maintain the groundwater lower on the mountain. Furthermore, the mountain's unusual size and appearance have evoked environmentalist, aesthetic, and spiritual attention since the late nineteenth century as well. These aspects have also drawn people to the region, and some have settled there permanently. These strands persist today, as witnessed by the new forms of tourism and the water bottling firms. Their views engage with each other, stimulating debates that are often quite open, since much of the land around Mt. Shasta is publicly owned and land use decisions consequently are made by public agencies that are open to scrutiny and that invite citizen participation.

At present, the residents and visitors value the mountain and its environs for a variety of reasons—spiritual, recreational, economic, and ecological. They report intermediate or high levels of concern for loss of snow and

FIGURE 4.8. Both the city and business owners utilize the image of Mt. Shasta on signage and advertising. (Photo Barbara F. Wolf.)

glaciers. Most concretely, they notice changes in the mountain primarily in terms of snow cover: How brown is the mountain? The length of the snow season has economic importance for the ski and climbing businesses, which are the main economic activities during winter, and for the safety of those participating. The aesthetics of the mountain is important as well, though many also appreciate its beauty as it changes through the year. Summer is the major tourist season, and visitors welcome the view of the mountain at this time, though few pause to consider whether the white matter on its summit is snow or ice. Shopkeepers and other business owners recognize the importance of this white cap and depict it prominently on their signage and advertising (Figure 4.8). The glaciers have an important visual and symbolic appeal, keeping the summit white throughout the year, though they do not directly influence visitors' activities or their choice to come to the Mt. Shasta region. Recreational assets and the quality of life in this beautiful, rural, safe environment are more often cited as the reasons visitors come to Mt. Shasta and the reasons many who come to visit never leave. In sum, Mt. Shasta's glaciers, the largest in the state and among the few in the world that are advancing, make a genuine contribution to the resources and risks that shape local economies and to the valued elements that influence local perceptions. Though it is difficult to measure, it is likely that glaciers offer a smaller contribution to these resources, risks, and perceptions than rainfall, groundwater, and snow that does not turn into glacial ice.

ACKNOWLEDGMENTS

We would like to thank Dennis Freeman, the library director and archivist of the Mount Shasta Collection at the College of the Siskiyous, Weed, California, Jenn Carr of Shasta Mountain Guides, Michelle Berditschevsky of the Mt. Shasta Bioregional Ecology Center, the people at the Blue House, and all the individuals who gave generously of their time for interviews.

REFERENCES CITED

Arnold, K. 2004. Sisson. Siskiyou County Sesquicentennial Committee. http://ww.siskiyouhistory.org (accessed September 27, 2004).

Benda, D. 2003. Residents don't bottle criticism. *Redding Record Searchlight*, October 1.

———. 2004. Nestlé OKs deal for plant. *Redding Record Searchlight*, July 24.

Berditschevsky, M. 1998. "No-ski" decision holds on Mount Shasta. Mount Shasta Bioregionai Ecology Center. http://www.mountshastaecology.org/12medicinelake01chron.html (accessed January 9, 2005).

Berditschevsky, M., and J. McCoy. 2004. Mountain thin project approved. *Eco Echo*, Summer/Fall, 5–6.

Biles, F. 1989. *Glaciers of Mount Shasta*. CSU Chico Regional Information Circular 89-1. Chico: California State University.

Blodgett, J.C., K.R. Poeschel, and W.R. Osterkamp. 1996. Characteristics of debris flows of noneruptive origin on Mount Shasta, northern California. Open-File Report 96-144 USGS. http://ca.water.usgs.gov/shasta (accessed June 1, 2004).

Bolender, E. 2003. Concerns voiced about Dunsmuir bottling. *Herald Press News*, March 12.

Breitler, A. 2004. Bottlers tap Mt. Shasta for a river of wealth. *Redding Record Searchlight*, March 28.

Callaghan, C.J. 2000. Debris flow initiation conditions on Mount Shasta, California. Master's thesis, University of Nevada.

Carlton, J. 2005. Can a water bottler invigorate one town? *Wall Street Journal*, June 9.

Dreidger, C.L., and P.M. Kennard. 1986. Ice volumes on Cascade volcanoes: Mount Rainier, Mount Hood, Three Sisters, and Mount Shasta. *U.S. Geological Survey Professional Paper* 1365:1–25.

Dunsmuir Chamber of Commerce. 2004. *Come to Dunsmuir . . . Come home*. Dunsmuir, CA.

———. n.d. *Welcome to Dunsmuir*. Dunsmuir, CA.

Eargle, D.H., Jr. 2000. *Native California guide*. Vol. 1. San Francisco: Trees Company Press.

Frank, Emille A. 1977a. Mud Creek: On the rampage again? *Dunsmuir News*, August 3.

———. 1977*b*. Wintun Glacier is acting up too. *Dunsmuir News*, August 17.

———. 1990. Notorious, rampageous Mud Creek. *Voice of the Mountain*, March 21.

Gabrukiewicz, T. 1999. River of life: River's origin is a source of debate. *Redding Record Searchlight*, September 12.

Guyton, B. 1998. *Glaciers of California: Modern glaciers, Ice Age glaciers, origin of Yosemite Valley, and a glacier tour in the Sierra Nevada*. Berkeley: University of California Press.

Hearden, T. 2002. Bottling plant seeks OK for discharge plan. *Redding Record Searchlight*, July 8.

Hirt, W. n.d. *Mt. Shasta, the legendary volcano*. Mt. Shasta, CA: Mount Shasta Visitors' Bureau.

Howat, I., S. Tulaczyk, P. Rhodes, K. Israel, and M. Snyder. 2006. A precipitation-dominated, mid-latitude glacier system: Mount Shasta, California. *Climate Dynamics*, doi:10.1007/s00382-006-0178-9.

Huntsinger, L., and M. Fernández-Giménez. 2000. Spiritual pilgrims at Mount Shasta, California. *Geographical Review* 90:536–58.

Keith, D., and E. Niemann. 2002. Northwest Economic Adjustment Initiative Assessment: McCloud, Siskiyou County, California. Sierra Institute for Community and Environment (Formerly Forest Community Research). http://www.sierrainstitute.us/neai/CA_case_studies/McCloud_CA.pdf (accessed November 19, 2006).

Kennedy, K. 2003. McCloud approves water bottling contract. *Herald Press News*, October 1.

Kennedy, K., and S. Gerace. 2001. CEQA document, permit approved for Danone (September 12). Siskiyou County Economic Development Council, Inc. http://www.siskiyoucounty.org/articles/sept12.htm (accessed May 6, 2004).

Le Guellec, M. 2004. Nestlé addresses water rights. *Mount Shasta Herald*, April 14.

Leviton, A.E., and M.L. Aldrich, eds. 1997. *Theodore Henry Hittell's California Academy of Sciences: A narrative history 1853–1906*. San Francisco: California Academy of Sciences.

Little, C.E. 2001. Keeping a place for prayer. In *Sacred lands of Indian America*, ed. J. Page, 48–53. New York: Harry N. Abrams.

Lowe, Diane. 2005. Update on Nestlé bottling plant threatening Mount Shasta's aquifer. *Eco Echo*, winter 2005–2006.

Miesse, William. 2005. Mount Shasta fact sheet. College of the Siskiyous. http://www.siskiyous.edu/library/shasta/factsheet/ (accessed November 18, 2006).

Mount Shasta Bioregional Ecology Center. 2004. *Eco Echo*, Summer/Fall 2004.

Mount Shasta Companion. 1888. Mt. Shasta. In *Picturesque California: The Rocky Mountains and the Pacific Slope: California, Oregon, Nevada, Washington, Alaska, Montana, Idaho, Arizona, Colorado, Utah, Wyoming*, ed. J. Muir, 231–32. New York: J. Dewing. Mount Shasta Companion, College of the Siskiyous Library. http://www.siskiyous.edu/library/Muirnationalparkproposal.htm (accessed August 5, 2007).

———. 1997. Glacial history of Mt. Shasta: Existing glaciers. Mount Shasta Companion, College of the Siskiyous. http://www.siskiyous.edu/shasta/env/glacial/exi.htm (accessed August 4, 2007).

Mount Shasta Visitors Bureau. n.d. *History of Mount Shasta City.* Mt. Shasta, CA.

Muir, J. 1877. Snow-storm on Mount Shasta. *Harper's New Monthly Magazine* 55:521–30.

———. 1918. A perilous night on Shasta's summit. In *Steep trails.* Sierra Club. http://www.sierraclub.org/john_muir_exhibit/frameindex.html?http://www.sierraclub.org/john_muir_exhibit/writings/steep_trails/ (accessed June 30, 2004).

Powell, J.W. 1885. *Fifth annual report of the United States Geological Survey to the Secretary of the Interior, 1883–'84.* Washington, DC: U.S. Government Printing Office.

Rhodes, P.T. 1977a. *Forest Service plans wilderness study.* Mt. Shasta, CA: Mount Shasta Resource Council.

———. 1977*b*. One last chance for Shasta. *Sierra Club Bulletin,* 49–51.

———. 1987. Historic glacier fluctuations on Mount Shasta. *California Geology* 40:205–11.

Sacred Land Film Project. 2002. The Wintu and Mt. Shasta http://www.sacredland.org (accessed June 4, 2004).

Sanchez, G. 2001. Siskiyou bottler gets permit: Opponents fear water quality will be affected. *Redding Record Searchlight,* September 1.

Save McCloud's Water Group. 2004. McCloud-Mount Shasta area spring water signed away. *Eco Echo,* Summer–Fall.

Sellers, T. 1988. Weed is lifting itself by its COS bootstraps. *Redding Record Searchlight,* August 4.

Shinn, C.H. 1889. The forest: Among the Siskiyou forests. *Garden and Forest* (2)94:598. http://www.siskiyous.edu/Library/Shasta/nationalparks/siskiyouforests.htm (accessed August 3, 2004).

Stuhl, E. 1924. The Mud Creek rampage of 1924. Ms, Mount Shasta Collection, College of the Siskiyous Library, Weed, CA.

Theodoratus, D.J., and N.H. Evans. 1991. *Statement of findings: Native American interview and data collection study of Mt. Shasta, California, USDA, Shasta-Trinity National Forests.* September. Redding, CA: USDA Forest Service.

U.S. Census Bureau. 2000. Profile of general demographic characteristics: McCloud CDP, CA. U.S. Census Bureau. http://censtats.census.gov/data/CA/1600644784.pdf (accessed July 2004).

USDA Forest Service. 1997. *Shasta-Trinity National Forests.* Redding, CA.

———. 2001. *A guide to the Mt. Shasta Wilderness and Castle Crags Wilderness.* Redding, CA.

U.S. Geological Survey. 1997. *The 1997 Whitney Creek debris flow.* Mount Shasta Collection, College of the Siskiyous, Weed, CA.

Vance, L. 2004. McCloud water forum provides education and invites public sentiment. *Eco Echo,* Winter.

Weed Chamber of Commerce. 2004. About the city. http://www.weedchamber.com (accessed September 27, 2004).

5

Glacial Hazards

PERCEIVING AND RESPONDING TO THREATS IN FOUR WORLD REGIONS

Christian Huggel, Wilfried Haeberli, and Andreas Kääb

Glacier-related hazards are common in high-mountain regions. Climate change and intensification of human activities in mountains have attracted increasing interest to the related risks, where risk is understood as a function of hazard (probability) and damage potential (e.g., Fell 1994). The management of these risks is a complex task involving analysis, evaluation, and communication as well as prevention and mitigation (Greminger 2003). In Switzerland, which has a comparatively long tradition of dealing with glacial hazards, research has focused on assessing these hazards and the related risks (Haeberli et al. 1989; Huggel et al. 2004) and includes risk evaluation approaches involving local stakeholders (Wegmann et al. 2004). In many places in the Swiss mountains, the glacial-risk debate represents an area of tension or conflict between different economic or political interests. How the risks are perceived is therefore often governed by different individual perspectives. Taking our experiences in Switzerland as a basis, we have contributed to several hazard assessment and management studies in other high-mountain regions of the world and have observed significant differences in the way glacial hazards and the related risks are dealt with.

This article is a physical scientist's view of glacial disasters and the responses of local populations and authorities in four high-mountain areas of the world. Local responses to hazards depend on hazard and risk perception by different actors. Although we cannot provide sufficient material for detailed insight into the local perception of hazards, it may be useful to refer briefly to some related issues. Risk and the perception of risk constitute a field in which physical or engineering science meets social science. Although this may represent an opportunity to bridge the two cultures, it has been an area of controversy (Renn 1998). The approaches to risk perception adopted by technical experts have often contrasted with those of social scientists. Many social scientists have argued that technical, probability-based risk assessments overlook the fact that risk is a social construction rather than a representation of real hazards (e.g., Jasanoff 1993). Technical risk analyses have furthermore drawn criticism for having failed to take sufficient account of the multitude of socioeconomic consequences that people associate with risks (Freudenberg 1989). Technical experts, for their part, have argued that the priorities of risk management should not be governed by the possible misperception or ignorance of the public (Sapolsky 1990). Individuals respond according to their own perceptions of risk and not according to an objective risk scientifically assessed. Risk management strategies that are based only on technical assessments and do not consider individual perceptions are likely to fail. More recently, therefore, efforts have been made to combine technical risk assessment with considerations of risk perception, social adequacy, and fairness (Renn 1998).

In view of the complexity of this highly interdisciplinary field, it is clearly beyond the scope of this chapter to provide a comprehensive study of the technical assessment or perception of glacial risks. The primary objective is rather a comparison of four case studies with regard to the nature of the hazards, the responses to them, and the management approaches adopted. The comparative character of this study prevents us from providing a full account of each case, and the picture that emerges may therefore be incomplete from the viewpoint of either physical or social scientists. It is, however, an attempt to narrow the gap between technical and social approaches to risk management in mountain areas by documenting the problems prevailing in different regions not only in terms of physical hazards but also in terms of their interaction with local populations and authorities. It is based on hazard assessment studies performed in recent years and experience in managing these hazards and risks in cooperation with the local, regional and national authorities. The cases involved the glacier-clad volcanoes of Central Mexico, the glacial lakes of the Cordillera Blanca of Peru, the complex hazards related to glacier instability in the Italian Alps, and a large ice/rock avalanche disaster in the Caucasus.

GLACIER-CLAD VOLCANOES IN CENTRAL MEXICO

Mexico's glaciers are situated on three different volcanoes, of which Popocatépetl is currently the most active. Glaciers on active volcanoes can pose an especially serious hazard to the surrounding population when volcanic activity interacts with ice and generates the melt-induced volcanic debris flows known as lahars (Major and Newhall 1989). The terrible catastrophe of Nevado del Ruíz, Colombia, in 1985, with a death toll of over 20,000, demonstrated the destructive potential of glacier-clad volcanoes (Thouret 1990). In Mexico, Popocatépetl returned to an eruptive cycle in 1994 and since then has erupted several times, with ashfall events occasionally reaching as far as Mexico City, 70 km away (Julio Miranda et al. 2005; Figure 5.1). Tens of thousands of people live around the volcano and could be affected by major debris flows descending its flanks. Small glaciers are nestled on top of the north-northwest side of the volcano. In recent decades, but particularly since 1994, they have suffered from a strong retreat attributed to atmospheric warming and, more recently, predominantly to volcanic activity (Delgado 1997; Huggel and Delgado 2000; Julio and Delgado 2003). Major lahar events, both presumably related to ice melt, occurred in 1997 and 2001 (Julio et al. 2005). Although these lahars stopped just before they reached local settlements, historical records indicate that lahars have traveled tens of kilometers farther, reaching areas that are heavily populated today (Siebe et al. 1996).

The activity of Popocatépetl in recent years has provoked a number of crises requiring risk management: In 1994, 1998, and 2000 there were major evacuations of local populations. The problems associated with these evacuations (opposition by local people) indicated that the different actors involved had divergent perspectives

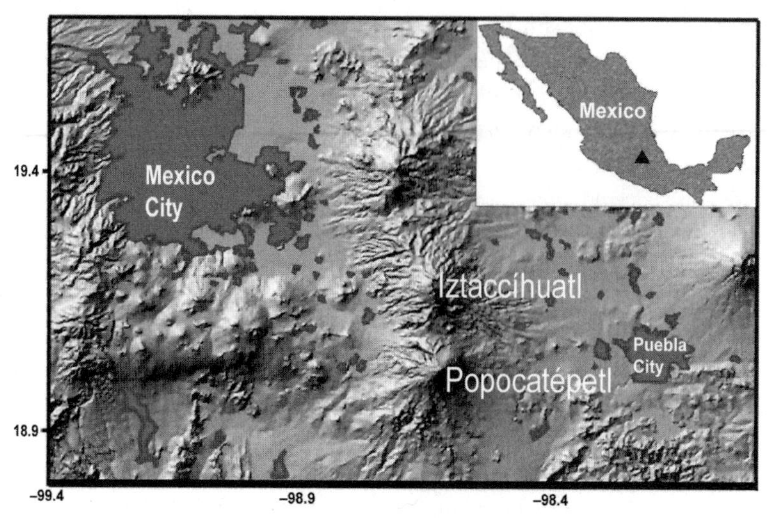

FIGURE 5.1. Shaded digital terrain model showing the area around the glacier-clad Popocatépetl and Iztaccíhuatl volcanoes, with the two major cities of Mexico and Puebla and several towns in the immediate vicinity (courtesy of P. Julio).

on hazards and risks. Our observations suggested that the main actors could be roughly grouped into three groups: the scientists, the authorities, and the local (indigenous) populations.

The scientists, mostly vulcanologists in this case, adopted a technical perspective on the professional level, although they may have developed more intimate personal relationships with "their" volcanoes. Hazards were assessed in terms of the available information regarding the volcano and its glaciers; different scenarios were developed and best estimates of uncertainties provided. Because risk assessments are based on information about hazards and potential damage, the scientists' role was communicating their findings to the authorities and providing well-considered and intelligible expert information to the public via the media.

The authorities were concerned with risk management, and their decisions depended heavily on the hazard assessments of the scientists. Evacuations of several tens of thousands of people are an enormous challenge in terms of logistics. They therefore needed the cooperation of residents. Obviously, at least partially similar perceptions of the hazards and possible risks would have facilitated evacuation efforts. It has long been recognized that knowledge plays a crucial role in risk perception (Johnson 1993; Okrent 1998), and therefore risk communication—the transmission of knowledge—was an important part of the authorities' risk-prevention efforts (De Marchi 1993). Risk communication has two components—education and warning (Voight 1996). Information on risks influences risk perception and eventually the public response during crises (e.g., Mileti and Fitzpatrick 1991). In Mexico, an important way of delivering a warning with regard to volcanic hazards is a three-level alert signal released by the media. The educational component includes elements such as simulation, maps of evacuation routes, and other training efforts (De la Cruz-Reyna 1999).

The risk perception of the indigenous and other people living around Popocatépetl cannot be adequately analyzed here, but our experience and that of others suggests that the major influences on it are ethno-religious and poverty-related. The indigenous people of Central Mexico view volcanoes as titans. They call Popocatépetl ("Smoking Mountain" in Nahuatl) "Don Goyo" and consider it sacred. The often-used diminutive "Don Goyito" points to the affection they have for it, and the offerings made in an attempt to calm it are another expression of this relationship. The

FIGURE 5.2. Scientific experts preparing for a glaciological field campaign in an indigenous village at the toe of the glacier-clad volcano Pico de Orizaba, Mexico. (Photo C. Huggel.)

Catholic Church has further influenced the people by preaching divine control of one's destiny.

The effect of poverty or lack of resources on risk perception is most clearly reflected in the evacuations of the past (Macías 1999). Many rural people lack the resources to leave their property (Figure 5.2). Further, historical distrust of the authorities makes them reluctant to leave their domestic animals with people they do not know (e.g., military) in the event of evacuation. The risk associated with leaving seems to be perceived as greater than the risk associated with staying. Social scientists have emphasized that the authorities' imposing their own risk perception upon residents is problematic (Macías 1999). Differences in the perceived benefits (e.g., fruitful volcanic soil) and harm associated with volcanic risk have often been identified by risk perception research (Slovic 1987). Whereas local people value staying in the location of their vital economic assets more than they fear the adverse effects of volcanic activity, the potential losses dominate the view of the authorities.

GLACIAL LAKES IN THE CORDILLERA BLANCA

The Peruvian Cordillera Blanca has been strongly affected by glacial disasters. In fact, some of the greatest glacial catastrophes ever documented roared down the Santa River valley in the Cordillera Blanca in the twentieth century (Figure 5.3). In 1941 the moraine that dammed the glacial lake Palcacocha suddenly burst, triggering an outburst flood that destroyed about one-third of the city of Huaraz more than 20 km away and killed some 6,000 people (Reynolds 1992). Even more destructive were the ice avalanches from Mt. Huascarán Norte in 1962 and 1970. The 1970 avalanche completely buried the settlement of Yungay, and more than 20,000 people were killed (Plafker and Ericksen 1978). The 1941 Palcacocha disaster had alerted the Peruvian authorities to the problem, and they had established the Comisión de Control de las Lagunas de la Cordillera Blanca, which subsequently became the División de Glaciología y Seguridad de Lagunas and today is the Unidad de Glaciología y Recursos Hídricos (Silverio 2003; Carey 2005). Mitigation measures such as lake level lowering, dam construction, and the creation of drainage tunnels were undertaken in the 1970s and 1980s, chiefly financed by the hydropower company ElectroPerú and its subsidiary in Huaraz, Hidrandina S.A. Mitigation activities have decreased in recent years because of a lack of resources, political changes, and the privatization of ElectroPerú, even though strong glacier retreat and the related formation or growth of numerous glacial lakes pose persistent hazards to the downstream communities. As is typical for glacial hazards, the prevailing hazards are characterized by a low probability of occurrence but high potential magnitude (Huggel et al. 2004). During glacial hazard studies performed in the Cordillera Blanca we observed a number of factors that played a role in hazard management and the response of the local population.

In contrast to the situation in many other regions, the memory of historical disasters is vivid. This is probably because of their particular severity and reminders such as the memorial at the place where the town of Yungay was buried by the avalanche. However, the oral tradition of the disasters is often not very accurate and tends to generate fear.

The poverty of many of the people in the Cordillera Blanca, together with a lack of knowledge about the hazards and limited available

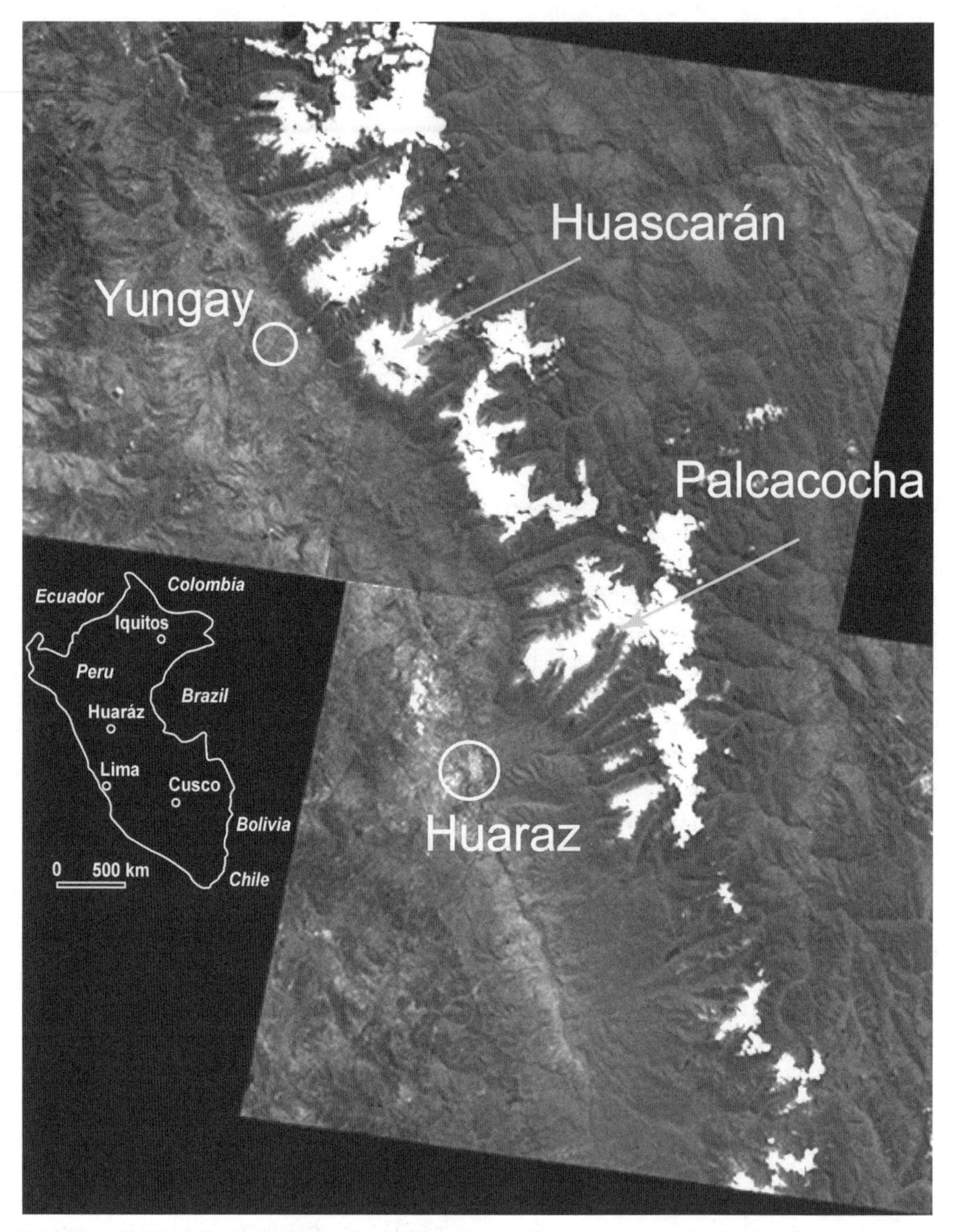

FIGURE 5.3. ASTER satellite image showing the Cordillera Blanca, with the Santa River valley, the city of Huaraz, and the former town of Yungay (destroyed by the 1970 ice avalanche).

land, forces them to settle in places susceptible to hazards—for example, close to or even within river channels. This is a common problem in less developed countries (Alcántara-Ayala 2002). Further, even if the hazards are known, the problems associated with daily survival are usually rated higher than the risk of infrequent glacial disasters.

The severely limited resources of the federal and regional agencies responsible for hazard mitigation and their sometimes limited efficiency make it difficult to reduce risks significantly. Carey (2005) furthermore points out that distrust of government and of scientific experts and past failures of the authorities to follow scientists' warnings have contributed to glacial disasters in the Cordillera Blanca. A lack of communication and trust has thus increased the vulnerability to glacial hazards.

An instructive instance of the role of communication and trust occurred in April 2003, when remote-sensing experts from the NASA Jet Propulsion Laboratory (JPL) misinterpreted an ASTER satellite image and released a warning on the JPL Web site about the threat of an ice avalanche's falling into Lake Palcacocha and producing an outburst flood that could seriously affect Huaraz (Steitz and Buis 2003). The report was taken up by the Peruvian media and provoked great concern and even panic in Huaraz. By unfortunate coincidence, it was released just 10 days after an actual accident at Lake Palcacocha, when one of the steep moraine slopes containing the lake had failed and generated an impact wave in the lake (Figure 5.4). The wave had destroyed one of the overflow dams and produced a small flood. Huaraz had been affected by an increased sediment load in the river, causing severe freshwater problems for a week. Thus, people were already worried when they received the news of a potential disaster. Although government officials asserted that there was no danger from Lake Palcacocha, radio stations in Huaraz urged residents to pack their valuables to be prepared for a quick getaway (Carey 2005). The population's distrust of government caused some to prefer to believe the NASA report. The most serious impact of the report was on the tourism industry and thus the local and regional economy. People from other parts of Peru who had intended to visit the Huaraz region for the Easter holidays canceled their trips in great numbers. The economic damage was estimated to be on the order of US$20 million.

Subsequent analysis of the satellite image and field studies at Palcacocha by one of us (C. Huggel) and others showed that the "crevasse" identified by NASA was a bedrock outcrop exposed by the normal process of glacier thinning. Further investigations of the situation indicated that the NASA report had been a serious misinterpretation. The situation at Palcacocha is not, however, without hazards. Spontaneous or earthquake-induced slope failures of the moraine are a major problem and could generate another impact wave in the lake. Fortunately, the lake level has been artificially lowered to its present volume of ca. 3.5 million m³ of water. It is possible that an outburst flood would be attenuated to some degree downstream, but in the worst case adverse effects on Huaraz could certainly not be excluded. The situation therefore needs monitoring, and this is currently the task of the Unidad de Glaciología y Recursos Hídricos in Huaraz.

FIGURE 5.4. Lake Palcacocha below the mountain massif of Palcaraju (6,274 m a.s.l.), July 2003. The moraine failure of March 2003 occurred on the right (*arrow*). The destroyed overflow dam is marked at the lower left. (Photo C. Huggel.)

On the one hand, this unfortunate incident shows the psychological and economic vulnerability of this area not only to glacial hazards but also to failed risk communication and emphasizes the importance of appropriate and responsible risk management. On the other hand, it perfectly demonstrates the increasing globalization of hazard assessment and related risk communication, including the growing possibility of looking from everywhere into anyone's backyard with modern satellite technology and thus interfering with governmental activities in foreign countries.

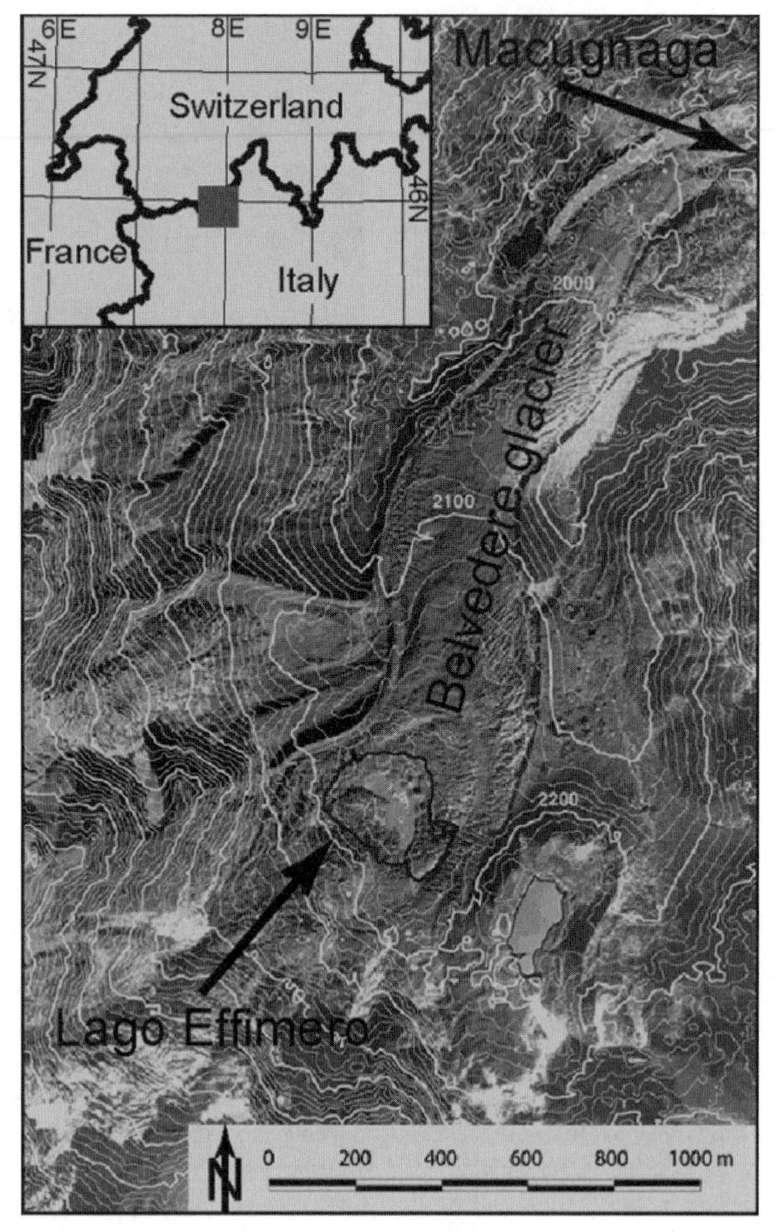

FIGURE 5.5. Overview of the Belvedere Glacier, October 2001 (courtesy of the Glaciorisk Project), showing the initial formation of the supraglacial lake (Lago Effimero) and indicating the lake area by 2002.

GLACIER INSTABILITY IN MACUGNAGA

Macugnaga is located on the southeastern flank of Monte Rosa (4,634 m a.s.l.), a major glaciated mountain massif in the Alps and the border region between Switzerland and Italy (Figure 5.5). Macugnaga has one of the most spectacular high-mountain landscapes in the Alps: the 2,500-m-high Monte Rosa east face, with the Belvedere Glacier at its base descending to about 1,700 m a.s.l. A moraine- and ice-dammed glacial lake, Lago delle Locce, adjacent to the Belvedere Glacier had an outburst in

FIGURE 5.6. The mayor of Macugnaga and a glaciologist at the margin of the Belvedere Glacier (*right*), July 2002. (Photo C. Huggel.)

FIGURE 5.7. Efforts to lower the level of Lago Effimero by the installation of pumps, August 2002. (Photo C. Huggel.)

1979 that caused massive damage in the village of Macugnaga and led to the initiation of mitigation measures (Haeberli and Epifani 1986; Dutto and Mortara 1992).

In 2001, an extraordinary surgelike process started at the Belvedere Glacier, with ice flow velocities increasing by almost one order of magnitude and a glacier surface uplift of up to 20–30 m. These developments threatened ski runs with ice/rock fall and obstructed hiking trails and cableway transport facilities (Figure 5.6 [Haeberli et al. 2002]). In 2002 a much more serious hazard developed: A large supraglacial lake formed on the Belvedere Glacier and reached a volume of 3 million m^3 of water (Tamburini et al. 2003), a uniquely large example of this phenomenon in the Alps (Kääb et al. 2004). A lake with such an enormous quantity of water towering above Macugnaga represented a serious threat to the mountain community. When scientists and community authorities observed a lake-level rise of 1 m per day by the end of June (due to intense melt conditions on the east face), the Ministry of Civil Protection in Rome was alerted. Within 24 hours a massive administrative, technical, and logistic organization was set in motion. The civil protection authorities assumed control over the military, police, and rescue services in accordance with Italian law. The charming mountain village of Macugnaga became an extremely busy disaster zone, with several hundred people from civil protection, the military, and other local, regional, and federal services, as well as the national and international media, just at the beginning of the summer tourism season. The heliport installed in Macugnaga became for a time the third-largest airport in Italy in terms of air traffic. The authorities installed pumps to lower the lake level (Figure 5.7), established a monitoring and alarm system, developed scenarios based on the scientists' assessments, and defined alarm thresholds accordingly (Kääb et al. 2004).

Our function as scientific advisers before, during, and after the crisis focused on the assessment and communication of the physical hazards. In practice, however, risk management and economic constraints inevitably interfere with scientific activities. In Macugnaga as in Central Mexico, we found three main groups involved in the management of hazards and the related risks: the community, the civil protection officials, and the scientists. These actors were characterized by different interests and perspectives and therefore different priorities in managing the hazards related to the threatening glacial lake.

For the community, a major determinant of its attitude toward the risks was its strong dependence on tourism. Emergency and preventive measures such as the closing of the cable-car facility due to the danger of an outburst flood were perceived as a threat to the local economy. However, driven by an interesting mixture of economic necessity and innovative ability, the community quickly responded to the new situation: As soon as the highest risk had decreased, the spectacular glacial lake was exploited as tourist attraction. The name given to the lake by the community, "Lago Effimero" (Ephemeral Lake), pretended that this was a unique phenomenon and only a temporary threat. Important hiking trails across the glacier were rapidly relocated, endangered ski runs were abandoned and new ones reconstructed, and a dam was constructed to prevent rock and ice fall from the glacier tongue. Beyond their economy-driven attitudes and actions, the local people took a relatively relaxed approach to the glacial risks, perhaps because of their long experience with mountain hazards.

The civil protection authorities' priority was primarily avoiding disaster. The management of the risk was highly professional and based on procedures that considered the scientific, technical, administrative, and communication aspects of the situation. Because the crisis attracted the mass media, including the regional, national, and international press and national television with regular broadcasting during the 8 p.m. news over several weeks, the authorities had an ideal platform for promoting a positive image of their service to the nation.

The scientists, finally, had a predominantly process-based perception of the hazards, striving to understand the complex processes and interactions. They were the most aware of the serious hazards from the enormous supraglacial Lago Effimero. Drawing on their experience and knowledge of glacial lake outbursts, they could anticipate the potential effects on Macugnaga—essentially the core task of scientists in such crises.

Although the people of Macugnaga had feared a serious loss of tourism due to the image of Macugnaga transmitted by the media, the massive presence of civil protection and military forces, and the closing of particular areas, the year 2003 turned out to be a very good year. The presence of Macugnaga in the newspapers and on television for weeks seems to have been a perfect marketing campaign for attracting an unexpectedly large number of tourists. Interestingly enough, tourists did not think that they were exposing themselves to a high risk in Macugnaga even though by this time the lake had formed again. It appears that in their individual benefit-versus-potential-damage evaluations, eagerness to see the spectacular location of a near-disaster was rated higher than any potential danger to be encountered there. Information dissemination by the authorities through the media, with statements such as "Lago Effimero is under control" and "Fear of the lake is over," likely influenced the perception of the public.

ICE/ROCK AVALANCHE DISASTER IN THE NORTH CAUCASUS

In the evening of September 20, 2002, a large ice/rock avalanche took place on the northern slope of the Kazbek Massif in North Ossetia, the Russian Caucasus (Kotlyakov et al. 2004; Haeberli et al. 2004). It started as a slope failure on the north-northeast face of Dzhimarai-khokh (4,780 m a.s.l.) below the summit and involved massive volumes of rock and ice (from hanging glaciers). The slide impacted the Kolka Glacier, largely entraining it (Huggel et al. 2005), and an ice/rock avalanche with a volume of about 100 million m^3 traveled down the Genaldon Valley for 19 km before being stopped at the entrance of the Karmadon Gorge (Figure 5.8). A mudflow continued downvalley for another 15 km and stopped 4 km short of the town of Gisel. Both the avalanche and the mudflow were devastating, causing the death of about 140 people and destroying important traffic routes, dwellings, and infrastructure. The ice dam that was formed by the ice/debris deposits at Karmadon created several marginal lakes of up to 5 million m^3 of water. Potential floods from these lakes were an imminent threat to the downstream areas after the disaster (Kääb et al. 2003).

An event entraining virtually a complete valley glacier had not been documented before and therefore had important implications for worldwide glacial hazard assessment (Huggel et al. 2005). The erosion of an entire valley glacier actually amounts to the unthinkable becoming reality. The disaster also acquired a political

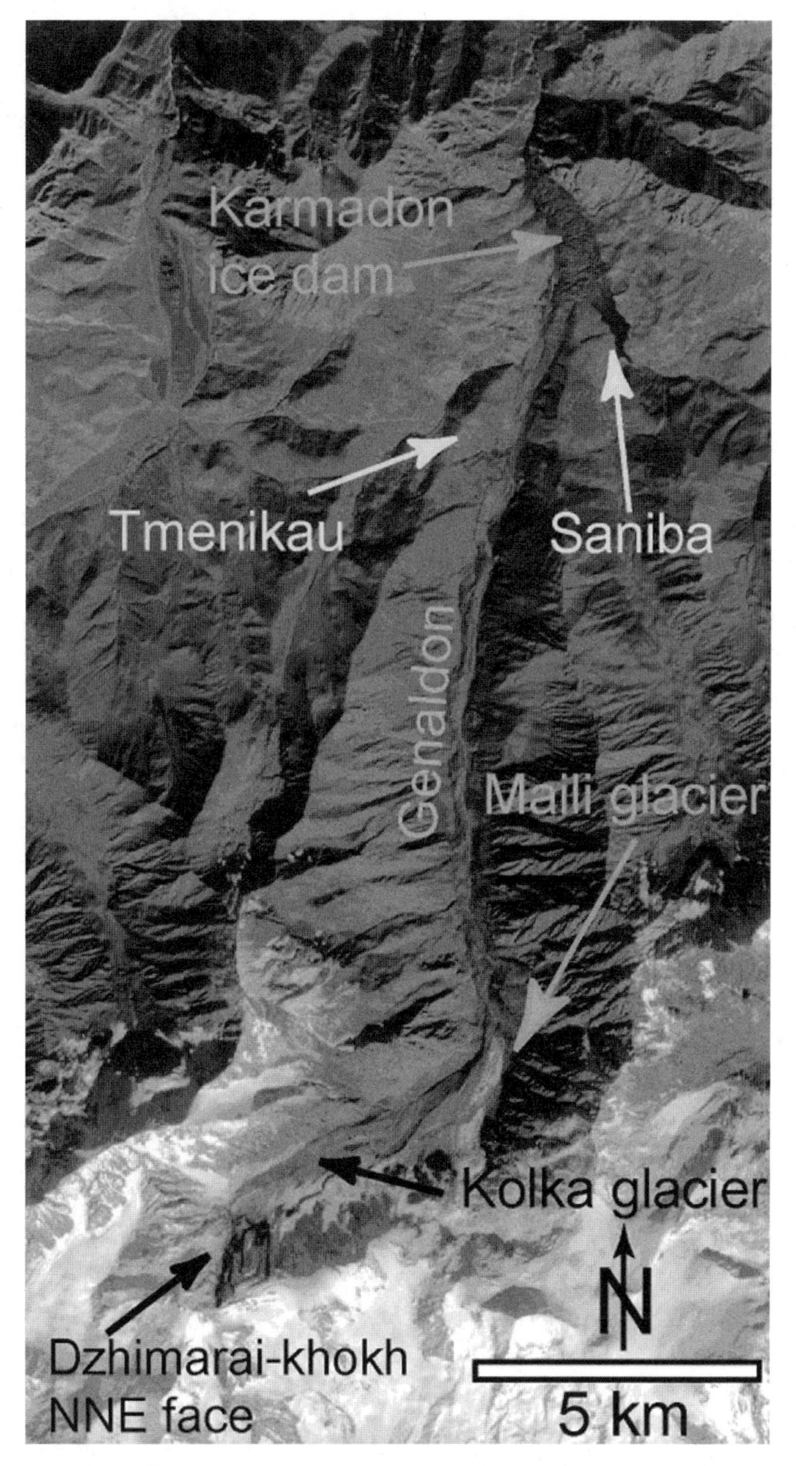

FIGURE 5.8. ASTER satellite image, October 6, 2002, showing the track of the September 2002 avalanche from Dzhimarai-khokh to Karmadon.

FIGURE 5.9. A traditional Caucasian village unaffected by the avalanche (*left*) and modern dwellings two days after the avalanche disaster with the beginning of lake formation (*right*). All of the buildings in the center of the picture were drowned when the lake reached its maximum level. (Photo C. Huggel and Department of Natural Resources, Republic of North Ossetia.)

dimension. Whether this event might have been predicted provoked intense debate affecting the government authorities of North Ossetia. As part of a team assembled by the Swiss Humanitarian Aid Unit to help them cope with the disaster and manage the subsequent hazards, we participated in determining that the timing of the disaster and its dimensions could not have been realistically predicted. It was, however, apparent that research and monitoring activities in this area had sharply declined after the breakdown of the Soviet Union because of the lack of resources. A project in cooperation with the authorities of North Ossetia initiated after the disaster and funded by the Swiss Agency for Development and Cooperation therefore established a new monitoring and alarm system for glacier-related and postdisaster hazards.

Observations made during the studies in Kolka/Karmadon indicated that attitudes of the local population toward glacial hazards had changed in recent decades. The traditional Caucasian villages of the region are situated not, as is commonly found elsewhere, in the valley bottom but high above the river channel on the valley flank, even though water has to be brought up from the river (Figure 5.9). It appears that people have long known about the occurrence of destructive processes in the Genaldon Valley. In fact, there had been comparable events in 1902 and probably in 1835. In 2002 the old village of Tmenikau remained unaffected by the avalanche. In recent years, however, a different settlement pattern had developed, with dwellings (often vacation homes) being constructed in Saniba on the valley bottom by wealthy people from the capital, Vladikavkas, who were unfamiliar with the mountain environment and lacked the traditional perspective regarding natural processes. These settlements were flooded by the water dammed by the massive ice/debris avalanche deposits (Figure 5.9). That poor people's settlements remained intact whereas rich people's property was destroyed contrasts with most situations in less developed countries, where poor people are often most exposed to hazards because they lack the resources to settle in safer areas or are ignorant of the potential danger.

CONCLUSIONS

Our engagement as physical scientists in glacial hazard studies and the management of crises in different regions of the world has shown that the social, cultural, economic, and historic context is a significant determinant of the local response to glacial hazards. This finding is intuitively reasonable and not a new insight considering the results from a body of literature dedicated to the perception of hazards. However, the gap between what has been recognized by social hazard or risk perception studies and what is applied in technical natural hazard studies is still considerable. We have found that factors such as lack of resources and poverty, ethno-religious tradition, distrust of the authorities, ignorance of risk, and economic dependence on tourism may influence the perception of risk in high-mountain regions. However, this list of factors may not present a complete picture of the conditions governing the response of the local population and the authorities to glacial hazards. Given the scarcity of studies of high-mountain hazards and the response to them, it is difficult to assess how representative these examples may be of conditions elsewhere. We emphasize the considerable differences in responses to glacial hazards among different regions and among the main actors involved at each location. We argue that, trivial as a consequence but profound as an implication in practice, such differences are fundamental and need to be recognized by experts concerned with hazard and risk management. Especially for experts unfamiliar with the local conditions, these sociocultural features can represent a challenge for efficient hazard prevention. Risk communication has been recognized as a major component in influencing risk perception and may help to harmonize the divergent perceptions of the actors involved to improve the response to glacial hazards in terms of hazard and risk prevention.

ACKNOWLEDGMENTS

We are grateful for the excellent and stimulating cooperation of colleagues, other experts, and government authorities at all the locations discussed here. Funding agencies such as the Swiss Agency for Development and Cooperation, Humanitarian Aid, and the Regione Piemonte and the Italian Ministry of Civil Protection are also acknowledged. ASTER data are a courtesy of NASA/GSFC/METI/ERSDAC/JAROS and the U.S./Japan ASTER Science Team. We furthermore thank Patricia Julio, Michael Kollmair, and Walter Silverio for critical review of the manuscript. Two anonymous reviewers helped to improve it.

REFERENCES CITED

Alcántara-Ayala, I. 2002. Geomorphology, natural hazards, vulnerability, and prevention of natural disasters in developing countries. *Geomorphology* 47:107–24.

Carey, M. 2005. Living and dying with glaciers: People's historical vulnerability to avalanches and outburst floods in Peru. *Global and Planetary Change* 47:122–34.

De la Cruz-Reyna, S. 1999. Escenarios de riesgo y percepción del peligro volcánico en México. In *Simposio Internacional "Volcán Popocatépetl," 22–24 March, 1999, Abstracts*, 11. Mexico City: UNAM/CENAPRED.

Delgado, H. 1997. The glaciers of Popocatépetl Volcano (Mexico): Changes and causes. *Quaternary International* 43:1–8.

De Marchi, B. 1993. Effective communication between the scientific community and the media. In *Prediction and perception of natural hazards*, ed. J. Nemec, J. M. Nigg, and F. Siccardi, 183–91. Dordrecht: Kluwer.

Dutto, F., and G. Mortara. 1992. Rischi connessi con la dinamica glaciale nelle Alpi italiani. *Geografia Fisica Dinamica Quaternaria* 15:85–99.

Fell, R. 1994. Landslide risk assessment and acceptable risk. *Canadian Geotechnical Journal* 31:261–72.

Freudenberg, W. R. 1989. Perceived risk, real risk: Social science and the art of probabilistic risk assessment. *Science* 242:44–49.

Greminger, P. 2003. Managing the risks of natural hazards. In *Debris-flow hazards mitigation: Mechanics, prediction, and assessment: Proceedings, 3rd International DFHM Conference, Davos, Switzerland, September 10–12, 2003*, ed. D. Rickenmann and C. L. Chen, 39–56. Rotterdam: Millpress.

Haeberli, W., J.-C. Alean, P. Müller, and M. Funk. 1989. Assessing risks from glacier hazards in high mountain regions: Some experiences in the Swiss Alps. *Annals of Glaciology* 13:77–101.

Haeberli, W., and F. Epifani. 1986. Mapping the distribution of buried glacier ice: An example from Lago delle Locce, Monte Rosa, Italian Alps. *Annals of Glaciology* 8:78–81.

Haeberli, W., C. Huggel, A. Kääb, S. Oswald, A. Polkvoj, I. Zotikov, and N. Osokin. 2004. The Kolka-Karmadon rock/ice slide of 20 September 2002: An extraordinary event of historical dimensions in North Ossetia (Russian Caucasus). *Journal of Glaciology* 50:533–46.

Haeberli, W., A. Kääb, F. Paul, M. Chiarle, G. Mortara, A. Mazza, P. Deline, and S. Richardson. 2002. A surge-type movement at Ghiacciaio del Belvedere and a developing slope instability in the east face of Monte Rosa, Macugnaga, Italian Alps. *Norwegian Journal of Geography* 56:104–111.

Huggel, C., and H. Delgado. 2000. Glacier monitoring at Popocatépetl Volcano: Glacier shrinkage and possible causes. In *Beiträge zur Geomorphologie: Proceedings der Fachtagung der Schweizerischen Geomorpho-logischen Gesellschaft vom 8. bis 10. Juli 1999, Bramois*, ed. C. Hegg and D. Vonder Mühll, Birmensdorf: WSL, 97–106.

Huggel, C., W. Haeberli, A. Kääb, D. Bieri, and S. Richardson. 2004. Assessment procedures for glacial hazards in the Swiss Alps. *Canadian Geotechnical Journal* 41:1068–83.

Huggel, C., S. Zgraggen-Oswald, W. Haeberli, A. Kääb, A. Polkvoj, I. Galushkin, and S.G. Evans. 2005. The 2002 rock/ice avalanche at Kolka/Karmadon, Russian Caucasus: Assessment of extraordinary avalanche formation and mobility and application of QuickBird satellite imagery. *Natural Hazards and Earth System Sciences* 5:173–87.

Jasanoff, S. 1993. Bridging the two cultures of risk analysis. *Risk Analysis* 13:123–30.

Johnson, B.P. 1993. Advancing understanding of knowledge's role in lay risk perception. *Risk-Issues in Health and Safety* 4:189–211.

Julio, P., and H. Delgado. 2003. Fast hazard evaluation employing digital photogrammetry: Popocatépetl glaciers, Mexico. *Geofísica Internacional* 42:275–83.

Julio Miranda, P., A. E. González-Huesca, H. Delgado Granados, and A. Kääb. 2005. Ice-fire interactions on Popocatépetl Volcano (Mexico): Case study of the January 22, 2001, eruption. *Zeitschrift für Geomorphologie* 140, suppl.:93–102.

Kääb, A., C. Huggel, B. Barbero, M. Chiarle, M. Cordola, F. Epifani, W. Haeberli, G. Mortara, P. Semino, A. Tamburini, and G. Viazzo. 2004. Glacier hazards at Belvedere Glacier and the Monte Rosa east face, Italian Alps: Processes and mitigation. In *Proceedings, 10th Interpraevent, Riva del Garda, May 24–27, 2004*, Klagenfurt: Interpraevent, 67–78.

Kääb, A., R. Wessels, W. Haeberli, C. Huggel, J.S. Kargel, and S.J.S. Khalsa. 2003. Rapid ASTER imaging facilitates timely assessments of glacier hazards and disasters. *Eos: Transactions American Geophysical Union* 13(84):117, 121.

Kotlyakov, V.M., O.V. Rototaeva, and G.A. Nosenko. 2004. The September 2002 Kolka glacier catastrophe in North Ossetia, Russian Federation: Evidence and analysis. *Mountain Research and Development* 24(1):78–83.

Macías, J.M. 1999. *Desastres y protección civil: Problemas sociales, políticos y organizacionales*. Mexico City: Dirección General de Protección Civil del Gobierno del Distrito Federal/CIESAS.

Major, J.J., and C.G. Newhall. 1989. Snow and ice perturbation during historical volcanic eruptions and the formation of lahars and floods. *Bulletin of Volcanology* 52:1–27.

Mileti, D.S., and C. Fitzpatrick. 1991. Communication of public risk: Its theory and its application. *Sociological Practice Reviews* 2(1):20–28.

Okrent, D. 1998. Risk perception and risk management: On knowledge, resource allocation, and equity. *Reliability Engineering and System Safety* 59:17–25.

Plafker, G., and F.E. Ericksen. 1978. Nevados Huascaran avalanches, Peru. In *Rockslides and avalanches, 1, Natural phenomena*, ed. B. Voight, 277–314. Amsterdam: Elsevier.

Renn, O. 1998. The role of risk perception for risk management. *Reliability Engineering and System Safety* 59:49–62.

Reynolds, J.M. 1992. The identification and mitigation of glacier-related hazards: Examples from the Cordillera Blanca, Peru. In *Geohazards natural and man-made*, ed. G.J.H. McCall, D.J.C. Laming, and S.C. Scott, 143–57. London: Chapman and Hall.

Sapolsky, H.M. 1990. The politics of risk. *Daedalus* 119(4):83–96.

Siebe, C., M. Abrams, J. M. Macías, and J. Obenholzner. 1996. Repeated volcanic disasters in Prehispanic time at Popocatépetl, Central Mexico: Past key to the future? *Geology* 24:399–402.

Silverio, W. 2003. *Atlas del Parque Nacional Huascarán, Cordillera Blanca, Perú*. Lima: Nova Graf.

Slovic, P. 1987. Perception of risk. *Science* 236: 280–85.

Steitz, D.E., and A. Buis. 2003. Peril in Peru? NASA takes a look at menacing glacier. Press release, April 11. http://www.nasa.gov/home/hqnews/2003/apr/HP_News_03138.html (accessed January 7, 2005).

Tamburini, A., G. Mortara, M. Belotti, and P. Federici. 2003. The emergency caused by the "short-lived lake" of the Belvedere Glacier in the summer 2002 (Macugnaga, Monte Rosa, Italy). *Terra Glacialis* 6:51–54.

Thouret, J. -C. 1990. Effects of the November 13, 1985, eruption on the snow pack and ice cap of Nevado del Ruiz volcano, Colombia. *Journal of Volcanology and Geothermal Research* 41:177–201.

Voight, B. 1996. The management of volcano emergencies: Nevado del Ruíz. In *Monitoring and mitigation of volcano hazards*, ed. S. Scarpa and R. Tilling, 719–69. Berlin and Heidelberg: Springer.

Wegmann, M., A. Bruderer, M. Funk, and C. Wuilloud. 2004. Natural hazard risk management: A participatory approach applied for glacier hazards. In *Proceedings, 10th Interpraevent, Riva del Garda, May 24–27, 2004*, Klagenfurt: Interpraevent, 297–308.

PART TWO

Scientific Observations

Measurement, Monitoring, and Modeling

6

Two Alpine Glaciers over the Past Two Centuries

A SCIENTIFIC VIEW BASED ON PICTORIAL SOURCES

Daniel Steiner, Heinz J. Zumbühl, and Andreas Bauder

For centuries high mountains and glaciers have been a source of both paralyzing fear and strange fascination. Natural scientists first began to show interest in Alpine glaciers at the beginning of the eighteenth century, but although the first simple observations of glacier movements and moraines were made, no systematic scientific investigations were carried out (Zumbühl 1980).

The nineteenth century witnessed the Ice Age hypothesis and the beginning of modern experimental glaciology. Initial detailed studies of glacier-related phenomena were undertaken on the Unteraar Glacier, in Switzerland, in the 1840s by Louis Agassiz (1807–73). His contributions to the newly established scientific field earned him the title of the "Father of Glaciology" (Lurie 1960). Impressive glacier advances affecting the majority of Alpine glaciers at about the same time can be recognized from archives and documentary evidence (Maisch et al. 1999; Nicolussi and Patzelt 2000). In addition, new kinds of documentary data that could be used to study glacier fluctuations appeared. The technical progress of photography made it possible to record a glacier's position without any artistic distortion (Gernsheim 1983), and the first geometrically exact topographic map of Switzerland, the so-called Dufour map, was published between 1844 and 1864 (Wolf 1879; Graf 1896; Locher 1954).

In general, all these historical records can give a very detailed picture of glacial fluctuations in the Swiss Alps during the Late Holocene. They allow the extension of the study of glacier history farther into the past than would be possible with the use of direct measurements alone (e.g., Holzhauser, Magny, and Zumbühl 2005). Empirical qualitative and/or quantitative data, mainly on the length but also on the area and volume of glaciers, can be derived from these sources.

With methods based on historical records, a temporal resolution of decades or, in some cases, even individual years can be achieved in the reconstruction of glacier time series. Probably the best example in this context is Switzerland's Unterer Grindelwald Glacier. Fluctuations in its length have been reconstructed with the aid of numerous texts and

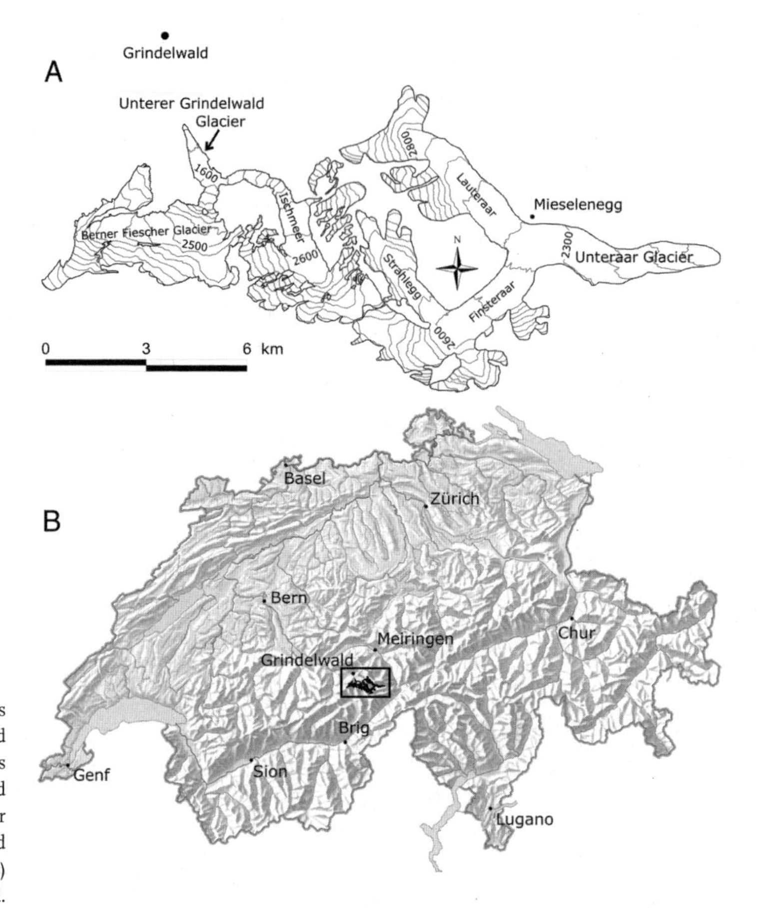

FIGURE 6.1. (A) The outlines of the Unterer Grindelwald and the Unteraar Glaciers and their main branches and tributaries and (B) the Unteraar Glacier and the Grindelwald region (*area with solid outline*) in Switzerland.

pictorial representations back to 1535 (Zumbühl 1980; Zumbühl, Messerli, and Pfister 1983; Holzhauser and Zumbühl 1996, 1999, 2003). For many other glaciers in the European Alps (e.g., the Mont Blanc region) the historical material appears exceptionally rich, but there may be untapped documentation for similar research in many other areas of the world.

The objective of this article is to bring together documentary data to quantify the fluctuations of the Unteraar and the Unterer Grindelwald Glaciers since their mid-nineteenth-century maximum extent. Whereas many other studies of glacier change focus on the regional or even global scale, our study deals with two exemplary glaciers in a relatively small area. In addition to having comparable climatic conditions and similar characteristics (e.g., area), the two glaciers are documented by a particularly rich sample of data. However, obtaining reliable results requires considerable effort in calibrating these data.

Finally, as a background to the beginning of modern glaciology, we also analyze the remarkable development of representation techniques from drawings and topographic surveys to the first scientific photographs within a few centuries.

THE LOCAL SETTING

The Unteraar and the Unterer Grindelwald Glaciers are located within a few kilometers of each other in the Bernese Alps (Figure 6.1).

The Unteraar (latitude 46°35′N, longitude 8°15′E) is a valley glacier 13.5 km long and covering 24.1 km². A prominent feature is its large

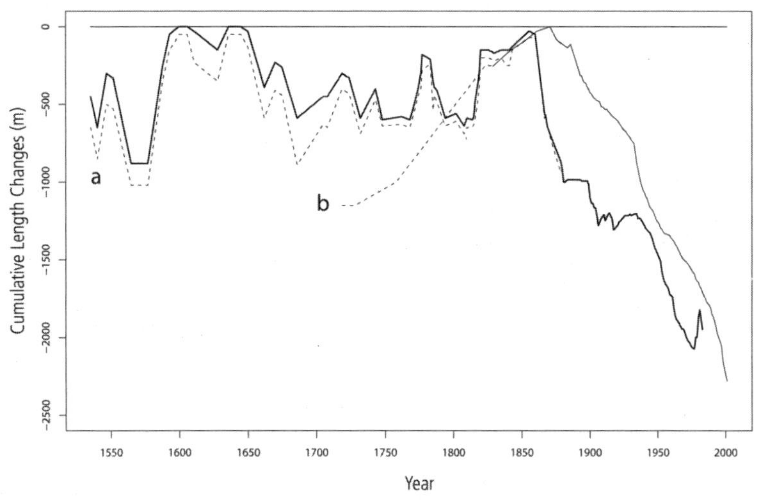

FIGURE 6.2. Cumulative length variations of (*a*) the Unterer Grindelwald Glacier from 1535 to 1983, relative to the 1600s maximum extent (=0), and (*b*) the Unteraar Glacier from 1719 to 2001, relative to 1871 (=0). Maximum extensions of the Unterer Grindelwald are represented by a thick line, minimum extensions by a dashed line. The Unteraar is represented by a solid line for the 1830–2001 period and a dashed line for the possible position between 1719 and 1829 (Zumbühl 1980; Zumbühl and Holzhauser 1988; Holzhauser and Zumbühl 1996, 2003; VAW/SANW 1881–2002).

ice-cored medial moraine and extensive debris cover, typically 5–15 cm thick (Bauder 2001; Schuler, Fischer, and Gudmundsson 2004). The tongue of the Unteraar is formed by the two main tributaries the Lauteraar and the Finsteraar. Mass balance measurements indicate that the present equilibrium line altitude of Unteraar is at 2,850 m a.s.l. (Bauder 2001). The present glacier terminus, 1.5 km from Lake Grimsel, is at an elevation of 1,950 m a.s.l. The cumulative length fluctuations of the Unteraar (Figure 6.2, *b*) show a continuous retreat of the glacier since the first observations in the 1880s (VAW/SANW 1881–2002; Zumbühl and Holzhauser 1988).

The Unterer Grindelwald Glacier (latitude 46°35'N, longitude 8°05'E) is a valley glacier 8.85 km long and covering 20.6 km^2. Ischmeer in the east and the Berner Fiescher Glacier in the west join to form its tongue. The main contribution of ice nowadays originates from the Berner Fiescher Glacier (Schmeits and Oerlemans 1997).

The approximate equilibrium line altitude (AAR [accumulation area ratio] = 0.67; Table 6.1), derived from digital elevation models and confirmed by the maximum elevation of lateral moraines (Gross, Kerschner, and Patzelt 1976; Maisch et al. 1999), is situated at 2,640 m a.s.l. in relatively flat areas. As a consequence of this specific hypsography, the relatively small ca. 40-m rise of the equilibrium line altitude in the past 140 years has resulted in a large increase of the ablation area and thus a large glacier retreat since the mid-nineteenth century (Nesje and Dahl 2000). Today the Unterer Grindelwald Glacier terminates at about 1,297 m a.s.l. in a narrow gorge, and reliable observations are difficult to obtain.

The last-published quantitative observation was made in 1983 (VAW/SANW 1881–2002). Because of the extraordinarily low position of its terminus and its easy accessibility, the Unterer Grindelwald is probably the best-documented glacier in the Swiss Alps. Figure 6.2, *a*, shows its cumulative length fluctuations, derived from documentary evidence, for the period 1535–1983 (Zumbühl 1980; Holzhauser and Zumbühl 1996, 2003).

The geometry of the two glaciers differs from that of the "model" glacier in having a wide variety of surface slopes and many basins that deliver ice to the main streams.

TABLE 6.1

Topographical Characteristics of the Unterer Grindelwald Glacier in 1860/61/72 and 2004

	1860/61/72	2004
Length (measured along the longest flowline 1; see Figure 6.10)	10.8 km	8.85 km
Elevation of head (Mönch)	4,107 m a.s.l.	4,107 m a.s.l.
Elevation of terminus	972 m a.s.l.	1,297 m a.s.l.
Surface area[a]	26.1 km^2	20.6 km^2
Estimation of equilibrium line altitude (accumulation area ratio = 0.67)	2,600 m a.s.l.	2,640 m a.s.l.
Average slope in %	29%	31.8%
Average slope in degrees	16.1°	17.6°
Exposure	N-NW	N-NW
Absolute ice volume change 1860/61/72–2004	$-1.56\ km^3$	
Average thickness change rate 1860/61/72–2004	$-0.42\ m\ a^{-1}$	

NOTE: The calculations are based on digital elevation models for 1861 and 2004 (Table 6.2, *a*).
[a]Including all subglaciers connected with major streams in 1860/61/72.

FROM FIRST OBSERVATIONS TO DETAILED STUDIES

Systematic observations on the Unteraar Glacier began with the field work of the naturalist Franz Josef Hugi (1796–1855) between 1827 and 1831. It was he who made the first observations on the surface velocity of the glacier (Hugi 1830). When his successor, Louis Agassiz (1807–73), visited the glacier in 1839, he found to his surprise that Hugi's hut on the medial moraine had moved since 1827, an important indication of glacier movement (Agassiz 1885; Portmann 1975). Agassiz's observations were also aimed at supporting his theory of ice ages. The idea that glaciers transported erratic material was not new and probably originated with the mountain farmer Jean-Pierre Perraudin (1767–1858), who explained it in 1815 to the geologist Jean de Charpentier (1786–1855) (Haeberli and Zumbühl 2003). Charpentier's argument that the Swiss glaciers had once been much more extensive was taken up by Agassiz, who further postulated that the whole Northern Hemisphere had once been covered with a gigantic glacier. This Ice Age hypothesis was rejected by many people because it challenged the prevailing Bible-based Christian worldview, which does not include huge glaciers. In addition to this, comparable glaciers or ice sheets had not yet been discovered (Bolles 1999; Haeberli, this volume).

Between 1840 and 1845 Agassiz conducted a research program on the glacier that constituted the beginning of modern experimental glaciology. He initially acted as a leader and program manager of his interdisciplinary team and was also responsible for the climatic data. His secretary and friend, the naturalist Jean Édouard Desor (1811–82), conducted the glaciological and geomorphological research. The results of their numerous observations were comprehensively documented in Agassiz (1847). Besides these extensive glaciological studies, the artists and engineers on the team produced the first outstanding representations of the glacier, among them the panorama of Jacques Bourkhardt (1811–67) and the glacier

FIGURE 6.3. (A) Bourkhardt's drawing (24.4 × 126 cm; pencil, pen, ink, watercolor, gouache) of the chromolithograph "Panorama de la mer de glace du Lauteraar et du Finsteraar–Hôtel des Neuchâtelois" from the Mieselenegg (Swiss national coordinates approx. 657′500/158′700, altitude approximately 2,620 m a.s.l.) in 1842 (*detail*). Private collection. (B) Recent panorama taken east from the Mieselenegg (Swiss national coordinates approximately 657′600/158′600, altitude approximately 2,520 m a.s.l.) on August 22, 2004. (Photos Daniel Steiner and Heinz J. Zumbühl.)

maps of Johannes Wild (1814–94) and Johann Rudolf Stengel (1824–57).

BOURKHARDT'S (1842) PANORAMA

Jacques Bourkhardt's drawing of the chromolithograph "Panorama de la mer de glace du Lauteraar et du Finsteraar–Hôtel des Neuchâtelois," which had been missing for a long time (Figure 6.3, A) is probably the most beautiful and topographically richest panorama ever done of the Unteraar Glacier. In it the different mountain peaks were named for the first time. The partial panorama includes the main mountain peaks in the background and the confluence area of the Finsteraar and Lauteraar Glaciers. Desor (1847) praised the panorama and pointed to the time-consuming drawing process that had produced it. The significance of the panorama may be judged from its use as the background for the portrait (Figure 6.4) of Agassiz and Desor by Fritz Berthoud (1812–90) as a reminder of the great scientific work done by the two naturalists.

On the medial moraine, below the confluence area, the panorama shows the "Hôtel des Neuchâtelois," a huge metamorphic boulder that served Agassiz and his team as accommodation and as a shelter during the summer field seasons. It also shows the newly built tent (20 m long and 5 m high) constructed by the guides and used for sleeping, drawing, and study (see Haeberli, this volume).

FIRST PHOTOGRAPHS OF GLACIERS

Besides portraiture, early photographers paid special attention to architecture, travel views, and landscapes. One possible reason that scientific and public interest focused specifically on glacier photographs is probably the dramatic Greenland expedition of the Arctic explorer Elisha Kent Kane (1820–57) from 1853 to 1855. During his search for survivors of another expedition, Kane's ship became icebound and his objective changed to one of survival. After two years of hardship and a difficult journey over ice and open water to Greenland, he and his shipmates were finally rescued (Kane 1856). His reports of an "ice-ocean of boundless dimensions" confirmed the existence of huge ice masses and brought final acceptance of the Ice Age hypothesis after 40 years of discussion and opposition. This prompted curiosity and attracted more and more tourists to locations where they could admire even "little" glaciers such as those of the Swiss Alps.

FIGURE 6.4. Portrait (in oil) of Louis Agassiz (*seated*) and Jean Édouard Desor by Fritz Berthoud (1812–90), with Bourkhardt's panorama (Figure 6.3, A) in the background. Musée d'Art et d'Histoire Neuchâtel, Switzerland (Inv.-Nr. 762).

Probably the earliest photos of Swiss glaciers were made in summer 1849/50 by Jean-Gustave Dardel (1824–99) and Camille Bernabé (1808–?) (Lagoltière 1989; Morand and Kempf 1989). In September 1855, Auguste-Rosalie Bisson took a panoramic photo of the Unteraar Glacier of the remarkable width of 1.85 m that was praised as "a gigantic ensemble with a wonderful effect" (Chlumsky, Eskildsen, and Marbot 1999). The Bisson Brothers, Auguste-Rosalie (1826–1900) and Louis-Auguste (1814–76), were among the best-known European photographers of the 1850s and 1860s. Their most famous body of work is made up of the high-altitude photographs they made in the Alps, among them the first photographs from the peak of Mont Blanc. Photography opened up regions that were not usually accessible (Guichon 1984; de Decker Heftler 2002). It is probable that the photo of Unteraar Glacier was shown at the Exposition Universelle of 1855 in Paris, where the Bisson Brothers' huge photos were among the major attractions and received first prize (Bonaparte 1856; Chlumsky, Eskildsen, and Marbot 1999). The Bisson photo "Glacier

FIGURE 6.5. The Unterer Grindelwald Glacier on the valley floor, exactly at its mid-nineteenth-century maximum extent. The black ink stamp "Bisson frères" at the edge of the photo, the snow-free glacier front, and an advertisement of the Bisson Brothers in the journal *L' Artiste* (December 14, 1856) show that the photo was taken in summer/autumn 1855/56. (Photograph "Glacier inférieur du Grindelwald [Oberland bernois]," Bisson Brothers, Alpine Club Library, London.)

inférieur du Grindelwald (Oberland bernois)" (Figure 6.5) was probably the first photograph of the Unterer Grindelwald Glacier.

After Queen Victoria took a fancy to the stereoscope at the Crystal Palace Exhibition in 1851, stereo viewing became the rage (Gernsheim 1983). The stereoscope slides that were produced allowed people to sit in their own homes and tour the world. Furthermore, because of their small size they were not too expensive. The most popular slides showed the world from the countrysides of Europe to the pyramids and tombs of ancient Egypt (Schönfeld 2001). However, both travelogue slides and books such as Kane's expedition reports produced increased interest in natural phenomena that had previously been inaccessible. Thus, these historical sources of glacier data also provide cultural information on social attitudes toward glaciers over time. The perception of glaciers ranges from fear and romantic transfiguration to broad scientific interest and nonscientific curiosity at a safe distance. In a similar way, representations of glacier retreat today could be interpreted as a social danger signal in a changing climate.

The earliest stereo views of the Unterer Grindelwald Glacier were taken by Adolphe Braun (1812–77) in summer/autumn 1858 (Figure 6.6, A). A list of Swiss stereo views by Braun published in the journal *La Lumière* (May 21, 1859), including Figure 6.6, A, shows that Braun had begun to market his series of depictions of chic tourist destinations in Switzerland a year earlier than was assumed by Kempf (1994, and personal communication, December 19, 2003). Figure 6.6 is a good example of the changes that can be observed in the forefield of the glacier since the mid-nineteenth century.

Braun's stereo views were taken two to three years after the mid-nineteenth-century maximum extent of the glacier in 1855/56. Because of the incomplete stereographic effects in the background of the photographs, the stereo views could not yield detailed topographic data. Nevertheless, a qualitative comparison of the stereo views with the Bisson photo of 1855/56 shows no significant

FIGURE 6.6. (A) The Unterer Grindelwald Glacier on the valley floor in 1858, two to three years after the maximum extent in 1855/56. Photograph on visiting card by Adolphe Braun (1812–77) at Dornach, Haut-Rhin ("510. Glacier inf. de Grindelwald"). Private collection of Richard Wolf, Fribourg, Switzerland. (B) Recent view of the glacier gorge and the Fiescherhorn/Pfaffestecki from Upper Stotzhalten (Swiss national coordinates approximately 645′900/163′700, altitude approximately 1,030 m a.s.l.) on September 25, 2003. (Photo Daniel Steiner.)

differences in the extent of the glacier. In fact, Zumbühl (1980) computed a change in length of less than −10 m for the 1855/56–1858 period. In contrast, the following two to three years brought a slightly accelerated tendency toward retreat. This can be seen in a projection of the 1855/56 moraine ridges extracted from an ortho photo map (Zumbühl 1980) onto the original plane-table sheet "Grindelwald 1860/61" (Figure 6.7). According to this, the glacier had withdrawn 30–60 m from the outer 1855/56 moraines. Furthermore, the arrangement of the moraine ridges and the glacier extent of 1870, extracted from the Siegfried map of 1870 ("Grindelwald 1870"), suggest that the glacier had been retreating in a rapid and asymmetric way since 1860/61.

EARLY TOPOGRAPHIC MAPS AND DIGITAL ELEVATION MODELS

The Unteraar Glacier was the subject of the first topographic map of a glacier with scientific value, generated by Wild in 1842. The lithography (scale: 1:10,000), published by Agassiz (1847), shows the tongue of the glacier, which was more than 8 km long, east of the confluence area and designated by a system of hachures. At that time, the glacier ended in a steep, partially ice-covered ice front (Zumbühl and Holzhauser 1988; VAW/SANW 1881–2002; Haeberli and Zumbühl 2003). Because of the absence of contour lines, no reliable interpretation in terms of glacier volume changes can be deduced. The glacier map "Carte Orographique du Glacier de l'Aar Montrant les détails de la Stratification et l'origine des glaciers de second ordre," surveyed by Stengel in 1846 and also published by Agassiz (1847), shows the same glacier extent as on the Wild map, but in it for the first time we find contour lines of the glacier surface.

The leading role of Swiss cartography is mainly based on the first modern official map series of Switzerland, produced between 1832 and 1864 under the supervision of General Guillaume-Henri Dufour (1787–1875). The 25-sheet series

FIGURE 6.7. Forefield of the Unterer Grindelwald Glacier, extracted from the original plane-table sheet "Grindelwald 1860/61" by Wilhelm August Gottlieb Jacky (1833–1915), with a projection of the 1855/56 moraine ridges (from a 1:2,000 ortho photo map [Zumbühl 1980]). The glacier extents in 1860/61 (from "Grindelwald 1860/61"; contour lines) and 1870 (from "Grindelwald 1870"; dashed outline) are also shown.

(scale: 1:100,000) is an admirable specimen of cartography. After the Dufour map was published, there was demand for 1:25,000 or 1:50,000 maps. The depiction of some areas surveyed by regional topographic surveys was no longer considered satisfactory, and the areas had to be completely resurveyed. In contrast, most surveys carried out by the federal topographic survey met the requirements with minor revisions. These efforts resulted in the first official map series of high resolution and continuous scale, known as the Siegfried map and published from 1870 on (Oberli 1968). Two plane-table sheets of the Dufour map and two first editions of the Siegfried map, covering the Unterer Grindelwald Glacier, were available in the mid-nineteenth century. The Siegfried maps were often based on a revision of the Dufour surveys (Zölly 1944), and this was the case for the Siegfried edition "Jungfrau 1872" (Table 6.2; Bundesarchiv E 27 20040, Geschäftsberichte des Eidg. Stabsbureaus, 1872), which was used as a basis for the development of the digital elevation model instead of Stengel's original plane-table sheet from 1851 because of the latter's inaccurate glacier texture.

TABLE 6.2

Digital Elevation Models (DEMs) Used in this Study and their Sources, Unterer Grindelwald Glacier

DEM NAME	DEM1861	DEM1926	DEM1993	DEM2000	DEM2004
Data basis	Plane-table sheet, topographic map	Topographic maps	Digital photogrammetry	Digital photogrammetry	Digital photogrammetry
Name	Grindelwald, Jungfrau	Interlaken, Jungfrau	DHM25 (Level 2)		
Survey/revision date	Terrestrial survey 1860/61, 1872	Terrestrial photographs 1926/1934	Aerial photographs 1993	Aerial photographs 1999/2000	Aerial photographs 2004
Scale	1:50,000	1:50,000			
Contour interval	30 m	20 m			

Both the original plane-table sheet "Grindelwald 1860/61" of Wilhelm August Gottlieb Jacky (1833–1915) and the Siegfried map "Jungfrau 1872" have been geo-referenced on the basis of the current Swiss geodetic datum CH1903. Figure 6.8 shows the composite of the two topographic maps used. The detailed correspondence between the two maps at the junction is remarkable. The early topographic maps were based on the Schmidt's 1828 ellipsoid equivalent conical projection and have not been converted to the present geodetic reference system. The shifts in *xy*-direction between the two reference systems are negligible for our purposes (Bolliger 1967; Urs Marti, personal communication, June 5, 2002). Because the origin of the elevation measurements has changed from 376.2 m a.s.l. (Dufour map) or 376.86 m a.s.l. (Siegfried map) to 373.6 m a.s.l. (CH1903) the *z*-coordinates of the historical data have been corrected by −3 m to obtain a comparable database.

A comparison of a recent pixel map (PK50©swisstopo) and the geo-referenced maps shows that the accuracy lies within the expected range (15 m in *xy*-direction, 1 m in *z*-direction; Urs Marti, personal communications, November 25, 2002, and August 16, 2004). This result testifies to the extraordinary work of many topographers of the time. Jacky's sheet, for example, was judged faultless and exemplary by the cartographic commission in 1862 (Locher 1954). The digital elevation model 1861 was generated by digitalizing contour lines and reference points from the composite map of Figure 6.8 (Hoinkes 1970; Funk, Morelli, and Stahel 1997; Wipf 1999; Bauder 2001). The high quality of the early maps makes it possible to develop modern digital elevation models from them and allows very precise calculations of surface and volumetric glacier loss in the Alps.

Using the first and completely revised edition of the official Swiss maps (scale: 1:50,000), based on terrestrial stereo-photogrammetry in 1926 and additional field surveys in 1934, the digital elevation model 1926 was developed (Bundesarchiv E 27 20042, Geschäftsberichte der Eidg. Landestopographie, 1926 [Vol. 2], 1934 [Vol. 3]). Additional models of the current state of the glaciers have been extracted from a recent set of aerial photographs by applying digital stereo-photogrammetry (Kääb and Funk 1999; Kääb 2001). Digital color aerial photos of the Unterer Grindelwald Glacier were produced in 1999 (accumulation area), 2000 (ablation area), and 2004 (©swisstopo). The photogrammetric interpretation and automatic generation of the

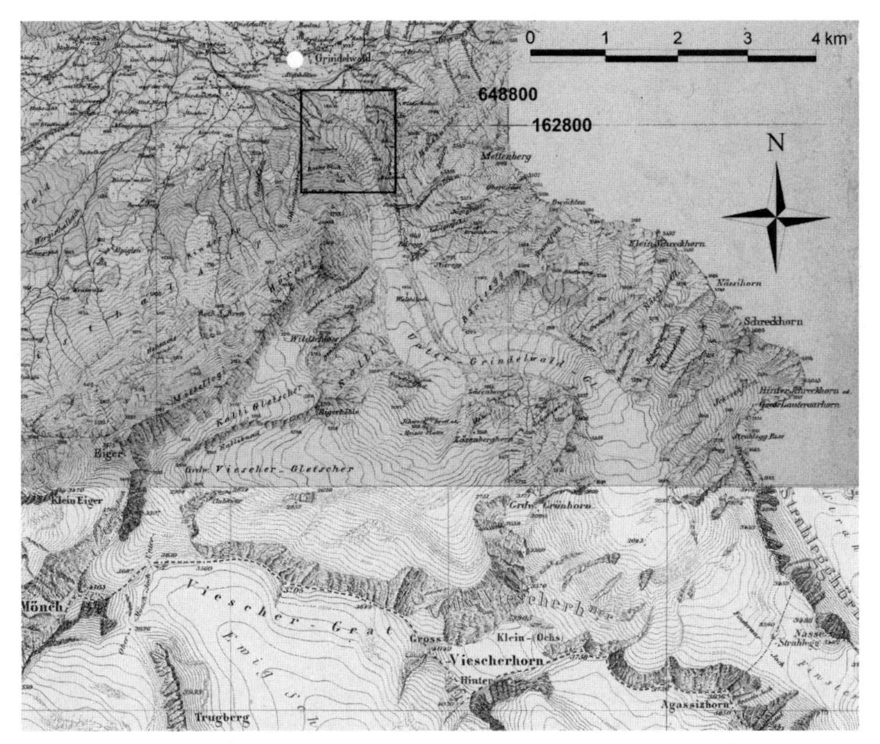

FIGURE 6.8. Composite map of the Unterer Grindelwald Glacier. *Above,* "Grindelwald 1860/61." *Below,* "Jungfrau 1872." The area outlined represents the region shown in Figure 6.7. The white dot indicates the approximate location of the photographer of Figure 6.6.

digital elevation model were performed with standard photogrammetric software. In order to improve the detail in the deglaciated glacier forefield, the digitalized contour lines of the aforementioned ortho photo map were merged with the models of 2000 and 2004. The automatic procedures for model generation are limited in areas with low texture, such as flat snowfields, and therefore points were checked manually and either deleted or corrected where necessary. The resulting high-resolution models 2000 and 2004 show an average grid width of ~10 m. Finally, the DHM25 (Level 2) of the Federal Office of Topography (DHM25©swisstopo) has been interpreted as the status of the Unterer Grindelwald Glacier in 1993 (Swisstopo 2004).

For the Unteraar Glacier there are five plane-table sheets of the Dufour map and four first editions of the Siegfried map. The quality of the maps covering the firn areas was inadequate for model construction, and we restricted our study to the ablation area below the confluence of Lauteraar and Finsteraar, where the major changes were expected (Table 6.3). Four additional digital elevation models, dating from 1927, 1947, 1961, and 1997, respectively (Bauder 2001), exist for this glacier. They are based on digitalized contours of two original photogrammetric analyses, a map of a regional cadastral survey, and recent digital photogrammetric analysis as presented before (Bauder 2001).

RESULTS

On the basis of these models, it is possible to calculate changes in glacier volume, area, and length (Wipf 1999; Maisch et al. 1999). The spatial distributions of volume and thickness changes have been calculated for both glaciers. They are based on the area of greatest extension, including areas of complete retreat and new advance. The results are summarized in Tables 6.4 and 6.5. After the well-documented maximum extents of the Unterer Grindelwald

TABLE 6.3

Digital Elevation Models (DEMs) Used in this Study and their Sources, Unteraar Glacier

DEM NAME	DEM1880	DEM1927	DEM1947	DEM1961	DEM1997
Data basis	Plane-table sheet	Photo-grammetrical contour analysis	Photo-grammetrical contour analysis	Cadastral map	Digital photogrammetry
Survey/ revision date	Terrestrial survey 1879/80	Terrestrial photographs 1927	Aerial photographs 1947	Aerial photographs 1961	Aerial photographs 1997
Scale	1:50,000	1:25,000	1:25,000	1:10,000	
Contour interval	30 m	20 m	20 m	10 m	

TABLE 6.4

Changes in Glacier Parameters, Unterer Grindelwald Glacier, 1860–2004

YEAR	AREA (KM^2)	LENGTH (KM)	VOLUME CHANGE (KM^3)	THICKNESS CHANGE (M)	AVERAGE THICKNESS CHANGE (M/YR)
1860/61/72	26.1	10.8	–	–	–
1926	24.3	9.8	−0.94	−36.0	−0.55
1993	22.3	9.0	−0.42	−17.3	−0.26
1999/2000	21.4	8.9	−0.14	−6.3	−0.90
2004	20.6	8.85	−0.06	−2.8	−0.70
1860–2004	–	–	−1.56	−59.8	−0.42

TABLE 6.5

Changes in Glacier Parameters, Unteraar Glacier, 1880–1997

YEAR	AREA (KM^2)	VOLUME CHANGE (KM^3)	THICKNESS CHANGE (M)	AVERAGE THICKNESS CHANGE (M/YR)
1880	28.4	–	–	–
1927	27.5	−0.32	−11.3	−0.24
1947	27.4	−0.42	−15.3	−0.76
1961	25.5	−0.34	−12.4	−0.89
1997	24.1	−0.51	−20.0	−0.56
1880–1997	–	−1.59	−56.0	−0.48

Glacier in 1855/56 and the Unteraar Glacier in 1871 (Zumbühl and Holzhauser 1988), we find a relatively high rate of thickness change of −0.55 m per year for the former and a lower rate of −0.24 m per year for the latter until the late 1920s. From then on we find a lower rate of thickness change for the Unterer Grindelwald Glacier and an increasing mass loss for the Unteraar Glacier.

A rapid volume loss of −0.90 m per year in the 1990s can be calculated for the Unterer Grindelwald Glacier. It is striking that this mass loss occurred almost without any substantial change in glacier length (Table 6.4). This is probably due to the fact that the glacier terminus is in a narrow gorge. Down-wasting of the glacier surface in the front basin of the glacier (Figure 6.9) is known from other glaciers (Paul et al. 2004) and primarily leads to a decrease in glacier width/area and not in length. The observed cumulative mass balance of three Swiss Alpine glaciers (Gries, Basòdino, Silvretta) amounts to −2 m for the period 1993–2000 and −2.6 m for the period 2001–2004, which includes the European heat wave of 2003 (VAW/SANW 1881–2002; unpublished data from Andreas Bauder). Therefore, the total decrease of the Unterer Grindelwald Glacier of −6.3 m for the period 1993–1999/2000 was significantly above average and the total thickness change of −2.8 m for the period 1999/2000–2004 was comparable to the mass balance of other Alpine glaciers.

The decrease of glacier area between 1860/61 (Unterer Grindelwald Glacer) or 1880 (Unteraar Glacier) and the end of the twentieth century amounts to approximately −20%. Furthermore, the overall mass loss is −1.56 km^3 ice (−1.4 km^3 water; assumed ice density 0.9 kg m^{-3}) for the Unterer Grindelwald Glacier and −1.59 km^3 ice (−1.43 km^3 water) for the Unteraar Glacier. The sum of this mass loss (2.83 km^3 water) is nearly equivalent to three years' water consumption (1.06 km^3 in the year 2000) in Switzerland (SVGW 2002). Both the glacier area and the volume loss correspond to studies based on inventory data (Zemp et al., this volume).

Finally, the calculated average thickness changes are similar to those revealed by investigations of other Alpine glaciers such as the Aletsch, the Rhône, and the Trift. Generally, an increased mass loss since the 1980s has been observed (see also Zemp et al., this volume). The rates and timing depend on the available digital elevation models.

Figure 6.10 shows the spatial distribution of the thickness change for the Unterer Grindelwald Glacier between DEM1861 and DEM2004. The average overall thickness change for this period amounts to −0.42 m per year. In general, the widespread decrease of glacier thickness expected since 1860 can be determined from Figure 6.10. Absolute negative thickness changes of up to −330 m can be seen in the ablation area, the upper Ischmeer, and south of the Berner Fiescher Glacier. The latter differences are much larger than we would expect in the accumulation areas (Bauder 2001). Nevertheless, there are some regions where a totally unexpected increase of glacier surface height is found. In some of these cases, such as the upper part of the Berner Fiescher Glacier, we can assume a lower quality of both DEM1861 and DEM2004 due to low snow/ice surface contrasts and/or complex terrain and perhaps the sparse point density of DEM1861 and DEM2004 in such areas.

This argument does not, however, apply to the lower part of the Berner Fiescher Glacier, which shows a large area of surface height increase of up to +100 m since the mid-nineteenth-century maximum extent (see also Figure 6.9, *b*, for the surface profile along flowline 2). According to the newly discovered field book, the topographer Jacky, one of the best topographers of his time, surveyed the Grindelwald region in July 1861 and worked very close to the area of increase, which is also a relatively flat region that is easy to survey. He measured a large number of trigonometric points with full spatial information on or close to the Unterer Grindelwald Glacier, including a profile crossing exactly the assumed area of increase. Because he measured these points in

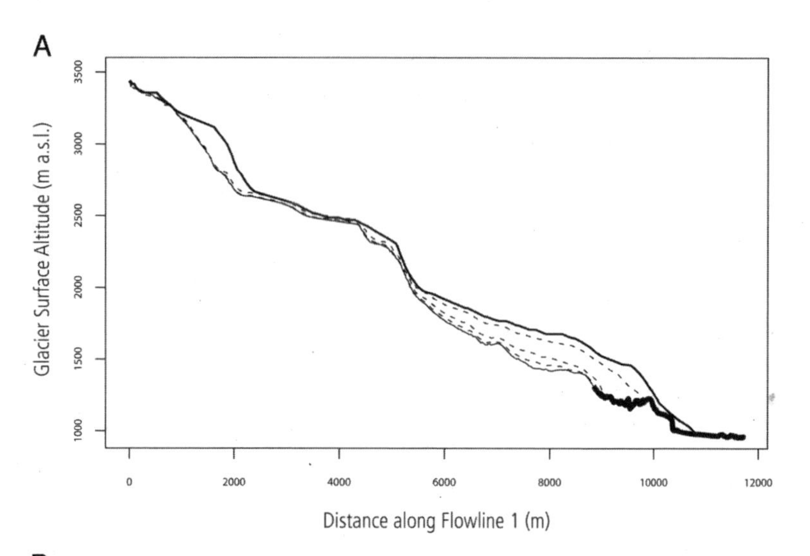

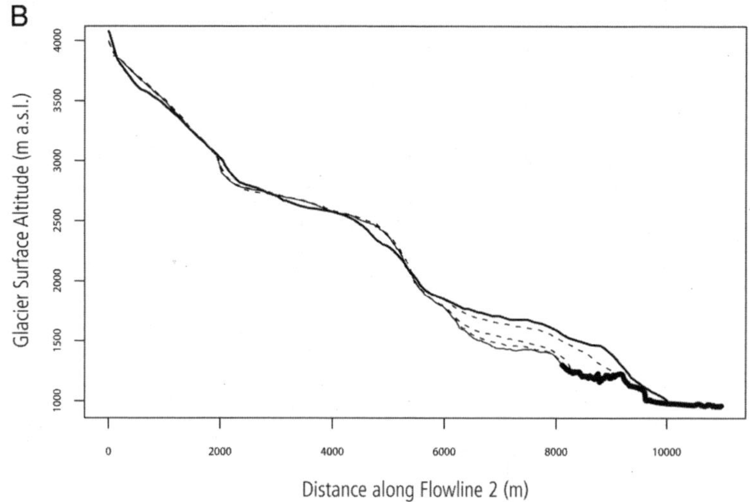

FIGURE 6.9. Surface profiles along (A) flowline 1 and (B) flowline 2 of the Unterer Grindelwald Glacier for 1860/61 (*thick line*), 1926 (*dashed line*), 1993 (*dashed line*), 1999/2000 (*dashed line*), and 2004 (*solid line*), derived from digital elevation models. The thick line downvalley indicates the exposed glacier forefield with the two pronounced "Schopf Rocks."

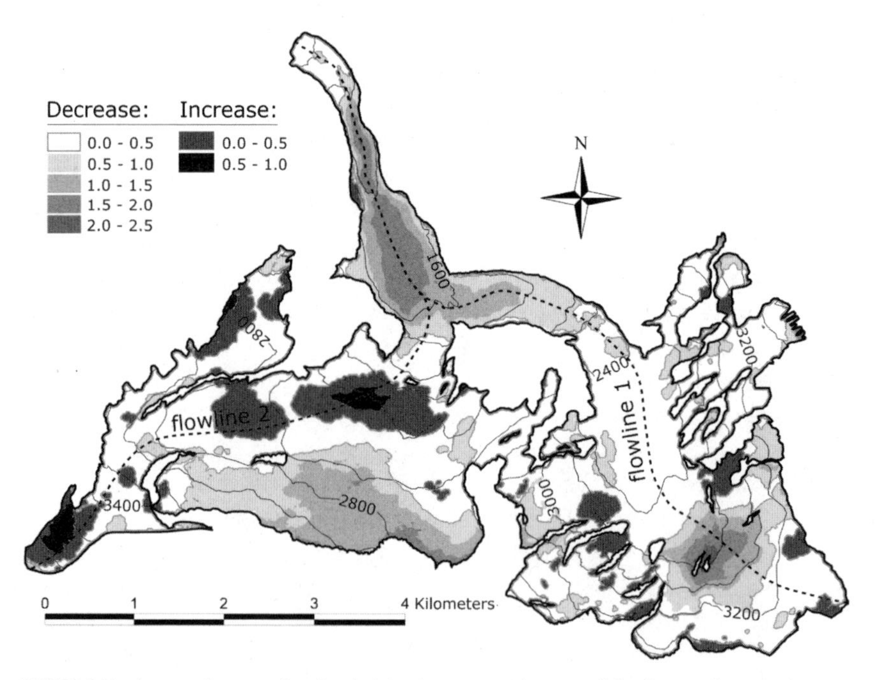

FIGURE 6.10. Average changes of surface heights (m per year) between 1860/61/72 and 2004 (DEM1861–DEM2004). The glacier outlines of 1860/61/72 and the contour lines of 2004 are also given.

situ to use them for drawing the contour lines (Martin Gurtner, personal communication, August 28, 2003), major systematic errors in his survey are very doubtful. It is possible that the area of increase could be dynamically linked with the decrease of surface heights in the previous mentioned southern part of the Berner Fiescher Glacier, but in the absence of detailed investigations of the dynamics of the Unterer Grindelwald Glacier this linkage cannot be investigated.

SUMMARY AND CONCLUSIONS

The first systematic investigations on the Unteraar Glacier marked the beginning of modern glacier research and represented an important proof of the Ice Age theory. Louis Agassiz was the motive force behind both the first systematic and scientific glaciological studies and the development of representations of glaciers in the form of drawings and topographic maps. A few years later, the first glacier photographs provided the means to show glacier changes in a relatively easy way. Thus, the photographic technique marked a new era of scientific representation of glaciers because even inaccessible areas could now be recorded quickly and easily. This change of scientific representation techniques from paintings and maps to photographs within only 15 years can be studied against the background of the inception of modern glaciology.

During the same period an increased glacier cover can be detected in many regions all over the world. This last glacier maximum extent is well documented for a few glaciers, among them the Unterer Grindelwald and the Unteraar. Therefore we were able to conduct this unique case study of research on two of the best-known glaciers in Europe and to demonstrate its development from the very beginnings to the present. We have also noted that the quality of the documentary evidence from these two glaciers is a benchmark against which other glaciers and studies should be measured.

The maximum extents of the two glaciers around the mid-nineteenth century and their subsequent retreat have been studied by combining a critical discussion of documentary data with the application of recent technical possibilities in photogrammetry. New photographic material and the evaluation of old topographic maps confirm that the retreat of the Unterer Grindelwald Glacier was relatively slow in a first period after the mid-nineteenth-century maximum extent and then accelerated in the 1861–70 period. Digital elevation model comparisons show reliable rates of thickness change of −0.42 m per year for the Unterer Grindelwald Glacier and −0.48 m per year for the Unteraar Glacier since their mid-nineteenth-century maximum extent.

Finally, the Unterer Grindelwald Glacier's spatial distribution of thickness changes shows some surprising patterns. Either the basic data are much less accurate than has been assumed, or this glacier demonstrates some interesting dynamic behavior that affected only parts of the area. Because a mass shift could have taken place within less than a decade, additional information between the 1860 and 1926 models is needed for further investigation of this question.

ACKNOWLEDGMENTS

This work has been supported by the Swiss National Science Foundation through its National Center of Competence in Research on Climate (NCCR Climate) project PALVAREX. We are grateful to Hermann Bösch for his never-ending help with photogrammetric problems and to Pierre Gerber, Urs Marti, Martin Gurtner, and Martin Rickenbacher for their competent help concerning problems in historical cartography. We thank the Federal Office of Topography for providing its topographical material and Heinz Wanner, Ben Orlove, Ellen Wiegandt, and Brian Luckman for their fruitful suggestions on various aspects of this article.

REFERENCES CITED

Agassiz, E. C. 1885. *Louis Agassiz: His life and correspondence.* Boston: Houghton, Mifflin.

Agassiz, L. 1847. *Système glaciaire ou recherche sur les glaciers, leur mécanisme, leur ancienne extension et le rôle qu'ils ont joué dans l'histoire de la terre.* Pt. 1. *Nouvelles études et expériences sur les glaciers actuels: Leur structure, leur progression et leur action physique sur le sol.* Paris: V. Masson/Leipzig: L. Voss.

Bauder, A. 2001. *Bestimmung der Massenbilanz von Gletschern mit Fernerkundungsmethoden und Fliessmodellierungen: Eine Sensitivitätsstudie auf dem Unteraargletscher.* Zurich: Versuchsanstalt für Wasserbau, Hydrologie und Glaziologie der Eidgenössischen Technischen Hochschule Zürich.

Bolles, E. B. 1999. *The ice finders: How a poet, a professor, and a politician discovered the Ice Age.* Washington, DC: Counterpoint Press.

Bolliger, J. 1967. *Die Projektionen der Schweizer Plan- und Kartenwerke.* Winterthur: Druckerei Winterthur.

Bonaparte, Napoléon-Joseph-Charles-Paul, Prince. 1856. *Exposition Universelle de 1855: Rapports du jury mixte international.* Paris: Imprimerie Impériale.

Chlumsky, M., U. Eskildsen, and B. Marbot. 1999. *Die Brüder Bisson. Aufstieg und Fall eines Fotografenunternehmens im 19. Jahrhundert: Katalog zur gleichnamigen Ausstellung, Museum Folkwang, Essen, 7.2.–28.3.1999; Fotomuseum im Münchner Stadtmuseum, 11.4.–30.5.1999; Bibliothèque nationale de France, Paris, 15.6.–15.8.1999.* Amsterdam: Verlag der Kunst.

de Decker Heftler, S. 2002. *Photographier Le Mont Blanc: Les pionniers.* Collection Sophie et Jérôme Seydoux. Chamonix: Éditions Guérin.

Desor, J. E. 1847. *Agassiz' und seiner Freunde geologische Alpenreisen in der Schweiz, Savoyen und Piemont,* ed. C. Vogt. Frankfurt am Main: Literarische Anstalt.

Funk, M., R. Morelli, and W. Stahel. 1997. Mass balance of Griesgletscher 1961–1994: Different methods of determination. *Zeitschrift für Gletscherkunde und Glazialgeologie* 33:41–56.

Gernsheim, H. 1983. *Geschichte der Photographie: Die ersten 100 Jahre.* Frankfurt am Main: Ullstein, Propyläen Verlag.

Graf, J. H. 1896. *Die Schweizerische Landesvermessung 1832–1864: Geschichte der Dufourkarte.* Bern: Buchdruckerei Stämpfli & Cie.

Gross, G., H. Kerschner, and G. Patzelt. 1976. Methodische Untersuchungen über die Schneegrenze in alpinen Gletschergebieten. *Zeitschrift für Gletscherkunde und Glazialgeologie* 12:223–51.

Guichon, F. 1984. *Montagne: Photographies de 1845 à 1914.* Paris: Denoël.

Haeberli, W., and H. J. Zumbühl. 2003. Schwankungen der Alpengletscher im Wandel von Klima und Perzeption. In *Welt der Alpen–Gebirge der Welt: Ressourcen, Akteure, Perspektiven,* ed. F. Jeanneret, D. Wastl-Walter, and U. Wiesmann, 77–92. Bern: Jahrbuch der Geographischen Gesellschaft Bern.

Hoinkes, H. 1970. Methoden und Möglichkeiten von Massenhaushaltsstudien auf Gletschern. *Zeitschrift für Gletscherkunde und Glazialgeologie* 6:37–90.

Holzhauser, H., M. Magny, and H. J. Zumbühl. 2005. Glacier and lake-level variations in west-central Europe over the last 3500 years. *The Holocene* 15:791–803.

Holzhauser, H., and H. J. Zumbühl. 1996. To the history of the Lower Grindelwald Glacier during the last 2800 years–palaeosols, fossil wood, and historical pictorial records–new results. *Zeitschrift für Geomorphologie,* n.s., suppl. 104:95–127.

———. 1999. Glacier fluctuations in the western Swiss and French Alps in the 16th century. *Climatic Change* 43: 223–37.

———. 2003. Nacheiszeitliche Gletscherschwankungen. In *Hydrologischer Atlas der Schweiz,* ed. R. Weingartner and M. Spreafico. Bern-Wabern: Bundesamt für Landestopographie.

Hugi, F. J. 1830. *Naturhistorische Alpenreise.* Solothurn: Amiet-Lutiger.

Jacky, W. A. G. 1861. *Feldbuch: Blatt Grindelwald, topographische Aufnahme 1:50000, 1860/61.* Bern-Wabern: Bundesamt für Landestopographie.

Kääb, A. 2001. Photogrammetric reconstruction of glacier mass balance using a kinematic ice-flow model: A 20-year time series on Grubengletscher, Swiss Alps. *Annals of Glaciology* 31: 45–52.

Kääb, A., and M. Funk. 1999. Modelling mass balance using photogrammetric and geophysical data: A pilot study at Griesgletscher, Swiss Alps. *Journal of Glaciology* 45:575–83.

Kane, E. K. 1856. *Arctic explorations: The Second Grinnell Expedition in Search of Sir John Franklin, 1853, '54, '55.* Philadelphia: Childs and Peterson.

Kempf, C. 1994. *Adolphe Braun et la photographie 1812–1877.* Illkirch: Éditions Lucigraphie/Valblor.

Lagoltière, R. M. 1989. Mulhouse et la conquête photographique des Alpes et du Mont-Blanc. *Annuaire Historique de la Ville de Mulhouse* 2:39–63.

Locher, T. 1954. Bernische Kartierung zur Zeit der Dufourkarte und Vorarbeiten zum bernischen Kataster. Ph.D. diss., University of Bern.

Lurie, E. 1960. *Louis Agassiz: A life in science.* Chicago: University of Chicago Press.

Maisch, M., A. Wipf, B. Denneler, J. Battaglia, and C. Benz. 1999. *Die Gletscher der Schweizer Alpen: Gletscherhochstand 1850, Aktuelle Vergletscherung, Gletscherschwund-Szenarien.* Zurich: vdf Hochschulverlag.

Morand, S., and C. Kempf. 1989. *Le temps suspendu: Le daguerréotype en Alsace au XIXe siècle, 150e anniversaire de la divulgation de la photographie.* Strasbourg: Éditions Oberlin.

Nesje, A., and S.O. Dahl. 2000. *Glaciers and environmental change.* London: Arnold.

Nicolussi, K., and G. Patzelt. 2000. Untersuchungen zur holozänen Gletscherentwicklung von Pasterze und Gepatschferner (Ostalpen). *Zeitschrift für Gletscherkunde und Glazialgeologie* 36:1–87.

Oberli, A. 1968. Vor 100 Jahren: Wie es zur Herausgabe der Siegfriedkarte kam. *Hauszeitung der Eidg. Landestopographie (Wabern)* 23:7–22.

Paul, F., A. Kääb, M. Maisch, T. Kellenberger, and W. Haeberli. 2004. Rapid disintegration of Alpine glaciers observed with satellite data. *Geophysical Research Letters* 31:L21402.

Portmann, J.P. 1975. Louis Agassiz (1807–1873) et l'étude des glaciers. *Denkschriften der Schweizerischen Naturforschenden Gesellschaft* 89:113–42.

Schmeits, M.J., and J. Oerlemans. 1997. Simulation of the historical variations in length of Unterer Grindelwaldgletscher, Switzerland. *Journal of Glaciology* 43:152–64.

Schönfeld, J. 2001. Die Stereoskopie: Zu ihrer Geschichte und ihrem medialen Kontext. Master's thesis, University of Tübingen.

Schuler, T., U.H. Fischer, and G. H. Gudmundsson. 2004. Diurnal variability of subglacial drainage conditions as revealed by tracer experiments. *Journal of Geophysical Research* 109:F02008.

SVGW (Schweizerischer Verein des Gas- und Wasserfaches). 2002. Der Trinkwasserkonsum in der Schweiz sinkt weiter. *Trinkwasser Informationsblatt* 9:1–2.

Swisstopo. 2004. *DHM25: The digital height model of Switzerland: Product information.* Bern-Wabern.

VAW/SANW (Versuchsanstalt für Wasserbau, Hydrologie und Glaziologie der Eidgenössischen Technischen Hochschule Zürich/Schweizerischen Akademie der Naturwissenschaften). 1881–2002. *Die Gletscher der Schweizer Alpen.* Jahrbücher der Glaziologischen Kommission der Schweizerischen Akademie der Naturwissenschaftern 1–122.

Wipf, A. 1999. *Die Gletscher der Berner, Waadtländer und nördlichen Walliser Alpen: Eine regionale Studie über die Vergletscherung im Zeitraum "Vergangenheit" (Hochstand 1850), "Gegenwart" (Ausdehnung 1973) und "Zukunft" (Gletscherschwundszenarien, 21. Jhdt.).* Zurich: Geographisches Institut der Universität Zürich.

Wolf, J.R. 1879. *Geschichte der Vermessungen in der Schweiz.* Zurich: Commission von S. Höhr.

Zölly, H. 1944. *Geodätische Grundlagen der Vermessungen im Kanton Bern: Geschichtlicher Überblick.* Schweizerische Zeitschrift für Vermessungswesen und Kulturtechnik, special issue.

Zumbühl, H.J. 1980. *Die Schwankungen der Grindelwaldgletscher in den historischen Bild- und Schriftquellen des 12. bis 19. Jahrhunderts: Ein Beitrag zur Gletschergeschichte und Erforschung des Alpenraumes.* Denkschriften der Schweizerischen Naturforschenden Gesellschaft 92. Basel, Boston, and Stuttgart: Birkhäuser.

Zumbühl, H.J., and H.P. Holzhauser. 1988. Alpengletscher in der Kleinen Eiszeit: Sonderheft zum 125jährigen Jubiläum des SAC. *Die Alpen* 64:129–322.

Zumbühl, H.J., B. Messerli, and C. Pfister. 1983. *Die Kleine Eiszeit. Gletschergeschichte im Spiegel der Kunst*: Katalog zur Sonderauss tellung, Schweizerisches Alpines Museum, Bern, 24.8.–16.10.1983; Gletschergarten-Museum, Luzern, 9.6.– 14.8.1983.

7

Long-Term Observations of Glaciers in Norway

Liss M. Andreassen, Hallgeir Elvehøy, Bjarne Kjøllmoen, Miriam Jackson, and Rune Engeset

Glaciers are an element of Norwegian life that is both fascinating and dangerous. Events such as icefalls, outburst floods, and glacier front advances have been recorded as far back as the 1700s; the total of 86 such events recorded in Norway over nearly three centuries includes outburst floods from Hardangerjøkulen (six events between 1736 and 1938 [Liestøl 1956; Elvehøy et al. 2002]) and Blåmannsisen (Engeset, Schuler, and Jackson 2005) and an icefall from Baklibreen in 1986 that killed three tourists (Kjøllmoen 2005). The popularity of glaciers with tourists has led to a long photographic and pictorial record of certain glaciers as visitors attempting to take their memories of the glaciers home with them recorded the extent and conditions of many glaciers over the past 150 years. Norwegian polar exploration, including the first visit to the South Pole and extensive early exploration on Svalbard, generally included a research component. Much activity in Norwegian glaciology has been carried out in the polar regions, especially in Svalbard but also in Antarctica. In the following, however, we focus on the glaciers of mainland Norway.

Glaciers cover about 1% of the land area in Norway, and many of them are situated in regions with considerable hydropower potential. In Norway, 98% of electricity is generated by hydropower production, and about 15% of the runoff used comes from glaciated basins. A good understanding of glaciers is therefore important to a coherent strategy of hydropower production. Glaciers are also considered a primary indicator of climate change (Houghton et al. 2001). While the mass balance reflects annual weather directly, records of length change (also termed front-position change) are considered as proxies for climate change on a decadal-to-century time scale (Oerlemans 2000; Hoelzle et al. 2003) and can be used to extract a temperature record (Oerlemans 2005).

The Norwegian glacier monitoring program is part of the Global Terrestrial Network for Glaciers coordinated by the World Glacier Monitoring Service (see Haeberli, Maisch, and Paul 2002). Norwegian mass balance and length change data sets are used extensively by researchers worldwide (e.g., Dyurgerov and Meier 1999; Rasmussen 2004). In this article we present an overview of the long-term monitoring program

FIGURE 7.1. Glaciers studied for length change (*left*) and mass balance (*right*). Glaciers are shaded in gray. Letters indicate glaciers mentioned in Figures 7.3 and 7.4. *Au*, Austdalsbreen; *Bu*, Buerbreen; *Br*, Briksdalsbreen; *E*, Engabreen; *H*, Hellstugubreen; *L*, Langfjordjøkelen; *M*, Midtdalsbreen; *N*, Nigardsbreen; *R*, Rembesdalsskåka; *Ste*, Stegholtbreen; *Sto*, Storbreen.

of Norwegian glaciers and report its main results on glacier change.

GLACIER MEASUREMENTS IN NORWAY

The first reliable observations of glaciers date from the first half of the eighteenth century (Hoel and Werenskiold 1962). In the period 1700–1750 glaciers in Norway advanced, often as much as several kilometers. The advancing glaciers caused serious damage to farms, some of which had to be struck off the tax register. Both Nigardsbreen and Engabreen (Figure 7.1) completely destroyed farms during their advance. During the nineteenth century several reports on Norwegian glaciers were produced by scientists and tourists who were traveling through Norway. (For a thorough review of early glacier work in Norway, see Hoel and Werenskiold 1962; for a chronological glaciological bibliography until 1961, see Hoel and Norvik 1962.)

The first systematic observations of Norwegian glaciers started around 1900, when glacier length change measurements were initiated and sketches, maps, and photographs were produced (e.g., Rekstad 1902; Øyen 1906). This pioneering work was carried out by individuals from various institutions rather than organized by the government. In 1927 the Norges Svalbard og Ishavsundersøkelser was established to investigate Svalbard and the polar regions, but investigations of glaciers in Jotunheimen were also undertaken. In 1948 glaciological work in Norway was taken over by the Norwegian Polar Institute (NPI), and Olav Liestøl was appointed glaciologist. Liestøl began the first mass balance measurements in mainland Norway at Storbreen in spring 1949 (Liestøl 1967). Then, in the 1960s, to investigate the contribution of glaciers to runoff in connection with the planning of hydropower production, systematic mass balance studies were initiated at selected glaciers and a glacier division was established by Gunnar Østrem in the Norwegian Water Resources and Energy Directorate (NVE). Through legislation, glacier mass balance measurements (short- and long-term programs) have been included in the licensing terms for hydropower production plants in glacierized basins. Although the bulk of glaciological work in Norway has been conducted by the NVE since the 1960s, considerable work has also been carried out by the NPI, the Universities of Oslo and Bergen, and researchers from other

TABLE 7.1

Organizations Involved in Glacier Monitoring in Norway (Excluding Foreign Researchers)

ORGANIZATION	COMMENT
Norges Svalbard og Ishavsundersøkelser (NSIU)	Established 1927. Survey, length change. Succeeded by NPI.
Norwegian Polar Institute (NPI)	Established in 1948. Initiated long-term mass balance program of Storbreen and Hardangerjøkulen, continued length change measurements. Activities transferred to NVE when NPI moved to Tromsø in 1994.
Norwegian Water Resources and Energy Directorate (NVE)	Glacier division established in 1962. Mass balance, length change; glacier inventories, glacier hydrology, ice dynamics, subglacial observatory. Contract work. Publishes annual report, *Glaciological Investigations in Norway*.
University of Oslo	Glaciological activity since the 1930s. Research, student projects, glacial geomorphology.
University of Bergen	Research, student projects, length change measurement of Midtdalsbreen since 1982, glacial geomorphology, reconstructions of Holocene glacier variations.

countries (Table 7.1). While traditional mass balance measurements have been the dominant activity of the NVE's glaciological work, many other studies have been conducted, often under contract to hydropower companies. Studies have included investigations in two unique subglacial laboratories (e.g., Hagen et al. 1993; Lappegard et al. 2005) and studies of ice dynamics (e.g., Jackson, Brown, and Elvehøy 2005).

The resulting mass balance and length change measurements as well as other glaciological investigations have been published annually or biannually since 1963 (e.g., Kjøllmoen 2005). The data are also reported to the World Glacier Monitoring Service and published in their series *Fluctuations of Glaciers* (e.g., Haeberli et al. 2005*b*) and *Glacier Mass Balance Bulletin* (e.g., Haeberli et al. 2005*a*).

MONITORING GLACIERS

SURVEY AND GLACIER INVENTORIES

Aerial photography and photogrammetric methods have long been popular ways of monitoring glaciers. The first commercial aerial photographs showing a Norwegian glacier were taken in 1937 and show part of the Folgefonni ice cap. Since then, most glacial areas in Norway have been photographed several times. A comprehensive aerial photographic survey of glaciers in Norway, covering 120 glacier units, was performed in 1997 and 1998 (Sorteberg 1998; Andreassen 2000). The photographs were compared with photos from the 1950s, 1960s, and 1970s and with topographical maps and have been used for detailed mapping of many glacier units. Repeated, detailed glacier mapping has been performed to calculate changes in glacier elevation, area, and volume. At the glaciers studied for mass balance, volume changes have been compared with cumulative mass balance results to evaluate the latter (Andreassen, Elvehøy, and Kjøllmoen 2002). However, the unknown accuracy of the maps lessens the significance of these comparisons. In recent years, high-accuracy satellite-based positioning systems (GPS) have been used to detect multiannual elevation changes at specific locations or along profiles, reflecting multiannual glacier volume changes. New technologies such as airborne laser scanning (altimetry) have shown promising results for mapping glacier areas at high resolution and accuracy (e.g., Geist et al. 2005).

The first detailed list of the numbers and areas of glaciers in Norway was made by Olav Liestøl in 1958 and published by Hoel and Werenskiold (1962). The list was based on topographic maps from the Norwegian Geographical Survey (Norsk Geografisk Oppmåling) at a scale of 1:100,000 and aerial photographs from the 1940s and 1950s for some areas. Subsequently, detailed glacier inventories based on vertical aerial photographs were made of southern Norway (Østrem and Ziegler 1969) and northern Scandinavia (Østrem, Haakensen, and Melander 1973). A second glacier inventory for southern Norway was completed in the late 1980s (Østrem, Selvig, and Tandberg 1988).

Methods used to measure mass balance have changed little over the years, and the amount of field work on the topic has been reduced on the basis of this experience. Further details on these measurements and calculations can be found in Østrem and Brugman (1991) and Andreassen et al. (2005). Since measurements started at Storbreen in 1949, mass balance has been measured on 42 glaciers, producing 535 observation years up to and including 2004 (Figure 7.1). To put this in perspective, the World Glacier Monitoring Service has a total of about 3,000 observation years from more than 200 glaciers worldwide, including the Norwegian glaciers. In Norway the total area of glaciers measured for mass balance is 483 km^2, nearly 20% of the glacierized area of mainland Norway ($\sim$2,600 km^2). In southern Norway, 6 glaciers have been measured since 1963 or before. They constitute a west-east profile extending from the maritime Ålfotbreen to the continental Gråsubreen. Half of the glaciers observed have been monitored for only six years or less. The number of glaciers being observed has varied through the years; since the 1960s it has been around 10.

Mass balance models are required to understand spatial and temporal variations in mass balance and the climate sensitivity of the glaciers. Traditionally, the degree-day model has been a popular choice, since in it ablation is related to temperature only (see, e.g., Laumann and Reeh 1993; Engeset et al. 2000), and a degree-day model has also been linked to a dynamic model to assess the possibilities of jökulhlaups (Elvehøy et al. 2002). Although degree-day models have the advantage of being computationally cheap and requiring only temperature as input data, a physical energy balance model is considered more physically correct (Greuell and Genthon 2003). Since October 2000 and September 2001, automatic weather stations have been operating on the tongues of Midtdalsbreen and Storbreen, respectively. These stations were erected and are operated by the Institute of Marine and Atmospheric Research, University of Utrecht, the Netherlands. The data are used to monitor the local climate of the glaciers and to calibrate and validate energy balance models (e.g., Andreassen and Oerlemans 2005).

The first satellite images of glaciers in northern Norway were taken in 1972 (e.g., Østrem 1975). These Landsat images showed good suitability for mapping of the transient snow line, and the transient snow line at the end of the melt season could be used as an estimate of the net balance (Østrem 1975; Østrem and Haakensen 1993). However, the weather conditions in Norway can make it difficult to find suitable cloud-free images at the end of the season.

NVE participated in the EU-funded OMEGA (Operational Monitoring System for European Glacial Areas) project, which considered several different methods for monitoring glaciers, including laser scanning, Landsat images, aerial photography, and airborne and satellite-based radar (e.g., Pellikka et al. 2001; Heiskanen et al. 2002). Several of these proved successful as a technique for measuring annual changes in glacier extent and volume, and one technique, laser scanning, had high enough resolution that it was successfully used to study seasonal changes in snow volume on the glacier (Geist et al. 2005).

GLACIER LENGTH CHANGE

Glacier length change is derived from annual measurements of distance between the glacier terminus and fixed landmarks. The glacier

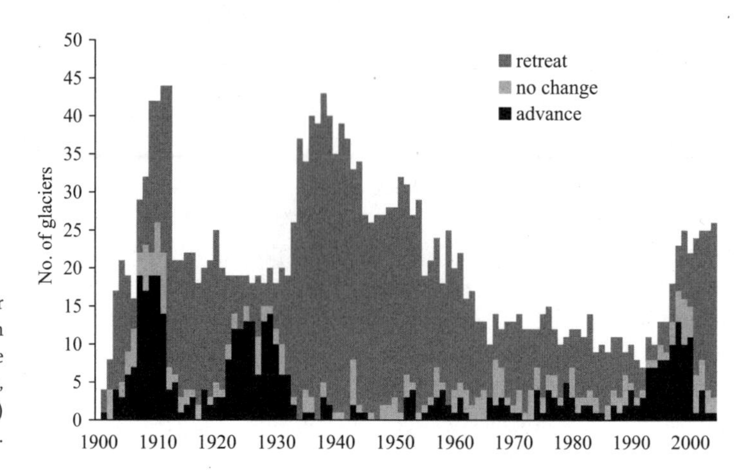

FIGURE 7.2. Annual number of measured Norwegian glaciers that retreated (more than 2 m), showed no change, and advanced (more than 2 m) between 1900 and 2004.

length measurements provide valuable information on glacier fluctuations and regional tendencies and variations when considering long time periods and patterns displayed by a number of glaciers in a single area. Since the first systematic observations were obtained around 1900, about 2,200 length change observation years from about 60 glaciers have been recorded (Figures 7.1 and 7.2). The size of the monitoring program has varied according to factors such as funding available. The observation program was revitalized in the 1990s in response to recent glacier advances. Eleven glaciers have a continuous record since initiation around 1900. The distribution of observed glaciers has been biased toward southern Norway, reflecting the costs in time and money of travel at the beginning of the twentieth century.

GLACIER MASS BALANCE

Glacier mass balance studies in Norway include measurements of accumulated snow (winter balance, b_w) during the winter season, and measurements of snow and ice removed by ablation (summer balance, b_s) during the summer season. The net balance is the sum of these two components: $b_n = b_w + b_s$, where b_s is negative. The methods used to measure mass balance have changed little over the years, whereas the amount of field work has been reduced on the basis of experience. Further details of measurements and calculations can be found in the works by Østrem and Brugman (1991) and Andreassen et al. (2005).

OBSERVED GLACIER CHANGES

Results from glacier length change records, surveys, and studies of aerial photographs and maps show that the Norwegian glaciers have been retreating throughout the twentieth century. The net retreat has been up to 2.5 km. However, several periods of advance and recession have been recorded since measurements started in 1899 (Figure 7.2). Many of the outlets from maritime ice caps in southern Norway had major advances culminating around 1910 and 1930. From the 1930s until 1990 a pronounced retreat took place for most glaciers in Norway. During this period many outlet glaciers from the coastal ice caps retreated 1–2 km, while many continental valley glaciers retreated 0.5–1 km. The most recent glacier advance in Norway started in the late 1980s and culminated in the late 1990s (Figures 7.2 and 7.3). This advance was mainly experienced by short and medium-length outlet glaciers from the coastal ice caps in western Norway. Since 2000 many glaciers have retreated rapidly, exceeding 90 m in a single year at Briksdalsbreen and Buerbreen (Figure 7.3).

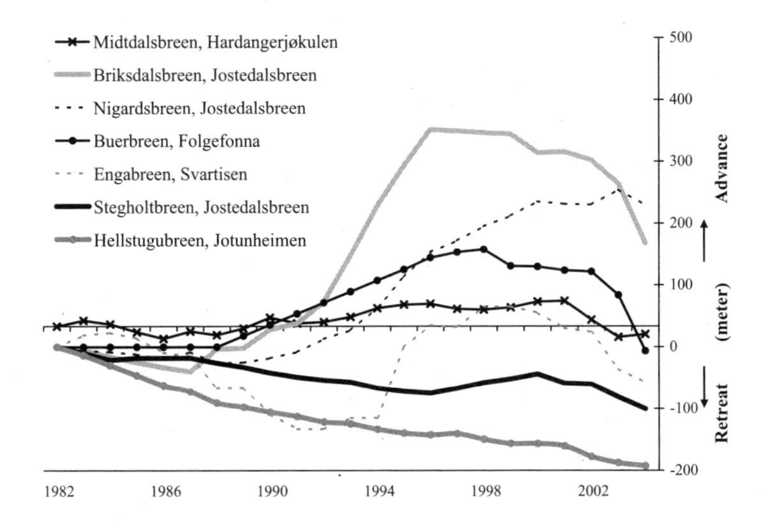

FIGURE 7.3. Cumulative length change for the period 1982–2004 for seven Norwegian glaciers.

Very few comparative ground-based observations of length changes are available from continental valley glaciers and ice caps in northern Norway. However, studies of aerial photographs and maps indicate that most glaciers in northern Norway have retreated considerably since 1900 (Andreassen 2000). Most glaciers were in a state of retreat until the 1960s, but in the period that followed up till 1998 the behavior of glaciers varied, with some advancing and some retreating even within the same ice cap (e.g., Svartisen and Blåmannsisen). However, glaciers in the northernmost part of the country (north of latitude 69°N) showed no sign of any advance from the 1960s to 1998.

The observed net balance profiles for nine of the glaciers observed for the period 1989–2003 reveal both the great difference in altitudinal range of the glaciers and the variation in mass turnover (Figure 7.4). The mass balance turnover at the tongues of Engabreen and Nigardsbreen is large in comparison with that for the glaciers at high elevations in the interior (Storbreen, Hellstugubreen, and Gråsubreen). The long-term recorded mass balance for the period 1963–2004 along a west-east profile in southern Norway reveals a strong gradient in both summer and winter values (Table 7.2). The glaciers located near the western coast had a much higher mass turnover than those located inland. The mean winter and summer balance values at Gråsubreen are only one-fifth and one-third, respectively, of those of Ålfotbreen.

The cumulative net balance series for glaciers in southern Norway for the period 1962(3)–2004 vary significantly (Figure 7.5, Table 7.2). While the coastal glaciers gained considerably in total mass, the continental glaciers had a total mass loss during the period 1962–2004. The period 1989–95 was exceptionally positive because of snow-rich winters, and the continental glaciers had a transient surplus in this period. However, the northernmost glacier, Langfjordjøkelen, where measurements started in 1989, had a net deficit in the same period. Although it is located near the coast, this glacier has had a cumulative deficit of 12-m water equivalent since 1989. This is in sharp contrast to the mass surplus of Engabreen, 500 km south of Langfjordjøkelen, and the coastal glaciers in southern Norway (Ålfotbreen and Nigardsbreen). However, since the mass balance year 2000–2001 all glaciers monitored for mass balance in Norway have shown a remarkable and rather uniform decrease in volume.

DISCUSSION

The Norwegian mass balance data set is comprehensive and reflects the importance of hydropower in Norway. Mass balance investigations

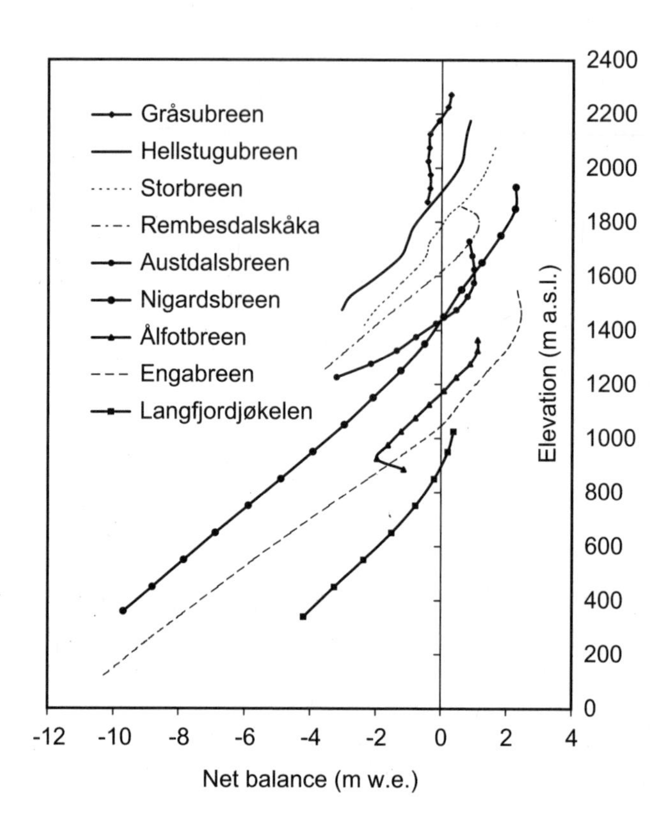

FIGURE 7.4. Mean net balance (m water equivalent) profiles by altitude for nine of the glaciers studied for mass balance in 1989.

TABLE 7.2

Results of Long-Term Mass Balance Investigations of Nine Glaciers in Norway Up to and Including 2004

GLACIER	AREA (KM^2)	PERIOD	NO. OF YEARS	b_w (M W.E.)	b_s (M W.E.)	b_n (M W.E.)	$\Sigma b_{n\ 63\text{-}04}$ (M W.E.)	$\Sigma b_{n\ 89\text{-}04}$ (M W.E.)
Ålfotbreen	4.5	1963–	42	3.7	−3.5	0.22	8.9	5.3
Nigardsbreen	47.8	1962–	43	2.4	−2.0	0.41	15.0	10.6
Rembesdalskåka	17.1	1963–	42	2.1	−2.0	0.12	4.9	6.1
Storbreen	5.4	1949–	56	1.4	−1.7	−0.26	−9.9	−3.0
Hellstugubreen	3.0	1962–	43	1.1	−1.4	−0.33	−15.2	−5.5
Gråsubreen	2.3	1962–	43	0.8	−1.1	−0.30	−14.0	−5.3
Engabreen	38.0	1970–	35	2.9	−2.3	0.63	—	12.1
Langfjordjøkelen	3.7	1989–	14[a]	2.2	−3.0	−0.75	—	−12.1

NOTE: b_w, winter, b_s, summer, and b_n, net balance; Σb_n, cumulative balances for these periods.

[a]Mass balance was not monitored in 1994 and 1995; the mass balance for these years was modeled using a degree-day model (Kjøllmoen and Olsen 2002).

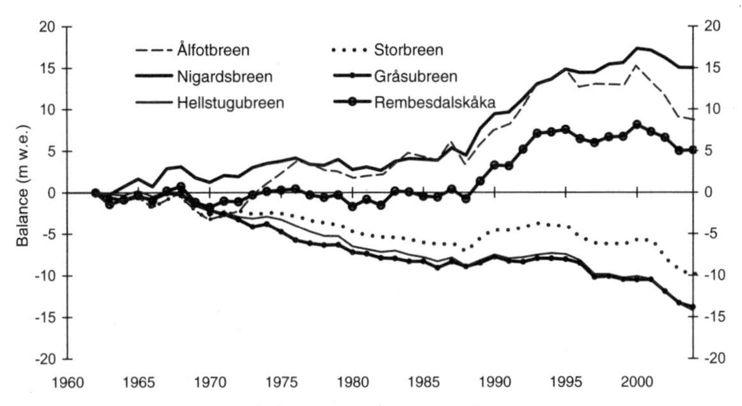

FIGURE 7.5. Cumulative net balance for six glaciers, 1963–2004.

are expensive and time-consuming, and future monitoring will depend on funding and unchanged licensing terms for hydropower companies as well as contributions from the government. However, although the data set is comprehensive, most glaciers remain unmeasured. Previous studies have shown that mass balance can vary widely with precipitation distribution, altitudinal range, and area within a region (Andreassen 2000; Engeset et al. 2000). Therefore, these factors should be taken into account before assuming that one glacier's mass balance record is representative for a whole area. This comment is even more applicable to length change records, since glaciers have different response times because of different steepness, length, and mass balance gradient (Jóhannesson, Raymond, and Waddington 1989). Therefore, monitoring of a number of glaciers in an area is needed to filter the influence of different glacier dynamics and geometries, as well as local meteorological conditions. The length change program is currently being extended to more glaciers in northern Norway to establish long-term series and gain more knowledge of glaciers in this part of the country. A detailed survey of the total glacierized area in Norway has not been performed since the glacier inventories of the mid-1980s for southern Norway (Østrem, Selvig, and Tandberg 1988) and the early 1970s for northern Norway (Østrem, Haakensen, and Melander 1973). Thus, a new updated survey of the Norwegian ice masses is planned to provide an overview of the present state of the glaciers and changes in them. As a Global Land Ice Measurements from Space (http://www.glims.org) regional center for Norway, NVE has started work on a new remote-sensing-based inventory.

All the glaciers monitored except the northernmost, Langfjordjøkelen, had a transient mass surplus for the period 1989–95. This mass surplus was caused mainly by the increase in winter precipitation for the maritime glaciers and a combination of a larger winter and a smaller summer balance than the average for the continental glaciers. This emphasizes the importance of winter precipitation for the net balance of maritime glaciers in Norway. The negative changes reported for Langfjordjøkelen reflect both its distance from the southern glaciers and variation in climate conditions from west to east and from south to north. The general trend of Norwegian glaciers in the twentieth century is mass deficit and reduction of area and length. This follows the trend in glacierized areas all over the world (e.g., Haeberli et al. 1999; Dyurgerov 2002; Arendt et al. 2002; Rignot, Rivera, and Casassa 2003). The mass deficit of Norwegian glaciers in 2001–04 can be explained in terms of a combination of less winter precipitation and higher summer ablation than normal. The 2002 summer was the warmest recorded in Norway since measurements began in 1876, while the 2003 summer was the fourth-warmest, causing record high ablation.

How will the predicted global warming (e.g., Førland et al. 2000; Engen-Skaugen et al. 2005) affect Norwegian glaciers? Studies of the possible effects of climate change on glacier mass balance at Ålfotbreen, Nigardsbreen, Hellstugubreen (e.g., Oerlemans 1992; Laumann and Reeh 1993; Roald et al. 2002), parts of the Svartisen ice caps including Engabreen (Engeset et al. 2000), and Rundvassbreen at the Blåmannsisen ice cap (Engeset, Schuler, and Jackson 2005) suggest a considerable mass deficit due to an increase in summer air temperature and consequently increased summer ablation. However, greater winter precipitation in the coastal areas might lessen the deficit. Glaciers with rather flat accumulation areas such as Hardangerjøkulen and narrow altitudinal ranges such as Ålfotbreen and Gråsubreen will be especially sensitive to the expected increase in equilibrium altitude. The impacts of climate change will be additional runoff to the reservoirs in the next few decades and a decrease in runoff when ice volume and extent are significantly reduced (Jóhannesson et al. 2004). The expected retreat and possible disappearance of glaciers will reduce or remove a fascinating and beautiful feature of the landscape and thus have an impact on the tourist industry.

CONCLUSIONS

Glaciers in Norway are important for hydropower production, climate research, local life, tourism, and safety. Glacier monitoring in Norway includes long-term records of mass balance and length, detailed glacier inventories, and mapping of changes by cartographic methods. Studies of ice dynamics, glacier hydrology, and a unique data series of ice (overburden) pressure measured at the ice-rock interface complete the monitoring program. Continuing long-term series are important both for the management of water resources in Norway and for national and international climate research.

The main findings of research on mass balance and length change in Norway are as follows: (1) There is a clear gradient in mean summer and winter balance along a west-east profile in southern Norway, the maritime glaciers located closer to the west coast having a much higher mass turnover than those located in drier continental regions. (2) The maritime glaciers, with a large annual mass turnover, had a large mass surplus between 1962 and 2000, while the continental glaciers, with small summer and winter balances, had a mass deficit over the same period. (3) In the period 1989–95 all the glaciers monitored except the northernmost, Langfjordjøkelen, had a transient mass surplus. The increase was mainly caused by higher winter precipitation. (4) For the period 2001–04, all the glaciers monitored had a considerable mass deficit. (5) Norwegian glaciers have retreated throughout the twentieth century. Continental glaciers have, with a few exceptions, retreated, while many maritime glaciers have been through periods of advance and recession. Since 2000 nearly all glaciers have retreated.

ACKNOWLEDGMENTS

We thank the numerous field workers who have been involved in collecting data from Norwegian glaciers. Special thanks go to the pioneers who initiated length change and mass balance measurements in Norway and provided us with such long time series. Tore Tonning made the map. Editor Ben Orlove and two anonymous reviewers provided useful comments.

REFERENCES CITED

Andreassen, L. M., ed. 2000. *Regional change of glaciers in northern Norway.* Report 1. Oslo: Norwegian Water Resources and Energy Directorate.

Andreassen, L. M., H. Elvehøy, and B. Kjøllmoen. 2002. Using aerial photography to study glacier changes in Norway. *Annals of Glaciology* 34: 343–47.

Andreassen, L. M., H. Elvehøy, B. Kjøllmoen, R. Engeset, and N. Haakensen. 2005. Glacier mass balance and length variation in Norway. *Annals of Glaciology* 42:317–25.

Andreassen, L. M., and J. Oerlemans. 2005. Measuring and modelling the energy and mass balance of Storbreen, Norway. *Geophysical Research Abstracts* 7:03923.

Arendt, A. A., K. A. Echelmeyer, W. D. Harrison, C. S. Lingle, and V. B. Valentine. 2002. Rapid wastage of Alaska glaciers and their contribution to rising sea level. *Science* 297:382–86.

Dyurgerov, M. 2002. Glacier mass balance and regime: Data of measurements and analysis. In *Institute of Arctic and Alpine Research Occasional Paper.* 55, ed. M. Meier and R. Armstrong. Boulder, CO: Institute of Arctic and Alpine Research, University of Colorado, distributed by National Snow and Ice Data Center, Boulder, CO.

Dyurgerov, M. B., and M. F. Meier. 1999. Analysis of winter and summer balances. *Geografiska Annaler* 81A:541–54.

Elvehøy, H., R. Engeset, L. M. Andreassen, J. Kohler, Y. Gjessing, and H. Björnson. 2002. Assessment of possible jökulhlaups from Lake Demmevatnet in Norway. *IAHS* 271:31–36.

Engen-Skaugen, T., L. A. Roald, S. Beldring, E. J. Førland, O. E. Tveito, K. Engeland, and R. Benestad. 2005. *Climate change impacts on water balance in Norway.* EMEP report 1/2005. Oslo: Norwegian Meteorological Institute.

Engeset, R. V., H. Elvehøy, L. M. Andreassen, N. Haakensen, B. Kjøllmoen, L. R. Roald, and E. Roland. 2000. Modelling of historic variations and future scenarios of the mass balance of the Svartisen ice cap, northern Norway. *Annals of Glaciology* 31:97–103.

Engeset, R. V., T. V. Schuler, and M. Jackson, M. 2005. Analysis of the first jökulhlaup at Blåmannsisen in northern Norway and implications for future events. *Annals of Glaciology* 42:35–41.

Førland, E., L. A. Roald, O. E. Tveito, and I. Hanssen-Bauer. 2000. *Past and future variations in climate and runoff in Norway.* DMNI Report. 19/00.

Geist, T., H. Elvehøy, M. Jackson, and J. Stötter. 2005. Investigations on intra-annual elevation changes using multitemporal airborne laser scanning data: Case study Engabreen, Norway. *Annals of Glaciology* 42:195–201.

Greull, W., and C. Genthon. 2004. Modelling land-ice surface mass balance. In *Mass balance of the cryosphere: Observations and modeling of contemporary and future changes,* ed. J. L. Bamber and A. J. Payne. Cambridge, UK: Cambridge University Press.

Haeberli, W., R. Frauenfelder, M. Hoelzle, and M. Maisch. 1999. On rates and acceleration trends of global glacier mass changes. *Geografiska Annaler* 81A:585–91.

Haeberli, W., M. Maisch, and F. Paul. 2002. Mountain glaciers in global climate-related observation networks. *WMO Bulletin* 51(1):18–25.

Haeberli, W., J. Noetzli, M. Zemp, S. Baumann, R. Frauenfelder, and M. Hoelzle, eds. 2005*a*. *Glacier mass balance bulletin 8 (2002–2003).* Zurich: World Glacier Monitoring Service/University of Zurich/IUGG (CCS)/UNEP/UNESCO/WMO.

Haeberli, W., M. Zemp, R. Frauenfelder, M. Hoelzle, and A. Kääb. 2005*b*. Fluctuations of glaciers 1995–2000. Vol. 8. Zurich: World Glacier Monitoring Service/IUGG(CCS)/UNEP/UNESCO.

Hagen, J. O., O. Liestøl, J. L. Sollid, B. Wold, and G. Østrem. 1993. Subglacial investigations of Bondhusbreen, Flgefonni. *Norsk Geografisk Tidsskrift* 47:117–62.

Heiskanen, J., K. Kajuutti, M. Jackson, H. Elvehøy, and P. Pellikka. 2003. Assessment of glaciological parameters using Landsat satellite data in Svartisen, Northern Norway. In *Proceedings of EARSeL Workshop on Observing Our Cryosphere from Space: Techniques and Methods for Monitoring Snow and Ice with Regard to Climate Change, 11–13 March 2002, Bern, Switzerland,* 34–42.

Hoel, A., and J. Norvik. 1962. *Glaciological bibliography of Norway.* Norsk Polarinstitutt Skrifter 126.

Hoel, A., and W. Werenskiold. 1962. *Glaciers and snowfields in Norway.* Norsk Polarinstitutt Skrifter 114.

Hoelzle, M., W. Haeberli, M. Dischl, and W. Pescke. 2003. Secular glacier mass balances derived from cumulative glacier length changes. *Global and Planetary Change* 36:295–306.

Houghton, J. T., et al., eds. 2001. *Climate change 2001: The scientific basis. Contributions of Working Group 1 to the third assessment report of the Intergovernmental Panel on Climate Change.* Cambridge, UK: Cambridge University Press.

Jackson, M., I. Brown, and H. Elvehøy. 2005. Velocity measurements on Engabreen. *Annals of Glaciology* 42:29–34.

Jóhannesson, T., C. F. Raymond, and E. D. Waddington. 1989. A simple method for determining the response time of glaciers. In *Glacier fluctuations and climate change,* ed. J. Oerlemans, 343–52. Dordrecht: Kluwer Academic Publishing.

Jóhannesson, T., G. Adalgeirsdóttir, H. Björnsson, C. E. Boggild, H. Elvehøy, S. Gudmundsson, R. Hock, P. Holmlund, P. Jansson, F. Pálsson, O. Sigurdsson, and T. Thorsteinsson. 2004. *The impact of climate change on glaciers in the Nordic countries.* Climate, Water, and Energy Glaciers Group Report 3. Reykjavik.

Kjøllmoen, B., ed. 2005. *Glaciological investigations in Norway in 2004.* NVE Report 2.

Kjøllmoen, B., and H. C. Olsen. 2002. *Langfjordjøkelen i Vest-Finnmark.* NVE Dokument 4.

Lappegard, G., J. Kohler, J. O. Hagen, and M. Jackson. 2005. Long time series with subglacial load cell data at Engabreen, Norway. *Journal of Glaciology* 52:137–48.

Laumann, T., and N. Reeh. 1993. Sensitivity to climate change of the mass balance of the glaciers in southern Norway. *Journal of Glaciology* 39:656–65.

Liestøl, O. 1956. Glacier dammed lakes in Norway. *Norsk Geografisk Tidsskrift* 15(3–4):122–49.

———. 1967. Storbreen glacier in Jotunheimen, Norway. *Norsk Polarinstitutt Skrifter* 141.

Oerlemans, J. 1992. Climate sensitivity of glaciers in southern Norway: Application of an energy-balance model to Nigardsbreen, Hellstugubreen, and Ålfotbreen. *Journal of Glaciology* 38:223–32.

———. 2000. Holocene glacier fluctuations: Is the current rate of retreat exceptional? *Annals of Glaciology* 31:39–44.

———. 2005. Extracting a climate signal from 169 glacier records. *Science* 308:675–77.

Østrem, G. 1975. ERTS data in glaciology: An effort to monitor glacier mass balance from satellite imagery. *Journal of Glaciology* 73:403–15.

Østrem, G., and M. Brugman. 1991. *Glacier mass-balance measurements: A manual for field office work.* Saskatchewan: Environment Canada. National Hydrology Research Institute Scientific Report 4.

Østrem, G., and N. Haakensen. 1993. Glaciers of Norway. In *Satellite image atlas of glaciers of the world,* ed. E. S. Williams and J. G. Ferrigno, 63–109. Professional Paper 1386–E. Washington, DC: United States Geological Survey.

Østrem, G., N. Haakensen, and O. Melander. 1973. *Atlas over breer i Nord-Skandinavia.* Meddelelse 22 fra Hydrylogisk Avdeling. Oslo: Norges Vassdrags- og Elektrisitetsvesen.

Østrem, G., K. D. Selvig, and K. Tandberg. 1988. *Atlas over breer i Sør-Norge.* Meddelelse 61 fra Hydrologisk Avdeling. Oslo: Norges Vassdrags- og Energiverk.

Østrem, G., and T. Ziegler. 1969. *Atlas over breer i Sør-Norge.* Meddelelse nr. 20 fra Hydrologisk Avdeling. Oslo: Norges Vassdrags- og Elektrisitetsvegen.

Øyen, P.A. 1906. Klima und Gletscherschwankungen in Norwegen. *Zeitschrift für Gletscherkunde* 1:46–61.

Pellikka, P., K. Kajuutti, R. Koskinen, M. Jackson, H. Stötter, H. Haggrén, K.-M. Luukkonen, T. Guneriussen, and A. Sharov. 2001. Development of an operational monitoring system for glaciers: Synthesis of earth observation data of the past, present, and future. In *Proceedings, International Workshop on Geo-spatial Knowledge Processing for Natural Resource Management, 28–29 June 2001, Varese, Italy,* 283–88.

Rasmussen, L. A. 2004. Altitude variation of glacier mass balance in Scandinavia. *Geophysical Research Letters* 31: L13401, doi: 10.1029/2004GL020273.

Rekstad, J. 1902. Iakttagelser fra bræer i Sogn og Nordfjord. *Norges Geologiske Underøkelse Aarbog* 1902(3): 1–48.

Rignot, E., A. Rivera, and G. Casassa. 2003. Contribution of the Patagonia icefields of South America to sea level rise. *Science* 302:434–37.

Roald, L. A., T. E. Skaugen, S. Beldring, T. Væringstad, R. Engeset, and E. Førland. 2002. *Scenarios of annual and seasonal runoff for Norway.* NVE Oppdragsrapport Series A 10. Oslo: Norwegian Meteorological Institute.

Sorteberg, H. K. 1998. *Regonal breovervåking i Sør-Norge 1997.* NVE Rapport 8. Oslo: Norwegian Water Resources and Energy Directorate.

8

Alpinewide Distributed Glacier Mass Balance Modeling

A TOOL FOR ASSESSING FUTURE GLACIER CHANGE?

Frank Paul, Horst Machguth, Martin Hoelzle, Nadine Salzmann, and Wilfried Haeberli

Glacier changes are the clearest natural signal of ongoing atmospheric warming (Houghton et al. 2001). The related temperature increase of about 1 °C since 1850 in the Alps has caused glacier retreats of up to 3 km with a mean volume loss of about 50% by the 1970s (Haeberli, Maisch, and Paul 2002). The glacier forefields, enclosed by lateral moraines more than 100 m high, are eloquent witnesses of the glaciers' former dimensions recently documented by Zängl and Hamberger (2004). Analyses of the latest satellite data (Paul et al. 2007) indicate that down-wasting (i.e., stationary thinning) was a dominant reaction of glaciers to the extraordinary warm decade of the 1990s. In particular, the summer 2003 heat wave in Central Europe (Beniston and Diaz 2004; Schär et al. 2004) caused record-breaking glacier thinning of about −2.5 m on average (Haeberli, Huggel, and Paul 2004), ten times the 1960–2000 annual mean (Hoelzle et al. 2003) and twice the melt of the former record year, 1998. Thus, changes in glacier thickness currently seem to have a major influence on further glacier evolution, even for comparatively large or fast-flowing glaciers. Because most glaciers in the Alps (81%) are smaller than 0.5 km^2 (Zemp et al., this volume) and such small glaciers show only a limited dynamic reaction to changes in climate (i.e., mass redistribution by glacier flow), thickness changes are currently the dominant reaction for most glaciers.

Changes in glacier thickness can be modeled with satisfying accuracy from what are called distributed glacier mass balance models based either on a temperature index method (e.g., Braithwaite 1989; Hock 2003) or on an energy balance approach (e.g., Arnold et al. 1996; Brock et al. 2000; Oerlemans 1991, 1992). The latter in general requires a large amount of meteorological input data (depending on the complexity of the model) and the calculation of the mass balance distributed over the terrain for all cells of an underlying digital elevation model (DEM) in the course of a year (e.g., Klok and Oerlemans 2002). Up to now, most studies have focused on a few individual glaciers only, and the full potential for large-area application has not been fully exploited. To apply mass balance models to large regions, some of the required climatic input data must be interpolated from point observations to continuous fields or obtained from gridded climatologies.

In this study we utilize such gridded data sets (i.e., radiation, albedo, precipitation) to apply a mass balance model to larger catchments. Melt of snow and ice (i.e., ablation) is calculated from an energy balance model developed by Klok and Oerlemans (2002), and accumulation is assimilated from interpolated precipitation data published by Schwarb et al. (2001). The main focus of our approach is whether mass balance models can produce realistic mass balance values and equilibrium line altitudes (ELAs) for all the glaciers of a larger catchment. If this is the case, it may be possible to explain the observed differences in the behavior of neighboring glaciers in response to the same climatic forcing. The validation of the model results will not account for measurements from specific years but will be compared with averaged long-term data such as the steady-state ELA, which is roughly approximated for most Alpine glaciers by a ratio of area of accumulation to ablation area of 2:1 (Maisch et al. 2000). Direct measurements have revealed that this ratio exhibits a large scatter among glaciers, from roughly 1:1 to 3:1 (Haeberli et al. 2005).

To speed up the processing, the gridded input data sets (in particular, the 365 grids of mean daily potential solar radiation) are prepared beforehand, including an interpolation to the cell spacing (25 m) of the DEM (e.g., for precipitation). While the mass balance model itself is based on a Fortran code, all related data processing is performed within a geographic information system (GIS), which also facilitates the final calculation of mass balance values and profiles of individual glaciers.

In this contribution we first give a brief overview of glacier mass balance modeling and then discuss the main components of a distributed mass balance model. Next the gridded data sets used and their application to a test region in the Swiss Alps are presented, and the results of the calculation are described and critically discussed. Finally, we offer some suggestions for future improvements.

PREVIOUS WORK ON GLACIER MASS BALANCE MODELING

A general characteristic of all mass balance models is that the more complex the model, the more input data it requires. Thus, if only a few input parameters (e.g., mean annual air temperature) are available for a specific region, there is no reason to run a physically complex model. Several field studies in recent years have, however, produced both parameterization schemes for several meteorological variables (e.g., Brutsaert 1975; Kimball, Idso, and Aase 1982; Munro 1989) and insight into the processes governing a glacier's mass balance (e.g., Greuell, Knap, and Smeets 1997; Oerlemans 2000; Strasser et al. 2004). This allows the application of physically more complex models if a suitable parameterization for an unmeasured but required variable can be found. The core of all distributed mass balance models is a DEM that allows the "distribution" of point measurements (e.g., data from climate stations or automatic weather stations) on the topography by means of elevation-dependent gradients such as the lapse rate for temperature. Generally, two types of models have evolved in recent years for the calculation of glacier melt (which is only one component of the mass balance): temperature index and energy balance models. While temperature index models are based on the sum of positive degree-days in a year fitted by a glacier-specific degree-day factor to observed glacier melt (e.g., Braithwaite and Zhang 2000; Hock 2003), energy balance models compute all terms of the energy balance (e.g., radiation balance and turbulent fluxes) explicitly and calculate melt from a positive energy balance (see review by Hock 2005). Degree-day models have proved to calculate runoff from snow and glacier melt quite accurately at well-calibrated sites (e.g., Hock and Noetzli 1997); they have also been applied in snow-melt runoff models for many years (e.g., Martinec and Rango 1986). Their success can be attributed mainly to incoming longwave radiation, which is strongly correlated with surface temperature (at screen height) and dominates the energy balance in the case of highly reflective surfaces such as snow and ice from polar glaciers (Ohmura 2001). Moreover, modified degree-day models have been used to obtain first-order estimates of future sea-level rise according to temperatures predicted by global climate models (e.g., Gregory and Oerlemans 1998; Van de Wal and Wild 2001), and they have recently been extended and improved by incorporating a factor for solar radiation calculated from digital elevation modeling (Hock 1999; Konya, Matsumoto, and Naruse 2004), getting somewhat closer to the energy balance approach.

Early mass balance models based on the calculation of the energy balance (e.g., Greuell and Oerlemans 1987; Oerlemans 1991, 1992) adopted a one-dimensional approach (elevation-dependent parameterizations) on a flat surface and climatic means of temperature (*T*) and precipitation (*P*). The mass balance profiles obtained (mean mass balance in specific elevation intervals) were in close agreement with measured profiles and clearly demonstrate the potential of such simple formulations for further applications. In particular, sensitivity studies for selected parameters (*T*, *P*) according to results from global climate models have been performed several times in an effort to assess the related changes in ELA for various glaciers (e.g., Cook et al. 2003; Kull, Grosjean, and Veit 2002; Schneeberger et al. 2003). Similar sensitivity studies have revealed that in the rough topography of the Alps the high temporal variation of incoming shortwave radiation and the high spatial variation of the glacier albedo are the most important parameters (Brock et al. 2000; Strasser et al. 2004). The forcing of energy balance models by direct measurements from automatic weather stations on the glacier or from nearby climate stations uses DEMs for the distribution of the point data on the terrain (e.g., Arnold et al. 2006; Brock et al. 2000; Escher-Vetter 2000). However, most researchers stress that some attention has to be paid to the spatial interpolation of point measurements because data from automatic weather stations are influenced by the microclimate of the glacier (e.g., wind direction [see Oerlemans 2000, 2001]) and data from appropriate climate stations (e.g., at a similar elevation) that are too far away may have no correlation to the conditions at the study site (e.g., because of clouds). If climatic means of temperature and precipitation are available on a monthly basis, the seasonal sensitivity characteristic will allow the determination of the temporal variation of mass balance sensitivity (Oerlemans and Reichert 2000). Such an approach can be used for reconstruction of the recent mass balance history of a glacier from measured climatic parameters (Wildt, Klok, and Oerlemans 2003).

In summary, the main advantage of degree-day models is the small amount of input data they require; their main disadvantage is the restriction to a specific glacier with a known degree-day factor, which can exhibit substantial variation from glacier to glacier (see Hock 2003). The main advantage of energy balance models is that their strict, process-based physical rules allow their application to large regions. Their major disadvantage is the large amount of (meteorological) input data they require, although these data can be parameterized from other known variables or held constant. With the advent of globally available reanalysis data (e.g., Kalnay et al. 1996; New, Hulme, and Jones 2000), new possibilities of forcing mass balance models through extended scales of time and space have emerged. Initial studies by Cook et al. (2003), Radick and Hock (2006), and Rasmussen and Conway (2003) have had promising results.

Up to now, energy balance models have been applied mostly to individual glaciers, with a strong focus on the exact physical formulation of the energy balance components (e.g., Hock 2005). The models have therefore not been applied to large regions with sparse data sets. In our approach we employ parts of the physically simpler model (e.g., the calculation of mean daily temperatures from a cosine wave) of Oerlemans (1992) with the more complex model of Klok and Oerlemans (2002) and combine the two with gridded data sets from parameters that show high spatial variation at local to regional scales (i.e., potential solar radiation, glacier albedo, precipitation) but do not change much from year to year. (Machguth et al. 2006)

COMPONENTS OF THE GLACIER MASS BALANCE MODEL

The mass balance of a glacier is a result of all the processes contributing to loss of glacier mass (ablation, calving) and gain of mass (solid precipitation) in the course of a balance year. For glaciers in the European Alps the dominant processes are ablation in summer due to a positive energy balance and accumulation of snow in winter due to solid precipitation (neglecting calving, refreezing of snow, etc.). Both components exhibit considerable spatial and temporal variation. However, our focus here is the calculation of mean potential glacier locations and sensitivity studies using climatic means of meteorological variables. The mass balance (MB) can thus be written (neglecting a cold reserve from winter) as MB = ablation + accumulation, with the energy balance (EB) composed of incoming and outgoing shortwave (SW_{in} and SW_{out}), and longwave (LW_{in} and LW_{out}) radiation terms and sensible (H_s) and latent (H_l) heat fluxes:

$$EB = SW_{in} - SW_{out} + LW_{in} - LW_{out} + H_s + H_l,$$
$$\text{with } SW_{in} - SW_{out} = (1 - \alpha) \cdot G.$$

Here α is the glacier albedo and G the global radiation (the sum of direct and diffuse radiation). While the radiation terms are mainly controlled by topography, cloud cover, and air temperature, the turbulent fluxes mainly depend on air temperature, wind speed, and humidity (see Oerlemans 2001).

TEMPERATURE

Air temperature influences energy balance in two major ways, affecting both incoming longwave radiation (emission from clouds at a specific elevation) and sensible heat flux (temperature difference between the air and the glacier surface). Temperature has an additional influence on glacier mass balance in that solid precipitation takes place only below a certain threshhold temperature (between 0 °C and 2 °C). In general, the temporal course of air temperature is governed by the annual and daily cycles, while its spatial variation depends mainly on terrain elevation. Two approaches for calculating temperature at a specific elevation have emerged. One uses a mean annual air temperature from climate station data and calculates the daily (hourly) value from an annual (daily) temperature range and a cosine function looped over 365 Julian days (24 hours). This approach is suitable for the calculation of mean mass balance profiles and allows studies of general sensitivity (change in ELA due to a change in climate) if daily temperature data are unavailable (e.g., Oerlemans 1992). It is also employed in this study with mean daily values. The other approach uses daily (or even hourly) climate station data and allows the calculation of mass balance for a specific year, thus permitting a comparison with measurements (Machguth et al. 2006). In both approaches temperature is calculated at a specific point of the DEM from a lapse rate and the elevation difference between the DEM and the climate station.

RADIATION

For the purpose of our model we calculated shortwave (solar) radiation by means of the computer code SRAD (e.g., Wilson and Gallant 2000), which gives the mean daily incoming potential solar radiation. This value includes all topographic effects (terrain slope and aspect, shadowing) related to direct and diffuse (i.e., global) radiation and assumes standard atmospheric profiles and zero cloud cover. Surface albedo and cloud cover are not accounted for at this stage. This allows an a priori calculation of the potential solar radiation for each day of a calendar year and the respective region. The shortwave radiation balance is then mainly modified by albedo and cloud cover. While the principal pattern of glacier albedo distribution is obtained from satellite data, we use an average climatic cloud factor of 0.5. Future refinements of the program can incorporate a time-dependent parameterization of cloud cover (e.g., from reanalysis data), but for present purposes cloud cover can be held constant. Outgoing longwave radiation is also held constant (assuming a 0 °C glacier surface temperature throughout the year), while incoming longwave radiation varies strongly with the temperature of the atmosphere and cloud cover. In our model we use the parameterization by Greuell and Oerlemans (1987) with constant cloud cover and cloud base height. In the Alps, incoming and outgoing longwave radiation are of the same order of magnitude, resulting in a longwave balance close to zero (e.g., Klok, Greuell, and Oerlemans 2003; Strasser et al. 2004). This is one reason for the domination of the shortwave balance over the entire radiation balance. The other reason is the low albedo of Alpine glaciers (due to soot and dust on the surface) and the long ablation period, which result in a high positive shortwave radiation balance. Extension of the model with a two-layer glacier surface would allow for the generation of a cold surface layer during winter that delays snow melt in spring (Greuell and Oerlemans 1987), but for present purposes this delay is neglected.

ALBEDO OF ICE AND SNOW

The shortwave radiation balance is strongly governed by the glacier (i.e., ice and snow) albedo. Several studies in recent years have shown that glacier albedo can be accurately mapped over large regions from 25-m-resolution Landsat Thematic Mapper (TM) data (e.g., Knap, Reijmer, and Oerlemans 1999; Klok, Greuell, and Oerlemans 2003). While there is not much spatial and temporal variation in glacier albedo in winter, in summer the retreat of the snowline (rapid temporal change) exposes the complex albedo pattern of the bare glacier ice (strong spatial variation). This complex pattern can be obtained from Landsat data acquired at the end of the ablation season in a year with a strongly negative mass balance. In our model the background bare-ice albedo is revealed in the course of the modeled snow line retreat. In the case of heavily debris-covered glaciers, a glacier map obtained from automatic mapping with TM data (using the TM4/TM5 ratio method (see Paul et al. 2002) could be used for delineation of the bare glacier ice, allowing the calculation of differential ablation due to debris cover (e.g., Pelto 2000; Takeuchi, Kayastha, and Nakawo 2000). This feature is not part of our current model.

The decrease of snow albedo due to aging (metamorphosis) is adjusted by an exponential decay rate related to the number of days since the last snowfall (Klok and Oerlemans 2002); other approaches use the sum of positive degree-days instead. Therefore the change in glacier albedo due to summer snowfall is considered in our model by starting with a constant value (0.72) for the albedo of fresh snow (snow depth >0) and then applying the aging function. The bidirectional reflectance characteristics of ice and snow result in a forward scattering of the incoming radiation at low solar elevations. This is not considered in the model because the change in albedo is only by a few percent (e.g., Greuell and de Ruyter de Wildt 1999; Knap and Reijmer 1998).

TURBULENT FLUXES

Several studies of the measurement and calculation of sensible and latent heat fluxes exist (e.g., Denby and Greuell 2000; Munro 1989; Suter, Hoelzle, and Ohmura 2004), but most of them are based on detailed field measurements of generally unknown parameters (e.g., wind speed at various heights above the surface, roughness length). Simpler approaches have yielded reliable results as well (see Oerlemans 2001). For instance, temperature, which mainly governs sensible and latent heat fluxes, can be interpolated from coarse gridded climatologies or nearby climate stations at the elevation of the DEM by using a specific lapse rate (e.g., Ishida and Kawashima 1993). Consideration of appropriate exchange coefficients for both sensible and latent fluxes as well as a mean relative humidity and atmospheric pressure for latent flux allows a sufficient description of turbulent fluxes (e.g., Oerlemans 2001) and is also used in our model. In other regions of the world, both fluxes may play a more dominant role for the energy balance and must be parameterized in more detail, and additional fluxes (e.g., from refreezing of rain or sublimation) have to be considered as well (e.g., Kaser and Osmaston 2002; Mölg and Hardy 2004; Wagnon et al. 1999). The inclusion of wind speed (e.g., from reanalysis data) in a future model would allow an improved formulation of turbulent fluxes.

PRECIPITATION

The modeling of glacier mass balance must also consider the amount of solid precipitation during the accumulation season. The normal way is to use a measured precipitation sum (e.g., monthly or yearly), extrapolate this value by means of a gradient (often dependent on elevation) to each cell of the DEM, and identify the solid part from a threshold temperature, normally assumed to be 0–2 °C (Oerlemans 2001). Because of the spatial variation of precipitation even in terms of climatic means (Frei and Schär 1998), this approach does not work well for large regions. Unfortunately, most of the available data on precipitation in high-mountain regions with rough topography are quite inaccurate, and the related interpolation schemes are very complex (see, e.g., Daly, Neilson, and Phillips 1994). One option is the use of the gridded climatologies that have recently become available (for an overview see Gyalistras 2003), but in general they have a too coarse cell size for direct use in a 25-m-resolution DEM, and questions of appropriate down-sampling strategies arise. In our model we use the highest-spatial-resolution data set available, one deduced from the data from more than 6,000 rain gauges by Frei and Schär (1998). It covers uncorrected annual and monthly climatic means for the period 1971–90 at 2-km spatial resolution for the entire perimeter of the Alps (Schwarb et al. 2001).

Another important issue is the redistribution of snow by wind and avalanches. Instead of complex three-dimensional transport models (e.g., Liston and Sturm 1998), some researchers have successfully employed rather simple schemes from GIS-based digital elevation modeling (Mittaz et al. 2002; Purves et al. 1998). Such schemes allow for calculation of snow erosion and deposition on stoss and lee sides of mountain ridges from a mean wind speed and direction obtained from climate station or reanalysis data. Similarly, snow redistribution by avalanches can be calculated from GIS-based modeling taking into account a critical slope threshold for snow deposition and rules of mass conservation (e.g., Mittaz et al. 2002). This allows the inclusion of snow-free rock walls and some additional accumulation of snow in glacier depressions and on lee sides. For the model results presented here we have not included snow redistribution by wind, but areas steeper than 60° are set to zero snow depth, and for slopes from 30° to 60° snow depth is multiplied by a linear factor from one to zero, respectively.

DATA SOURCES AND PROCESSING

TEST REGION

To assess the performance of the model we have selected a larger test site in the Mischabel group of mountains in the eastern Valaisian Alps, Switzerland, close to the Italian border (Figure 8.1, *inset*). The region is 23 by 27 km in size (equivalent to 680 by 930 pixels at 25-m spacing) and covers about 50 glaciers of various sizes, types, elevation ranges, exposures, and slopes. The elevation range is from 1,000 to 4,550 m a.s.l., with ELAs ranging from 2,800 to 3,400 m a.s.l. The mean annual air temperature at Zermatt (1,640 m a.s.l.) is 3.5 °C, and the cumulative annual precipitation ranges from 900 to 1,700 mm at 3,000 m a.s.l. (Schwarb et al. 2001). The region is shielded from precipitation (which normally comes from the northwest) by the Alpine main ridge, resulting in more continental-type glaciers. With a mean annual air temperature of about −5 °C at the

FIGURE 8.1. Location of the Mischabel test region (*inset, black rectangle*) and overview of the region modeled in a fused satellite image (using data from IRS-1C and Landsat TM). Image size is 17 km by 23 km; glaciers show up in light gray and snow in white. *F*, Findelen Glacier; *S*, Schwarzberg Glacier.

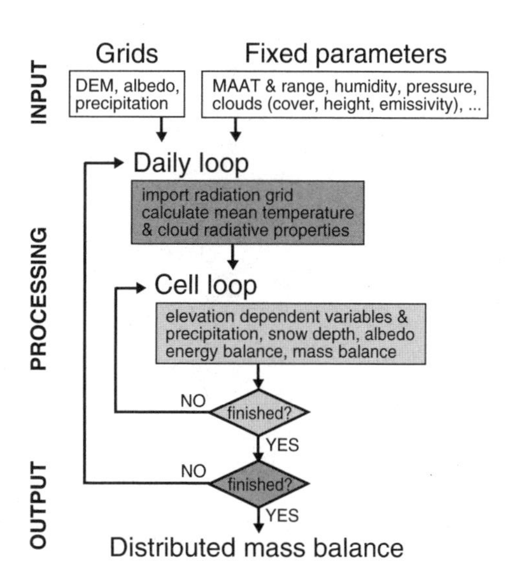

FIGURE 8.2. Schematic workflow of the mass balance model. The inner cell loop computes mass balance for each cell of the digital elevation model before the outer daily loop succeeds it on the following day. After all days are calculated, the final mass balance grid is exported.

ELA, glaciers are mostly temperate, with partly cold ablation areas surrounded by discontinuous permafrost.

INPUT DATA

A schematic workflow of the model is given in Figure 8.2. The program is initiated with a number of constant meteorological parameters (e.g., latent heat of melt, Stefan-Boltzmann constant, specific heat capacity of air) and several variables that are held constant but can later be taken from other sources (e.g., reanalysis data). For the reference simulation we use the following fixed variables: mean annual air temperature at 2,000 m a.s.l. 2 °C with an annual temperature range of 7.5 °C, snowfall threshold temperature 1.5 °C, cloud cover 0.5, cloud height 5,000 m, clear-sky emissivity 0.7, mean surface pressure (at sea level) 1,013 hPa, and relative humidity 80%. Furthermore, the following raster data sets are directly imported on a cell-by-cell basis from the hard disk at the beginning of each model run: the DEM25 (level 2) from swisstopo, the annual precipitation sum from the Schwarb et al. (2001) climatology, which is converted by bilinear interpolation to the 25-m cell size beforehand (Figure 8.3, A), and the albedo map derived from a Landsat TM scene (path-row 195-28) acquired on August, 31, 1998 (Figure 8.3, B). The mean daily potential solar radiation from the SRAD code is imported within the daily loop. Examples for March 15 and July 15 are presented in Figure 8.4. They clearly indicate the dependence of the available radiation energy on topography and exposure as well as the much higher radiation values (up to two times) obtained in July.

DATA PROCESSING

The program for the mass balance model is based on a Fortran code and organized as follows: Input data are assigned (meteorological variables) and imported (grids), and the program starts with a daily loop beginning on Julian day 271 (October 1). The respective radiation grid is imported and elevation-independent variables are calculated. Then a cell loop calculates the

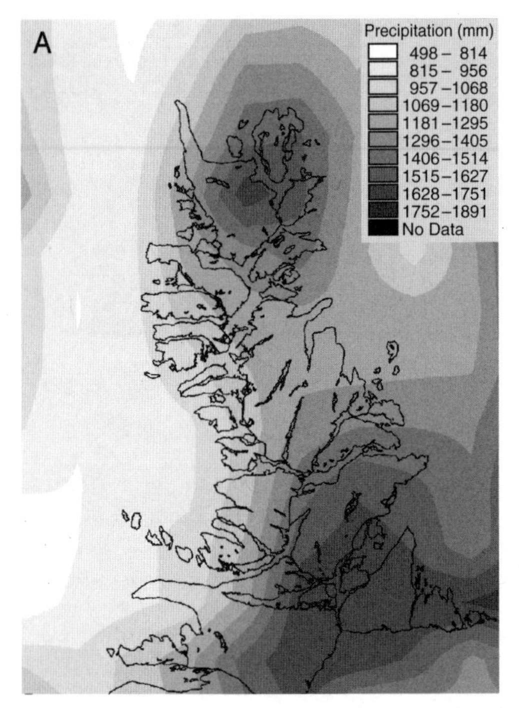

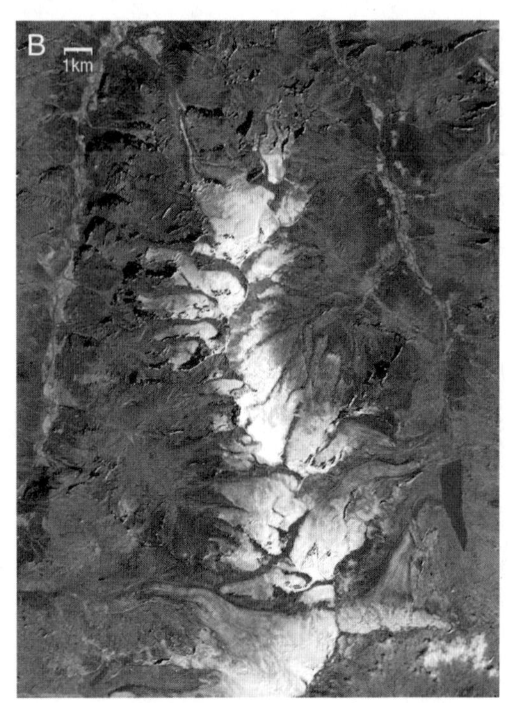

FIGURE 8.3. A, annual precipitation sum from the Schwarb et al. (2001) climatology after bilinear resampling to 25-m cell size; *black lines*, 1973 glacier extent. B, albedo map as derived from the Landsat TM scene 195–28 acquired on August 31, 1998. The effects of shading have been mostly removed, and the terrain appears quite flat. For better visibility, albedo values from 0.0 to 1.0 have been linearly rescaled from 0 (*black*) to 255 (*white*).

mass balance for each cell of the DEM (net energy balance, precipitation, melt, snow depth) and the specified day and proceeds with the following Julian day. In the current version of the program we have constant precipitation on every fifth day with 1/73 of the annual sum. After a one-year cycle is finished, the cumulative mass balance is exported to an ASCII file and transformed to the Arc/Info grid format for further analysis and visualization. A one-year simulation for the 632,400 cells of the test region is accomplished in about eight minutes on a 900-MHz Sunfire workstation. For comparison with other studies and evaluation of the model's performance, the sensitivity of the model is analyzed with respect to the reference simulation for changes in temperature, precipitation, and albedo. Temperature is changed by ±2 °C, precipitation by ±20%, and albedo by ±0.1. Mass balance values and profiles are obtained for each simulation for Findelen Glacier, including the adjacent Adler Glacier.

RESULTS

Selected results of the distributed mass balance model are presented in Figure 8.5. For better visibility of details we display only the southern part of the test site, with Findelen Glacier (area 16.7 km^2) in the lower center (Figure 8.1). Selected mass balance profiles of the model experiments are displayed in Figure 8.6. The reference simulation using the input data is depicted in Figure 8.5, A. The mass balance distribution pattern in the ablation area (light gray) results from the interaction of potential solar radiation (see Figure 8.4) with local variations in albedo (see Figure 8.3, B), which is also valid for the sensitivity studies discussed below. For the brighter parts of the Findelen Glacier tongue, minimum

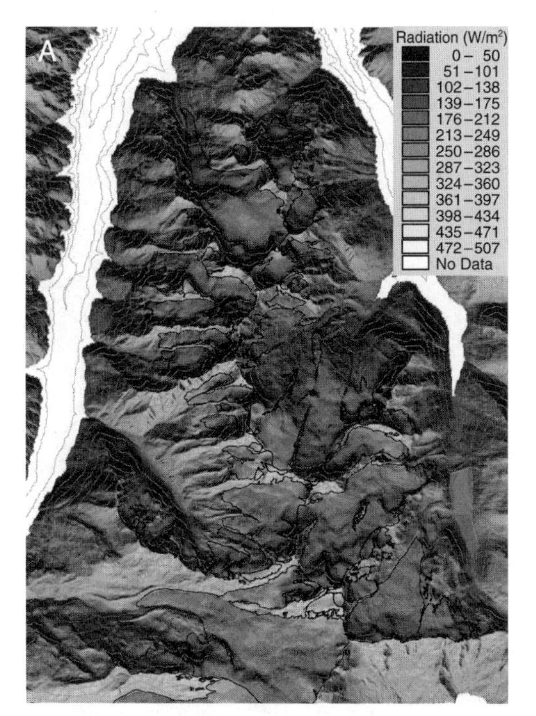

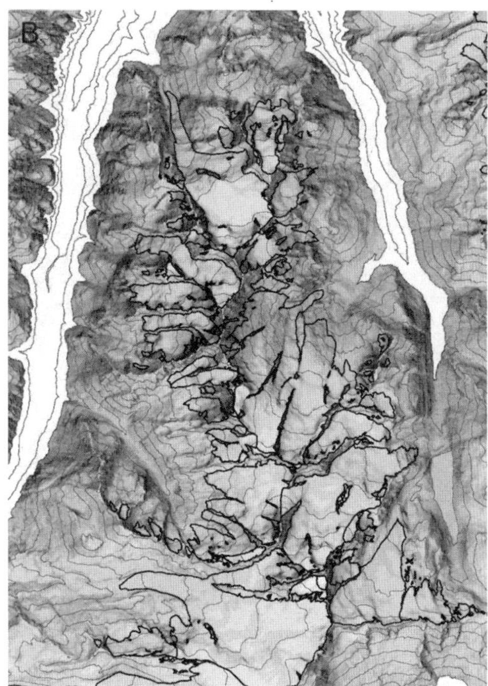

FIGURE 8.4. Mean daily potential solar radiation as computed for the Mischabel test region from the SRAD code for March 15 (A) and July 15 (B). *Thin white lines*, Elevation contours at 200 m equidistance; *thicker white lines and black lines*, glacier outlines. Reproduced by permission of swisstopo (BA057338).

mass balance values of −8 m water equivalent result. The accumulation pattern (*dark gray*) depends more on the elevation (see Figure 8.3, A), with locally some lower-reaching snow bands in more radiation-shielded parts of the terrain. Some snow has been removed in the model in the highest parts of some glaciers because of the steepness of the terrain. Some of the high, small glaciers are completely snow-covered, indicating highly positive mass balances.

The mass balance profile for Findelen Glacier (Figure 8.6) gives an ELA of 3,210 m with a gradient of 0.9 m per 100 m below the equilibrium line. The ELA is thus about 80 m higher than an assumed 2:1 ELA at 3,130 m. However, the mass balance for the reference simulation is −0.25-m water equivalent, indicating that the steady-state ELA is somewhat lower than 3,210 m (if the model is correct), and, of course, the 2:1 approximation may have a different ratio for this glacier as well.

In Figure 8.5, B, the mass balance distribution for a 2 °C increase in mean annual air temperature is shown. In this simulation all the glaciers except a few smaller ones (*white oval*) between the Findelen and Schwarzberg glaciers have negative mass balances. While both of these glaciers receive the highest precipitation amounts (Figure 8.3, A), the left one (to the west) also receives the highest amounts of radiation (Figure 8.3, B). In this case the positive balance can be explained by the high albedo as obtained from satellite data. Maximum ablation values on the tongue of the Findelen Glacier partially exceed −10-m water equivalent, which is in accord with observations reported from the Gries Glacier during the hot summer of 2003. The ELA increases by 250 m, which is in close agreement with other observations (reporting +100 to +150 m for a 1 °C temperature increase). The mass balance sensitivity is −0.9 m per degree temperature increase and +0.6 m per degree decrease.

A decrease in precipitation by 20% (Figure 8.5, C) does not change the mass balance distribution (also visible in Figure 8.6) but results in more negative balances for all glaciers.

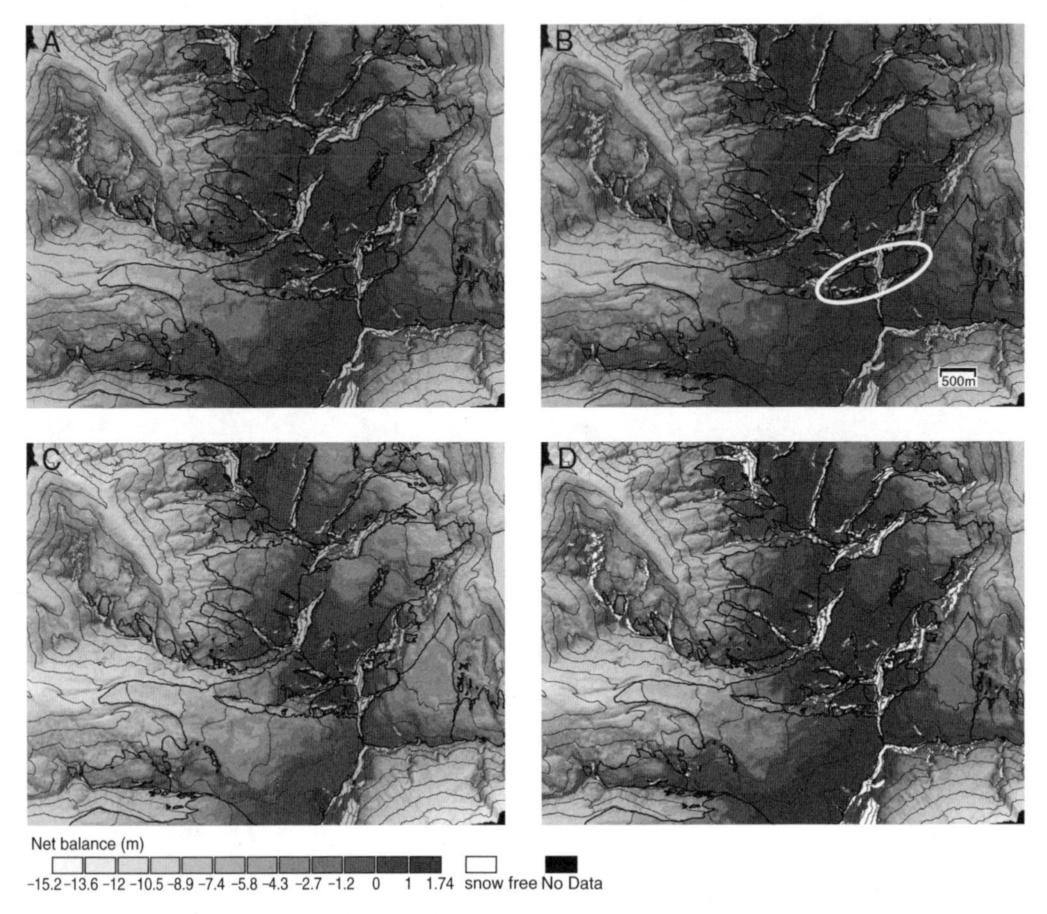

FIGURE 8.5. Mass balance distribution as derived from the model for (A) the reference simulation (B), a temperature increase of 2 °C (C), a precipitation decrease of 20%, and (D) an albedo decrease of 0.1. *White*, regions steeper than 60° (snow depth = 0); *thin black lines*, contours at 200-m equidistance; *thick black lines*, glacier outlines. Reproduced by permission of swisstopo (BA057338).

On the Findelen Glacier the ELA shifts 80 m upward (about 60 m downward) for a precipitation decrease (increase) of 20% (Figure 8.6). This indicates that a 1°C increase in temperature may be compensated for by at least a 40% increase (about 500 mm at the ELA) in precipitation. In fact, an even greater increase is required because of the positive feedback with the snowfall temperature. The mass balance for the Findelen Glacier is −0.9 m (+0.25 m) for a −20% (+20%) change in precipitation.

While a general decrease of the albedo by 0.1 (Figure 8.5, D) leads only to a slight shift in ELA (by about +30 m), it has a much stronger effect on the mass balance in the ablation area (sometimes reaching −10-m water equivalent) with a related steepening of the mass balance gradient (see Figure 8.6). Thus, the gradual decrease of glacier albedo due to the accumulation of soot and dust in dry, hot summers (as in 2003), when there is no cleaning by heavy precipitation events, produces a strong positive feedback for enhancing glacier melt. The mass balance is −0.56 m (+0.03 m) for an albedo change of −0.1 (+0.1). The digital subtraction of the respective mass balance grids (reference minus sensitivity study) reveals the greatest changes near the equilibrium line,

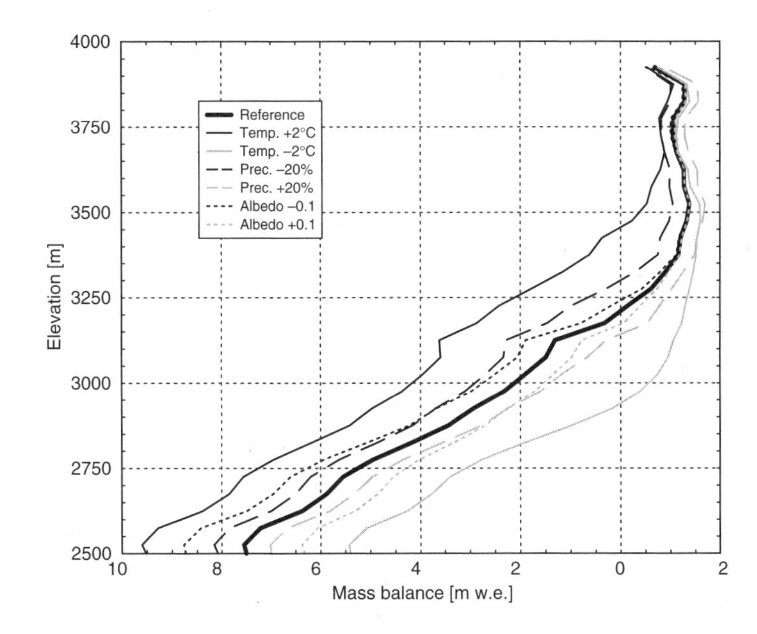

FIGURE 8.6. Mass balance profiles of Findelen Glacier for the six sensitivity studies performed. Average values are calculated for elevation intervals at 50-m equidistance.

moderate changes in the ablation zone, and small changes in the accumulation area (not shown).

DISCUSSION

Although we find a reasonable mass balance distribution and ELA for the Findelen Glacier, visual inspection of the reference simulation clearly indicates larger deviations on neighboring glaciers. While the Findelen Glacier has a slightly negative mass balance (−0.25-m water equivalent), other glaciers, in particular the smaller ones at higher elevations, are in balance or strongly positive. Of course, the hypsography of a glacier governs its local mass balance sensitivity and results in natural variation of the net balance from glacier to glacier under identical climate conditions (Haeberli et al. 2005). However, the strong variations over a small distance observed here cannot be fully explained by hypsography. Moreover, meteorological variables that exhibit variations only on a regional scale (e.g., temperature, lapse rate, pressure, humidity) cannot be responsible, and the high spatial and temporal variability of precipitation is present more on a daily time scale than in climatic means. Thus the change to a more realistic temporal precipitation pattern in our model (e.g., using monthly means and a more realistic forcing instead of every fifth day) will not alter the simulated pattern significantly. Albedo values that are too high (snow in the satellite image) cannot explain the positive mass balances of many glaciers either, as the snow line on the 1998 satellite image was much higher (see Figure 8.3, B) than the one resulting from the simulation. Therefore, more local processes that have not yet been considered, such as redistribution of snow by wind and avalanches or orographic clouds, may be responsible for the local deviations. Their modeling remains a major challenge for future studies.

In principle the model can be used to calculate the mass balance distribution for much larger regions (e.g., 100 km by 100 km) and for any other location on earth where DEM data are available and glaciers are present. As these data can now be obtained at the appropriate spatial resolution (e.g., from the Shuttle Radar Topography Mission or stereoscopic ASTER or SPOT satellite data) for nearly every place on earth, a remaining challenge would be the compilation of meteorological data for remote areas. While the forcing of a mass balance model by climatic mean values can be achieved even with data from printed climate atlases (e.g., mean annual temperature and precipitation sum), a forcing for a specific year requires data from climate stations or automatic weather stations near the location of interest. If no such data are available, an alternative might be the use of reanalysis data (e.g., Kalnay et al. 1996; New, Hulme, and Jones 2000), which can be further downscaled with regional climate models to better resolve topographic effects. Satellite data to account for glacier-specific albedo variation are available more or less worldwide, but an appropriate DEM and some image-processing software are required for orthorectification and atmospheric correction. In high-mountain topography the spatial interpolation of precipitation remains a major challenge. While in small catchments an elevation-dependent gradient might be applied, for larger catchments such a gradient might vary considerably (e.g., Frei and Schär 1998). However, Oerlemans (2001) has shown that even simple models provide realistic mass balance profiles if the major components are described and tuned in a realistic way. Therefore there is a good chance of replacing the degree-day approach with distributed mass balance models for an improved assessment of glacier mass balance in highly glacierized and climate-sensitive regions (temperate glaciers).

The availability of reanalysis data should also permit the inclusion of parameterization schemes for the redistribution of snow by wind (e.g., Mittaz et al. 2002; Purves et al. 1998) and nighttime cloud cover, which exert a major impact on regional mass balance distribution and energy balance, respectively, but are not yet accounted for in most models. The validation of the modeled distribution of snow water equivalent for glaciers at the end of winter would require expensive field surveys (e.g., Plattner, Bran, and Brenning 2006). Major improvements of the model presented here in the ablation area should focus on the calculation of differential ablation due to debris cover and the incorporation of a two-layer snow/ice surface.

CONCLUSIONS AND PERSPECTIVES

Glacier down-wasting has become an increasingly important part of glacier-climate feedback in the past two decades. Calculation of such changes by means of distributed melt modeling can already assess the related discharge with high accuracy. Of the two main approaches available, energy balance calculation is more suitable in the Alps than the degree-day approach because of the much lower and more variable albedos of Alpine glaciers compared with Arctic ones and the potential for large-area application. The latter requires the incorporation of gridded data sets (radiation, albedo, precipitation, reanalysis data) that have become available in recent years or can be computed from additional sources (satellite, DEMs). However, some processes operating at a local scale (especially snow redistribution by wind) must be included in future models to obtain more realistic results. Therefore, future research should focus on such processes.

The great potential of such distributed and calibrated models consists in their ability to be forced with meteorological parameters obtained from regional climate models. Although the current spatial resolution of these models (about 20 km) requires the application of appropriate downscaling techniques (e.g., Salzmann et al. 2007), their resolution is steadily improving. Moreover, some parameters (e.g., temperature, pressure) do not change much over large regions and can therefore be easily incorporated into current mass balance models. In particular, the availability of reanalysis data will improve current models with respect to regional processes (e.g., wind, clouds) that are not yet accounted for and may be responsible for the variation of mass balance from glacier to glacier observed at a catchment scale.

ACKNOWLEDGMENTS

This work has been supported by grants from COST Action 719 (BBW C001.0041) and the European Union–funded Fifth Framework Project ALP-IMP (EVK-CT-2002-00148).

REFERENCES CITED

Arnold, N.S., W.G. Rees, A.J. Hodson, and J. Kohler. 2006. Topographic controls on the surface energy balance of a high Arctic valley glacier. *Journal of Geophysical Research* 111(F2), F02011.

Beniston, M., and H.F. Diaz. 2004. The 2003 heat wave as an example of summers in a greenhouse climate? Observations and climate model simulations for Basel, Switzerland. *Global and Planetary Change* 44:73–81.

Braithwaite, R.J. 1989. Calculation of glacier ablation from air temperature. In *Glacier fluctuations and climatic change*, ed. J. Oerlemans, 219–33. Dordrecht: Kluwer.

Braithwaite, R.J., and Y. Zhang. 2000. Sensitivity of mass balances of five Swiss glaciers to temperature changes assessed by tuning a degree-day model. *Journal of Glaciology* 46(152):7–14.

Brock, B.W., I.C. Willis, M.J. Sharp, and N.S. Arnold. 2000. Modelling seasonal and spatial variations in the surface energy balance of Haut Glacier d'Arolla, Switzerland. *Annals of Glaciology* 31:53–62.

Brutsaert, W. 1975. On a derivable formula for long-wave radiation from clear skies. *Water Resources Research* 11:742–44.

Cook, K.H., X. Yang, C.M. Carter, and B.N. Belcher. 2003. A modeling system for studying climate controls on mountain glaciers with application to the Patagonian Icefields. *Climatic Change* 56:339–67.

Daly, C., R.P. Neilson, and D.L. Phillips. 1994. A statistical-topographic model for mapping climatological precipitation over mountainous terrain. *Journal of Applied Meteorology* 33:140–58.

Denby, B., and W. Greuell. 2000. The use of bulk and profile methods for determining surface heat fluxes in the presence of glacier winds. *Journal of Glaciology* 46(154):445–52.

Escher-Vetter, H. 2000. Modelling meltwater production with a distributed energy balance method and runoff using a linear reservoir approach: Results from Vernagtferner, Oetztal Alps, for the ablation seasons 1992 to 1995. *Zeitschrift für Gletscherkunde und Glazialgeologie* 36:119–50.

Frei, C., and C. Schär. 1998. A precipitation climatology of the Alps from high-resolution rain-gauge observations. *International Journal of Climatology* 18:873–900.

Gregory, J.M., and J. Oerlemans. 1998. Simulated future sea-level rise due to glacier melt based on regionally and seasonally resolved temperature changes. *Nature* 391:474–76.

Greuell, W., W.H. Knap, and P.C. Smeets. 1997. Elevational changes in meteorological variables along a mid-latitude glacier during summer. *Journal of Geophysical Research* 102 (D22):25941–54.

Greuell, W., and M.S. de Ruyter de Wildt. 1999. Anisotropic reflection by melting glacier ice: Measurements and parametrizations in Landsat TM bands 2 and 4. *Remote Sensing of Environment* 70:265–77.

Greuell, W., and J. Oerlemans. 1987. Sensitivity studies with a mass balance model including temperature profile calculations inside the glacier. *Zeitschrift für Gletscherkunde und Glazialgeologie* 22:101–24.

Gyalistras, D. 2003. Development and validation of a high-resolution monthly gridded temperature and precipitation data set for Switzerland (1951–2000). *Climate Research* 25:55–83.

Haeberli, W., C. Huggel, and F. Paul. 2004. Gletscherschwund im Hochgebirge: Eine Herausforderung für die Wissenschaft. In *Alpenvereinsjahrbuch Berg 2005*, ed. W. Theil, 25–31. Munich: Deutscher Alpenverein/Innsbruck: Österreichischer Alpenverein/Bozen: Alpenverein Südtirol.

Haeberli, W., M. Maisch, and F. Paul. 2002. Mountain glaciers in global climate-related observation networks. *WMO Bulletin* 51(1):18–25.

Haeberli, W., J. Noetzli, M. Zemp, S. Baumann, R. Frauenfelder, and M. Hoelzle, eds. 2005. *Glacier mass balance bulletin. 8 (2002–2003)*. Zurich: World Glacier Monitoring Service/University of Zurich/IUGG(CCS)/UNEP/UNESCO/WMO.

Hock, R. 1999. A distributed temperature index ice and snow melt model including potential direct solar radiation. *Journal of Glaciology* 45(149):101–11.

———. 2003. Temperature index melt modelling in mountain areas. *Journal of Hydrology* 282:104–15.

———. 2005. Glacier melt: A review of processes and their modelling. *Progress in Physical Geography* 29:362–91.

Hock, R., and C. Noetzli. 1997. Areal melt and discharge modelling of Storglaciären, Sweden. *Annals of Glaciology* 24:211–16.

Hoelzle, M., W. Haeberli, M. Dischl, and W. Peschke. 2003. Secular glacier mass balances derived from cumulative glacier length changes. *Global and Planetary Change* 36:295–306.

Houghton, J. T., et al., eds. 2001. *Climate change 2001: The scientific basis (Contribution of Working Group 1), Third assessment report of the Intergovernmental Panel on Climate Change.* Cambridge, UK: Cambridge University Press.

Ishida, T., and S. Kawashima. 1993. Use of cokriging to estimate surface air temperature from elevation. *Theoretical and Applied Climatology* 47:147–57.

Kalnay, E., M. Kanamitsu, R. Kistler, W. Collins, D. Deaven, L. Gandin, M. Iredell, S. Saha, G. White,

J. Woollen, Y. Zhu, A. Leetmaa, B. Reynolds, M. Chelliah, W. Ebisuzaki, W. Higgins, J. Janowiak, K. C. Mo, C. Ropelewski, J. Wang, R. Jenne, and D. Joseph. 1996. The NCEP/NCAR 40-year reanalysis project. *Bulletin of the American Meteorological Society* 77:437–71.

Kaser, G., and H. Osmaston. 2002. *Tropical glaciers.* Cambridge, UK: Cambridge University Press.

Kimball, B. A., S. B. Idso, and J. K. Aase. 1982. A model of thermal radiation from partly cloudy and overcast skies. *Water Resources Research* 18:931–36.

Klok, E. J., W. Greuell, and J. Oerlemans. 2003. Temporal and spatial variation of the surface albedo of Morteratschgletscher, Switzerland, as derived from 12 Landsat images. *Journal of Glaciology,* 48(163):491–502.

Klok, E. J., and J. Oerlemans. 2002. Model study of the spatial distribution of the energy and mass balance of Morteratschgletscher, Switzerland. *Journal of Glaciology* 49(167):505–18.

Knap, W. H., and C. H. Reijmer. 1998. Anisotropy of the reflected radiation field over melting glacier ice: Measurements in Landsat-TM bands 2 and 4. *Remote Sensing of Environment* 65:93–104.

Knap, W. H., C. H. Reijmer, and J. Oerlemans. 1999. Narrowband to broadband conversion of Landsat-TM glacier albedos. *International Journal of Remote Sensing* 20:2091–110.

Konya, K., T. Matsumoto, and R. Naruse. 2004. Surface heat balance and spatially distributed ablation modelling at Koryto Glacier, Kamchatka Peninsula, Russia. *Geografiska Annaler* 86A:337–48.

Kull, C., M. Grosjean, and H. Veit. 2002. Modelling modern and late Pleistocene glacio-climatological conditions in the North Chilean Andes (29°S–30°S). *Climatic Change* 52:359–81.

Liston, G. E., and M. Sturm. 1998. A snow-transport model for complex terrain. *Journal of Glaciology* 44:498–516.

Machguth, H., F. Paul, H. Hoelzle, and W. Haeberli. 2006. Distributed glacier mass balance modelling as an important component of modern multi-level glacier monitoring. *Annals of Glaciology* 43:335–43.

Maisch, M., A. Wipf, B. Denneler, J. Battaglia, and C. Benz. 2000. *Die Gletscher der Schweizer Alpen: Gletscherhochstand 1850, Aktuelle Vergletscherung, Gletscherschwund-Szenarien.* Zurich: vdf Hochschulverlag.

Martinec, J., and A. Rango. 1986. Parameter values for snowmelt runoff modeling. *Journal of Hydrology* 84:197–219.

Mittaz, C., M. Imhof, M. Hoelzle, and W. Haeberli. 2002. Snowmelt evolution mapping using an energy balance approach over an alpine terrain. *Arctic, Antarctic, and Alpine Research* 34:274–81.

Mölg, T., and D. R. Hardy. 2004. Ablation and associated energy balance of a horizontal glacier surface on Kilimanjaro. *Journal of Geophysical Research* 109:D16104.

Munro, D. S. 1989. Surface roughness and bulk heat transfer on a glacier: Comparison with eddy correlation. *Journal of Glaciology* 35(121):343–48.

New, M., M. Hulme, and P. D. Jones. 2000. Representing twentieth-century space-time climate variability. Pt. 2. Development of 1901–96 monthly grids of terrestrial surface climate. *Journal of Climate* 13:2217–38.

Oerlemans, J. 1991. A model for the surface balance of ice masses. Pt. 1. Alpine glaciers. *Zeitschrift für Gletscherkunde und Glazialgeologie* 27/28:63–83.

———. 1992. Climate sensitivity of glaciers in southern Norway: Application of an energy-balance model to Nigardsbreen, Hellstugubreen, and Ålfotbreen. *Journal of Glaciology* 38(155):223–32.

———. 2000. Analysis of a 3-year meteorological record from the ablation zone of Morteratschgletscher, Switzerland: Energy and mass balance. *Journal of Glaciology* 46:571–79.

———. 2001. *Glaciers and climate change.* Lisse: A. A. Balkema.

Oerlemans, J., and B. K. Reichert. 2000. Relating glacier mass balance to meteorological data using a seasonal sensitivity characteristic (SSC). *Journal of Glaciology* 46(152):1–6.

Ohmura, A. 2001. Physical basis for the temperature-based melt-index method. *Journal of Applied Meteorology* 40:753–61.

Paul, F., A. Kääb, and W. Haeberli. 2007. Recent glacier changes in the Alps observed from satellite: Consequences for future monitoring strategies. *Global and Planetary Change* 56: 111–22.

Paul, F., A. Kääb, M. Maisch, T. Kellenberger, and W. Haeberli. 2002. The new remote-sensing-derived Swiss glacier inventory: 1. Methods. *Annals of Glaciology* 34:355–61.

Paul, F., A. Kääb, M. Maisch, T. W. Kellenberger, and W. Haeberli. 2004. Rapid disintegration of Alpine glaciers observed with satellite data. *Geophysical Research Letters* 31:L21402.

Pelto, M. S. 2000. Mass balance of adjacent debris-covered and clean glacier ice in the North Cascades, Washington. *IAHS* 264:35–42.

Plattner, C., L. Braun, and A. Brenning. 2006. The spatial variability of snow accumulation at Vernagtferner, Austrian Alps, in winter 2003/2004. *Zeitschrift für Gletscherkunde und Glazialgeologie* 39:43–57.

Purves, R. S., J. S. Barton, W. A. Mackaness, and D. E. Sugden. 1998. The development of a rule-based spatial model of wind transport and deposition of snow. *Annals of Glaciology* 26:197–202.

Radić, V., and R. Hock. 2006. Modeling future glacier mass balance and volume changes using ERA −40 reanalysis and climate models: A sensitive study at Storglaciaren, Sweden. *Journal of Geophysical Research* 111(F3), F03003.

Rasmussen, L.A., and H.B. Conway. 2003. Using upper-air conditions to estimate South Cascade Glacier (Washington, U.S.A.) summer balance. *Journal of Glaciology* 49(166):465–62.

Salzmann, N., C. Frei, P.-L. Vidale, and M. Hoelzle. 2007. The application of regional climate model output for the Simulation of high-mountain permafrost scenarios. *Global and Planetary Change* 56:188–202.

Schär, C., P.L. Vidale, D. Lüthi, C. Frei, C. Häberli, M. Liniger, and C. Appenzeller. 2004. The role of increasing temperature variability in European summer heatwaves. *Nature* 427:332–36.

Schneeberger, C., H. Blatter, A. Abe-Ouchi, and M. Wild. 2003. Modelling changes in the mass balance of glaciers of the Northern Hemisphere for a transient $2CO_2$ scenario. *Journal of Hydrology* 282:145–63.

Schwarb, M., C. Daly, C. Frei, and C. Schär. 2001. Mean annual precipitation in the European Alps 1971–1990. In *Hydrological atlas of Switzerland,* ed. Swiss Federal Council, pl. 2.6. Bern: Landeshydrologie und Geologie.

Strasser, U., J. Corripio, F. Pellicciotti, P. Burlando, B. Brock, and M. Funk. 2004. Spatial and temporal variability of meteorological variables at Haut Glacier d'Arolla (Switzerland) during the ablation season 2001: Measurements and simulations. *Journal of Geophysical Research* 109(D3):D03103.

Suter, S., M. Hoelzle, and A. Ohmura. 2004. Energy balance at a cold Alpine firn saddle, Seserjoch, Monte Rosa. *International Journal of Climatology* 24:1423–42.

Takeuchi, Y., R.B. Kayastha, and M. Nakawo. 2000. Characteristics of ablation and heat balance in debris-free and debris-covered areas on Khumbu Glacier, Nepal Himalayas, in the pre-monsoon season. *IAHS* 264:35–42.

Van de Wal, R.S.W., and M. Wild. 2001. Modelling the response of glaciers to climate change by applying volume-area scaling in combination with a high-resolution GCM. *Climate Dynamics* 18:359–66.

Wagnon, P., P. Ribstein, B. Francou, and B. Pouyaud. 1999. Annual cycle of energy balance of Zongo Glacier, Cordillera Real, Bolivia. *Journal of Geophysical Research* 109(D4):3907.

Wildt, M.S., E.J. Klok, and J. Oerlemans. 2003. Reconstruction of the mean specific mass balance of Vatnajøkull (Iceland) with a seasonal sensitivity characteristic. *Geografiska Annaler* 85A:57–72.

Wilson, J.P., and J.C. Gallant. 2000. *Terrain analysis: Principles and applications.* New York: Wiley.

Zängl, W., and S. Hamberger. 2004. *Gletscher im Treibhaus: Eine fotografische Zeitreise in die alpine Eiswelt.* Steinfurth: Tecklenborg Verlag.

9

Modeling Climate-Change Impacts on Mountain Glaciers and Water Resources in the Central Dry Andes

Javier G. Corripio, Ross S. Purves, and Andrés Rivera

The Central Dry Andes form a high mountain barrier dividing Chile and Argentina between around latitude 31° and 35°S. They run north–south, reaching the highest elevations in the Southern Hemisphere: Aconcagua is 6,954 m a.s.l., while many other peaks rise over 6,000 m. Their slopes descend abruptly toward the Pacific coast to the west, sandwiching a narrow stretch of land that sustains some of Chile's richest agriculture and largest populations. To the east they descend more gradually toward the Argentine Pampas (Figure 9.1). On both sides of the range the precipitation regime has a marked seasonality, with most precipitation at high altitudes occurring during the austral winter, while summers are dry and sunny. Although the Chilean piedmont has annual precipitation in excess of 500 mm, the Argentine city of Mendoza receives only some 180 mm (Miller 1976). In both cases almost all crops are irrigated and represent an important element of the economy. The Mediterranean subtropical production is sold at high prices in the Northern Hemisphere because of inverted seasonality (Brignall et al. 1999). According to the Chilean wine producers' association, wine exports reached US$877 million in 2005 (Associación Viñas de Chile 2004). In Mendoza, according to the Mendoza Tourist Board, wine production is second in economic importance only to oil production. Santiago hosts the largest population in Chile, in excess of 5 million people, while Mendoza and surrounding towns are home to more than 1 million. Supporting this population and irrigated agriculture would be impossible without the water derived from melting snow and ice (Ribbe and Gaese 2002).

The importance of meltwaters is demonstrated in Table 9.1. Precipitation is very low in the summer months, just 1 mm in December, while runoff is at its maximum, over 42 m^3 s^{-1}, on the Río Aconcagua at Río Blanco (1,420 a.s.l.) during the same month (Legates and Willmott 1990; LBA-Hydronet 2002). This inverse pattern, with maximum discharge corresponding to minimum precipitation, is highly beneficial for human activities in the region in that the time of maximum heat stress and water demand coincides with the time of maximum availability. The mechanism for this

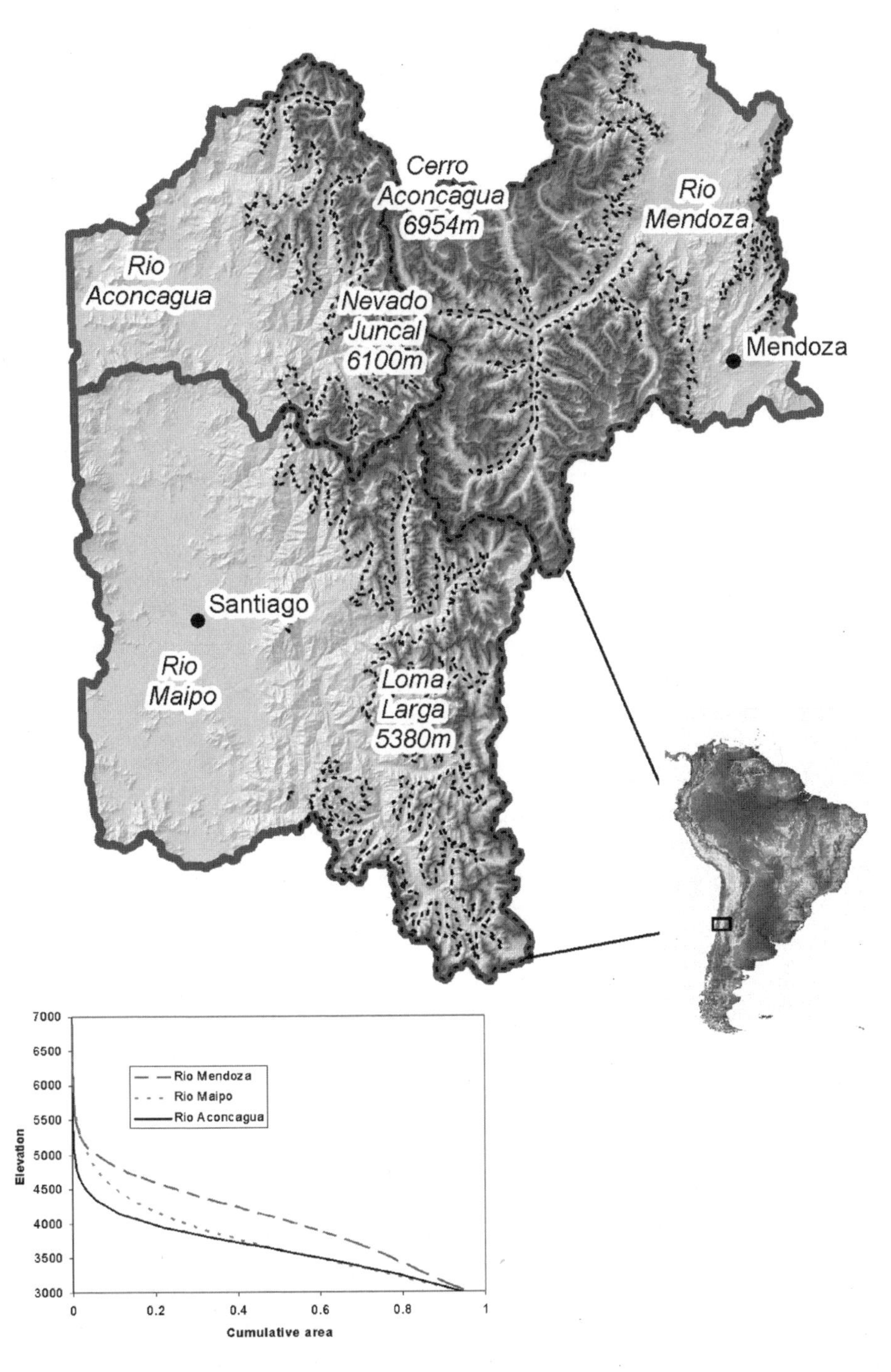

FIGURE 9.1. Map of the area of study, about latitude 33°S and longitude 70°W, showing the three main river catchments in the region and the hypsographic distribution of elevation above 3,000 m for each catchment. Regions used in calculation of hypsometry are delineated by a dashed black line. Note the more linear trend of the Mendoza basin and the rapid decline of surface area with altitude for the Aconcagua basin. Digital elevation model source, NASA SRTM (http://www2.jpl.nasa.gov/srtm/); map source, NOAA-NGDC Globe Project (http://www.ngdc.noaa.gov/mgg/topo/globe.html).

TABLE 9.1

Pattern of Monthly Discharge (m^3s^{-1}) and Precipitation (mm) in the Upper Aconcagua River Basin, Chile, at Río Blanco

MONTH	J	F	M	A	M	J	J	A	S	O	N	D
Discharge	22.04	13.88	8.53	3.90	3.12	2.98	3.06	3.10	4.16	6.62	14.36	22.72
Precipitation	3	15	10	31	109	102	79	94	41	22	11	1

asynchrony is the storage of water as snow and ice in the Cordillera during the winter months and its release through melt during the summer months. Whether the system is sustainable in the medium to long term depends on the balance between snow accumulation and ablation and the rate of melting of old water reserves in the form of glacier ice.

The Central Dry Andes are distinguished from glaciers and snow-covered surfaces at other latitudes by the presence of *penitentes* (Figure 9.2)—pinnacles of ice up to several meters in height that are sculpted by differential ablation of the snow surface; the peaks remain frozen because of sublimation, while the troughs act as solar traps, enhancing their deepening (see Lliboutry 1954). Their name comes from their perceived similarity to the penitents who march in Easter processions in Spain and Latin America wearing distinctive white conical headdresses. These ice pinnacles were first described by Darwin (1839) during his voyage in the *Beagle*. Corripio (2003) has demonstrated that most of the energy on penitentes is redirected to sublimation and temperature changes, which enhance the conservation of the snow cover and reduce ablation. The labyrinth of ice and snow pinnacles appears chaotic, but penitentes are invariably oriented toward the zenith angle of the sun at noon, with walls aligned from east to west. Because the incoming solar radiation is distributed over a large surface area while the peaks are cooled by sublimation and evaporation in the very dry atmosphere, they are an efficient mechanism for the preservation of the snow cover.

A full study of the response of the glaciers of the Central Dry Andes to future climate change is beyond the scope of this article and, we would argue, the available data. Instead, we present here a detailed study of potential changes in the ablation regime during the crucial melt season based on data from two glaciers and considering the influence of penitentes. We then hypothesize about the likely consequences of these results for the three catchments identified earlier through consideration of catchment hypsometry.

INVESTIGATING THE SENSITIVITY OF THE CENTRAL DRY ANDES TO CLIMATE CHANGE

In investigating the sensitivity of the Central Dry Andes to potential climate change, we wish to explore in particular the potential response of the system in terms of summer ablation—the most important source of meltwater (and thus water) in the region. Our approach to the problem of modeling this response is limited by both the complexity of potential feedbacks within the system and the scarcity of available data at appropriate scales (in terms of both past change and potential future responses). In general, climate data for the region are both temporally and spatially limited. Rösenbluth, Fuenzalida, and Aceituno (1997) produced a time series showing warming rates at latitude 33°S of around 2 °C per century between 1933 and 1992. Glacial response in the region appears to be correlated with this warming trend, with Leiva (1999) showing a general pattern of glacier retreat in the region in the twentieth century.

The region lies close to the interface between the influence of the southern westerlies (Kull,

FIGURE 9.2. Penitentes in the Central Dry Andes of Argentina. The pinnacles in the picture are 2 m in height and may be as high as 5 m elsewhere. The sun is in the west, and the penitentes are tilted about 12° north, toward the position of the sun at midday. (Photo © Javier G. Corripio.)

Grosjean, and Veit 2002), which are a strong driver of glacial systems farther south in Patagonia (Villalba et al. 2003), and the mainly anticyclonic climate system experienced to the north (Compagnucci and Vargas 1998). Changes in the magnitude or position of the westerlies would have significant effects on the regional climate through advection of moisture from the Pacific. For example, Kull, Grosjean, and Veit (2002) hypothesize that an intensification of the westerlies at or around the Last Glacial Maximum doubled precipitation at latitude 29°S. Furthermore, the region lies within the zone of influence of the El Niño Southern Oscillation. Compagnucci and Vargas (1998) have shown that runoff in the Mendoza basin increases in El Niño years, most likely as a result of increased accumulation.

Any projection of future glacier response to climate change must take into account the availability of data and models for characterizing the system (Haeberli, Hoelzle, and Suter 1998). Approaches to modeling the response of systems to future change vary with the available data and the nature of the questions being asked. A key consideration is the spatial and temporal scales at which change is to be examined. In this case, the question is how, given current scenarios for climate change in South America, the availability of runoff will change in the short to medium term within the catchments fed by ablation from the Central Dry Andes. Previous work has shown that penitentes significantly perturb the ablation regime with respect to more typical glaciers (Corripio and Purves 2005). Thus the question of how the distribution of penitentes will change with possible change in climate is also important. Finally, the influence of any change in ablation regime will be strongly dependent on catchment hypsometry. Therefore, a third question is whether differences in the hypsometry of catchments in the Central Dry Andes could produce significantly different responses to climate change.

Approaches to the modeling of ablation vary in complexity from point models utilizing either the temperature index (e.g., Hock 2003) or more physically based energy balance approaches (Greuell and Konzelmann 1994) to distributed mass and energy balance models that take into account differential ablation and accumulation (Klok and Oerlemans 2002). Here we apply a distributed energy balance model that has been validated at a point. We have no data regarding accumulation and therefore have considered only the ablation season.

MODELING ABLATION IN THE CENTRAL DRY ANDES

Portable, light automatic weather stations were installed on the surfaces of two glaciers in the region and the information logged every 10 minutes with a Campbell CR10 data logger. The glaciers were the Juncal Norte, at 3,335 m a.s.l. (latitude 32.986°S, longitude 69.956°W), and the Loma Larga, at 4,667 m a.s.l. (latitude 33.692°S, longitude 70.0°W). The data, collected from December 2000 to February 2001, were air and snow temperature, relative humidity, wind speed and direction, and incoming and outgoing shortwave radiation.

To model the ablation of snow and ice in the region we use the physically based energy balance model known as SnowDEM (Snow Distributed Energy Balance Model). This is a highly distributed multilayered snow energy balance model that takes full account of topographic influences and simulates incoming and outgoing shortwave radiation (direct, diffuse, and reflected), incoming and outgoing longwave radiation (atmospheric thermal radiation and radiation emitted from surrounding slopes), snow surface and subsurface temperature, and latent and sensible turbulent heat interchange with the atmosphere. The model is slightly modified from that described by Corripio (2003), which can be summarized as follows:

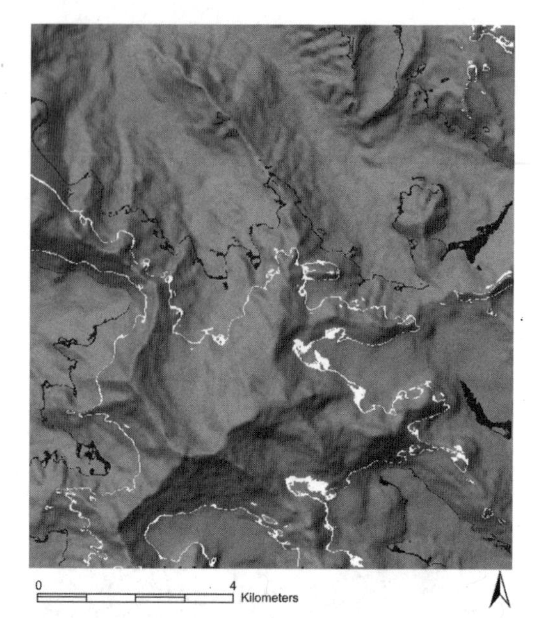

FIGURE 9.3. Penitentes migration for future climatic scenario. The black line is the observed and modeled line while the white one is the one forecast for $\Delta T = +4$ °C. The Juncal Sur and Río Plomo glaciers *(lower center right and bottom)* are today completely covered with penitentes, while these would be almost absent in future scenarios. The highest peak in the lower section of the figure is the Nevado Juncal, 6,110 m a.s.l. Digital elevation model source, Instituto Geográfico Militar, Chile.

$$I_G(1-\alpha) + L\downarrow + L\uparrow + H + L_vE + Q_S + Q_M = 0$$

where I_G is global shortwave radiation, α is albedo, $L\downarrow$ is downward flux of longwave radiation, $L\uparrow$ is upward flux of longwave radiation, H and L_vE are sensible and latent heat fluxes, Q_s is internal heat flux within the snowpack, and Q_M is available heat for melting. Advective heat to the snowpack is neglected in this formulation but could be computed if high-resolution snowcover information and data on the thermal properties of the bare ground were available.

A detailed study of the effect of penitentes with a high-resolution (1-cm grid cell) synthetic model of penitentes surface energy balance has been reported elsewhere (Corripio 2003; Corripio and Purves 2005). To account for this influence while running a computationally tractable model, areas covered by penitentes were parameterized as areas of increased roughness length. An additional parameter to account for the increased efficiency of radiative cooling by penitentes is currently being developed. Our observations show that in the initial stage of the formation of penitentes there is always a thin radiative crust on the snow surface. This crust is found when the uppermost skin surface layer of the snow has a zero or slightly negative energy balance while the subsurface layer has a positive net balance. This criterion has been confirmed by direct observation of the lowest limit of penitentes occurrence in the Central Dry Andes (~4,100 m a.s.l.), and it is used in this chapter to distinguish areas where the model uses increased roughness to simulate the effects of penitentes. Figure 9.3 shows where this modeled skin surface net energy balance of around zero is found on the Juncal Norte and surrounding glaciers.

SCENARIOS

In this study we apply to the region a very simple scenario for climate change derived from the HadCM3-coupled atmosphere-ocean

TABLE 9.2

Variables Used for the Two Model Runs

PRESENT CONDITIONS	ΔT = + 4 °C	ΔT = + 4 °C + Penitentes Migration
Air temperature (*Ta*)	*Ta* + 4	*Ta* + 4
Surface roughness	Surface roughness	Surface roughness
0.002 < 4,100 m;	0.002 < 4,100 m;	0.002 < 4,700 m;
0.20 > 4,200 m	0.20 > 4,200 m	0.20 > 4,800 m
RH measured	RH measured	RH measured (implies higher atmospheric water content)
Wind measured	Same wind	Same wind
SW↓ measured + topography	SW↓ + topography	SW↓ as today, topography corrected
SW↑ measured	SW↑ as today	SW↑ as today, implying same albedo
Ts for validation	—	—

NOTE: *Ta*, air temperature at 2-m screen level; *RH*, relative humidity; *SW*, shortwave radiation; *Ts*, snow temperature; *Sk*, sky view factor. Modeled melt is in millimeters (mm) of water equivalent.

general circulation model (Gordon et al. 2000; Pope et al. 2000). The model was run for present conditions and for two climate change scenarios based on a warming of 4 °K, as predicted by the above-mentioned general circulation model for the end of the twenty-first century, with an all-anthropogenic forcing integration scenario. This warming is at the upper end of estimates for likely warming: for example, Bradley, Keimig, and Diaz (2004) suggest mean warming on the order of 2.5 °K at this latitude on the basis of a scenario with 2 × CO_2 derived from seven general-circulation-model simulations. However, our experiments are aimed not at predicting actual change in the system but rather at exploring the linkages between penitentes, ablation, and orography.

In the first warming scenario, the model was run for present-day conditions, with increased temperature and relative humidity held constant (i.e., increased specific humidity), implying an increase in atmospheric water vapor content. The increase in humidity is based on the hypothesis that increased sea-surface temperatures would increase available moisture for transport. All other input variables are held constant in this simple scenario with, in particular, no changes in accumulation modeled. This is in line with our aim of exploring model sensitivities through a simple set of experiments and with the fact that predictions of change in precipitation in this region are uncertain and appear to be very low. In the second scenario, the position of zero net energy balance at the snow surface was calculated to determine the lowest elevation at which penitentes would form, and the surface was then reparameterized by changes in surface roughness according to the calculated upward migration of penitentes.

RESULTS

Table 9.2 shows the input variables for the different runs of the model. The daily average is considered representative of mean ablation during the ablation season. Because the climatic conditions for the year of the study are similar to those of the long-term mean for the area, according to reanalysis data from NCEP/NCAR (Earth System Research Laboratory 2007), it is not unreasonable to assume that these computed daily melts are representative of the average conditions in the region. Roughness length has an important effect on the evaporation/sublimation of the snow and on

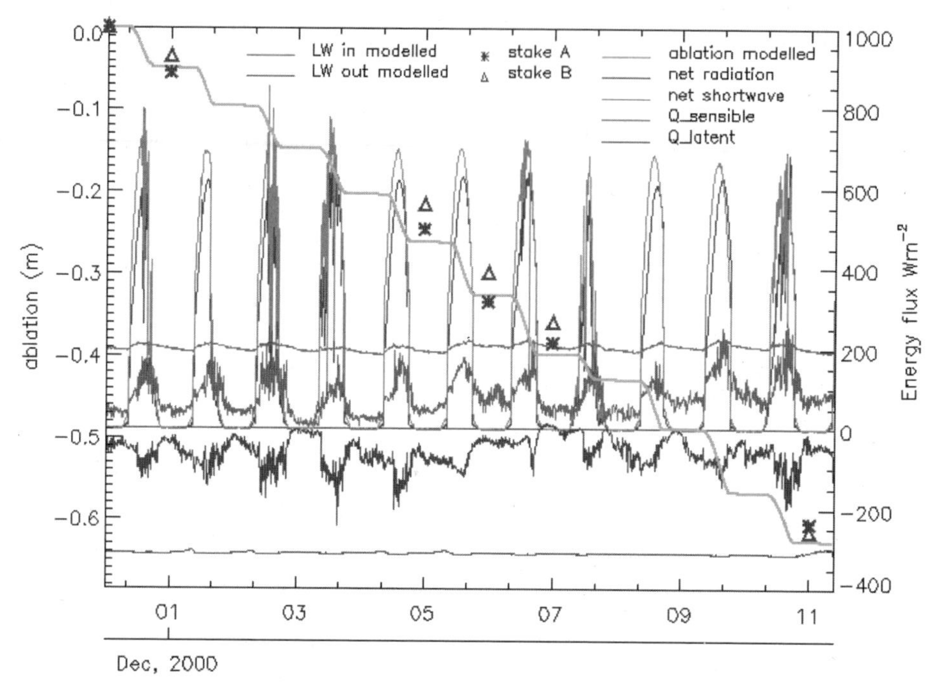

FIGURE 9.4. Plot of the different energy balance components during ten days of December for the weather station (3,305 m a.s.l.) on the Juncal Norte Glacier. Superimposed are the modeled melt *(stepped line)* and the measured melt at two nearby ablation stakes *(triangles and stars)*.

the sensible heat transfer with the atmosphere. The heat interchange with the atmosphere seems to be much higher on these snow covers than on those at higher altitudes (Corripio 2003). Model outputs for the present day were validated through the use of several ablation stakes in the vicinity of the weather station. The results were satisfactory, especially with regard to cumulative melt (see Figure 9.4).

The modeled melt outputs are, in fact, potential melt—the amount of melt that would be possible if the ground were covered by a snowpack of infinite depth. More realistic initial snow-cover conditions would be preferable, but direct measurements are not available. The differences between the results for present conditions and those for future scenarios are summarized in Figure 9.5. The results show a large increase at lower altitudes, of about 8-mm water equivalent or 17–22% of the present melt. The maximum increase in melt, 34–48%, occurs between 4,100 and 4,700 m a.s.l. because of the upward migration of penitentes. This value of up to 11-mm water equivalent is probably an underestimate, as some researchers have observed a dramatic decrease of melting on penitentes-covered areas (Kotlyakov and Lebedeva 1974). At higher altitudes temperatures remain relatively low and the slopes are steeper, leading to lower solar radiation interception and lower melt for all scenarios.

DISCUSSION

Analysis of these results together with the hypsographic distribution of land within the three catchments studied and the extent of the glaciers suggest a maximum increase in melt in the areas of greatest ice storage at present. This would imply a likely increase in future runoff generation in the ablation season. As general circulation models do not appear to forecast an increase in winter precipitation, it seems that stored ice is likely to be depleted rapidly. Some compensation may be brought about by an increase in El Niño events, which

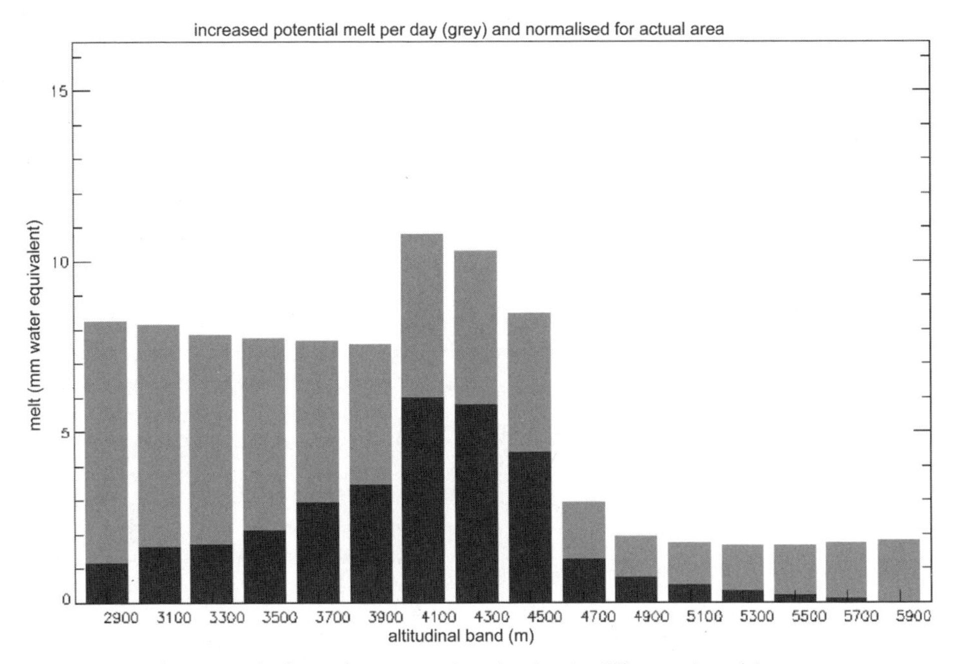

FIGURE 9.5. Melt variation for future climatic warming, showing the differences in melt between present modeled conditions and a future scenario of temperature increased by 4 °C and upward migration of the penitentes line *(gray bars)* and normalized differences in melt weighted for the actual surface area in every altitudinal band (nondimensional units) *(black bars)*.

are associated with increased winter precipitation and increased winter glacier mass balance (see, e.g., Leiva 1999). However, the depletion of ice storage will eventually lead to water scarcity and decreased runoff during the summer months. This situation will follow different patterns in different catchments depending on the altitudinal distribution of the land. Thus, the Mendoza River is likely to experience gradual decay as the altitude level of enhanced melt increases because of the almost linear distribution of land area with elevation. At the same time, in the Aconcagua catchment runoff is likely to decrease rapidly after a period of much enhanced runoff. Here the surface area of land distribution has a peak between 4,100 m and 4,700 m a.s.l., precisely the area of maximum melt increase under the two scenarios.

That increased temperatures will produce increased melt is not a surprising or novel result. However, what is especially significant in this region is that the melt regime is nonlinear because of the effect of the penitentes in mediating ablation. The fact that the lower limit of penitentes is now located at the peak of glacier surface distribution will have a positive feedback on future increases in ablation, but this is not likely to last long. As shown in Figure 9.5, the expected upward migration of the penitentes is in some cases rather dramatic. The fact that some glaciers whose main ice bodies are at present entirely above the penitentes line, such as the Juncal Sur, a very large glacier south of the highest peak in Figure 9.4, or the Río Plomo glaciers, at the lower right, part of the largest glacier system in the Argentine Central Dry Andes, will be entirely below it under the new scenarios shows that this increase in melt occurs precisely in the areas of maximum ice concentration in these glaciers.

It is important to sound a note of caution with respect to these results. The scenarios used and the models applied are relatively simple, and the uncertainties in many elements of the system are very significant. In particular, in the scenarios presented here we have assumed that precipitation (and thus accumulation) and wind speeds, an important influence on turbulent heat fluxes, remain unchanged. Nonetheless, at present precipitation is negligible during the ablation season, and the main driver of local wind circulation is katabatic forcing, which is likely to remain similar or decrease if glacier extent decreases.

CONCLUSIONS AND SUGGESTIONS FOR FURTHER WORK

Despite the scarcity of data in the region of study, our modeling results suggest that meltwater availability is likely to increase in the medium term, with faster depletion of glacier resources in the longer term. Because of local orography it appears that the rate of variation will be more pronounced in the Aconcagua and Maipo basins than in the Mendoza basin.

Given the importance of meltwater for the economy and populations of Chile and Argentina and the time required to adapt to new conditions, it would be desirable to be able to anticipate with greater precision the timing and magnitude of changes in water supplies. The tools presented here seem appropriate for gaining insight into such changes, but greater knowledge of the initial conditions is necessary, including (1) reliable data on precipitation and snow cover, (2) an updated inventory of glaciers that will permit the estimation of total runoff, (3) continuous meteorological data from a high-altitude weather station to permit modeling of the annual cycle of accumulation and ablation and approximation of the vertical distribution of meteorological variables, and (4) measurements of actual ablation in penitentes-covered areas and techniques for applying an energy balance model that fully represents fluxes within penitentes fields as opposed to the roughness-based parameterization employed here.

REFERENCES CITED

Associación Viñas de Chile. 2004. http://www.vinasdechile.cl/archivo_estadistica/18exp_anosano92-2005.xls (accessed July 26, 2007).

Bradley, R. S., F. T. Keimig, and H. F. Diaz. 2004. Projected temperature changes along the American cordillera and the planned GCOS network. *Geophysical Research Letters* 31:L16210+4, doi: 10.1029/2004GL020229.

Brignall, A. P., T. E. Downing, D. Favis-Mortlock, P. A. Harrison, and J. L. Orr. 1999. Agricultural drought in Europe: Site, regional, and national effects of climate change. In *Climate, change, and risk*, ed. T. E. Downing, A. J. Olsthorn, and R. S. J. Tol. New York: Routledge.

Compagnucci, R. H., and W. M. Vargas. 1998. Interannual variability of the Cuyo River's streamflow in the Argentinian Andean mountains and ENSO events. *International Journal of Climatology* 18:1593–1609.

Corripio, J. G. 2003. Modelling the energy balance of high altitude glacierised basins in the Central Andes. Ph.D. diss., University of Edinburgh. http://www.ihw.ethz.ch/staff/jcorripi/corripiophd.pdf.

Corripio, J. G., and R. S. Purves. 2005. Surface energy balance of high-altitude glaciers in the Central Andes: The effect of snow penitentes, In *Climate and hydrology in mountain areas*, ed. C. de Jong, D. Collins, and R. Ranzi. London: Wiley. http://www.ihw.ethz.ch/staff/jcorripi/corripiopurves04.pdf.

Darwin, C. 1839. *Journal of researches into the geology and natural history of the various countries visited by H.M.S.* Beagle, *under the command of Captain Fitz Roy, R. N., 1832 to 1836*. London: H. Colburn.

Earth System Research Laboratory. 2007. NOAA. http://www.cdc.noaa.gov/ (accessed July 26, 2007).

Gordon, C., C. Cooper, C. Senior, H. Banks, J. Gregory, T. Johns, J. Mitchell, and R. Wood. 2000. The simulation of SST, sea ice extents, and ocean heat transports in a coupled model without flux adjustments. *Climate Dynamics* 16:147–68.

Greuell, W., and T. Konzelmann. 1994. Numerical modelling of the energy balance and the englacial temperature of the Greenland Ice Sheet: Calculations for the ETH–Camp location (West Greenland, 1115 m a.s.l.). *Global and Planetary Change* 9:91–114.

Haeberli, W., M. Hoelzle, and S. Suter, eds. 1998. *Into the second century of worldwide glacier monitoring: Prospects and strategies.* Studies and Reports in Hydrology 56. Paris: UNESCO.

Hock, R. 2003. Temperature index melt modelling in mountain areas. *Journal of Hydrology* 282:104–15.

Klok, E. J., and J. Oerlemans. 2002. Model study of the spatial distribution of the energy and mass balance of Morteratschgletscher, Switzerland. *Journal of Glaciology* 48:505–18.

Kotlyakov, V. M., and I. M. Lebedeva. 1974. Nieve and ice penitentes: Their way of formation and indicative significance. *Zeitschrift für Gletscherkunde und Glazialgeologie* 10:111–27.

Kull, C., M. Grosjean, and H. Veit. 2002. Modelling modern and late Pleistocene glacio-climatological conditions in the North Chilean Andes. *Climatic Change* 52:359–81.

LBA-Hydronet. 2002. http://www.lba-hydronet.sr.unh.edu (accessed July 26, 2007).

Legates, D. R., and C. J. Willmott. 1990. Mean seasonal and spatial variability in gauge-corrected global precipitation. *International Journal of Climatology* 10:111–27.

Leiva, J. C. 1999. Recent fluctuations of the Argentinian glaciers. *Global and Planetary Change* 22:169–77.

Lliboutry, L. 1954. The origin of penitents. *Journal of Glaciology* 2:331–38.

Miller, A. 1976. The climate of Chile. In *World survey of climatology: Climates of Central and South America,* ed. W. Schwerdtfeger. Amsterdam: Elsevier.

Pope, V. D., M. L. Gallani, P. R. Rowntree, and R. A. Stratton. 2000. The impact of new physical parameterizations in the Hadley Centre climate model: HadAM3. *Climate Dynamics* 16:123–46.

Ribbe, L., and H. Gaese. 2002. Water management issues of the Aconcagua watershed, Chile. *Technology Resource Management and Development: Water Management* 2:86–108.

Rösenbluth, B., H. A. Fuenzalida, and P. Aceituno. 1997. Recent temperature variations in southern South America. *International Journal of Climatology* 17:67–85.

Villalba, R., A. Lara, J. A. Boninsegna, M. Masiokas, S. Delgado, J. C. Aravena, F. A. Roig, A. Schmelter, A. Wolodarsky, and A. Ripalta. 2003. Large-scale temperature changes across the southern Andes: 20th-century variations in the context of the past 400 years. *Climatic Change* 59:177–232.

PART THREE

Trends in Natural Landscapes

Climate Change

10

Glacier Mass Balance in the Northern U.S. and Canadian Rockies

PALEO-PERSPECTIVES AND TWENTIETH-CENTURY CHANGE

Emma Watson, Gregory T. Pederson,
Brian H. Luckman, and Daniel B. Fagre

Alpine glaciers in the U.S. and Canadian Rocky Mountains reached their maximum Holocene extent during the Little Ice Age (Luckman 2000; Carrara 1989). Subsequently, glaciers throughout the region underwent dynamic and sometimes rapid phases of frontal recession (Carrara 1989; Luckman 2000; Key, Fagre, and Menicke 2002). Recent glacier research has focused on developing detailed histories of glacier fluctuations throughout the Little Ice Age. These data, though sparse, indicate multiple periods of glacier advance and suggest that the timing of maximum advance may not have been synchronous (Luckman 2000). Moraine dates at several of the northernmost glaciers studied (e.g., in Jasper National Park) indicate maximum glacier extent between 1700 and 1750 (Luckman 2000) with a subsequent only slightly less extensive advance between 1800 and 1850. Farther south (e.g., in Kananaskis and Glacier National Park) the Little Ice Age maximum extent occurred between ca. 1800 and 1850 (Carrara 1989; Smith, McCarthy, and Colenutt 1995; Luckman 2000; Key et al. 2002).

Mass balance records for this region are limited to two records from the Canadian Rockies (Peyto Glacier 1965–present and Ram River Glacier 1965–75 [Demuth and Keller 2006; Young and Stanley 1977]). Watson and Luckman (2004*a*) and Pederson et al. (2004) have used tree-ring data to investigate the paleoclimatic drivers of glacier fluctuations for two sites located along the Continental Divide (Figure 10.1). Although these reconstructions used different paleoclimate and glacier data, the results produced some interesting similarities and differences in inferred glacial dynamics over the past 300 years. In this article we use instrumental and proxy climate data to investigate whether the differences between these proxy mass balance series reflect regional differences in mass balance over time or result from differences in the approaches and data used to develop the reconstructions. In doing so, we begin to explore differences in timing of the Little Ice Age maximum glacier advance and

Note: Emma Watson's contribution to this chapter is courtesy Environment Canada.

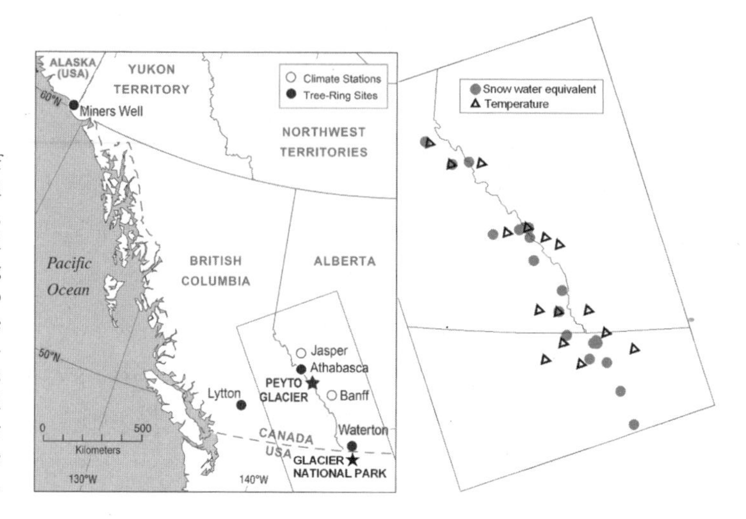

FIGURE 10.1. Location of Peyto Glacier, Alberta, and Glacier National Park, Montana, showing selected meteorological stations and tree-ring chronology sites used to develop the Peyto Glacier mass balance reconstructions. The larger-scale map shows the stations located along the Continental Divide from which snow water equivalent and temperature records were obtained.

their implications for future paleoglaciological research.

CONSTRUCTION OF GLACIER MASS BALANCE PROXY RECORDS

The variability in width (or density) of annual growth rings in many species of trees has a demonstrated sensitivity to climate variables such as precipitation and temperature (Fritts 1976). Tree-ring chronologies are routinely used as predictor variables in statistical models that provide valuable estimates of climate conditions before the advent of instrumental measurement. The mass balance of continental glaciers is sensitive to many of the same climate factors that influence tree growth. Therefore, tree-ring chronologies can be used to generate useful estimates of past glacier conditions.

In this chapter we compare tree-ring-based mass balance reconstructions for Peyto Glacier in the Canadian Rockies and for the glaciers of Glacier National Park in the northern U.S. Rockies. Approximately 35 years of mass balance data for Peyto Glacier allow direct calibration of mass balance estimates. Watson and Luckman (2004*a*) have developed 322 year-long reconstructions of winter, summer, and net mass balance for Peyto Glacier using temperature and precipitation-sensitive tree-ring chronologies from Canada and Alaska. The winter component was estimated using a *Tsuga mertensiana* (mountain hemlock) ring-width chronology from as Miners Well[1] and a July–June precipitation reconstruction from *Pinus ponderosa* (ponderosa pine) from Lytton, British Columbia. Summer mass balance predictors include the Columbia Icefield maximum temperature reconstruction from *Picea engelmannii* (Engelmann spruce) (Luckman and Wilson 2005) and a July–June precipitation reconstruction from *Pseudotsuga menziesii* (Douglas fir) for Waterton (Watson and Luckman 2004*b*). Both winter and summer models were calibrated against the measured mass balance records (1966–94), and they explain > 40% of the variance in these records and pass conventional verification tests conducted in dendroclimatological studies.

Pederson et al. (2004) used tree-ring-based proxy records to compare and explain measured fluctuations in the Agassiz and Jackson Glaciers (Carrara 1989; Key et al. 2002) over the past 300 years. Winter mass balance was estimated using a reconstruction of the Pacific Decadal Oscillation (PDO; D'Arrigo, Villalba, and Wiles 2001) because of the demonstrated strong linkage between this sea-surface-temperature anomaly, atmospheric circulation, and regional snowpack patterns (e.g., Selkowitz, Fagre, and Reardon 2002; Moore and McKendry 1996). Summer balance estimates were based on a summer drought (June–August; precipitation-potential evapotranspiration) reconstruction for Glacier National Park from mid-elevation *Pseudotsuga menziesii* (Douglas fir) and *Pinus flexilis* (limber pine) chronologies.

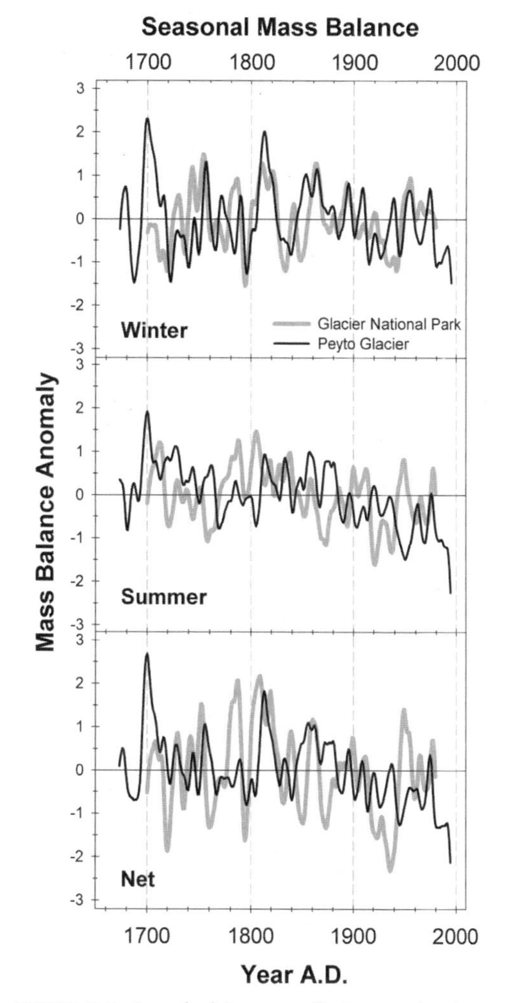

FIGURE 10.2. Smoothed (ten-year spline) seasonal and net mass balance reconstructions for Peyto Glacier (Watson and Luckman 2004*a*) and Glacier National Park (Pederson et al. 2004).

COMPARISON OF MASS BALANCE PROXY RECORDS

The methods and data used to infer and reconstruct glacier mass balance differ between the two studies, resulting in interesting similarities and differences in balance estimates. The seasonal balances in both regions exhibit strong decadal and multidecadal variation (Figure 10.2). The winter balance proxies often share periods of above- and below-average accumulation events, though the magnitude and intensity of individual events may vary. There is general correspondence between records from the 1770s to the 1790s throughout the nineteenth century and a common period of high accumulation from the mid-1940s to the late 1970s. There are fewer similarities between the records of summer balance. The two periods of highest summer balance correspondence are the favorable (cool) summer conditions of the early 1700s and the early to mid-1800s.

The many similarities between the seasonal mass balance records result in common periods of predicted positive and negative net balance, but differences between the net balance records point to some striking and perhaps important disparities between the seasonal proxies. For example, above-average accumulation is reconstructed for the early 1700s for Peyto Glacier, whereas the D'Arrigo, Villalba, and Wiles (2001) PDO reconstruction used to estimate winter balance in Glacier National Park indicates average to below-average accumulation. The proxy records also indicate major differences between snowpack levels for the early twentieth century. There are also significant differences between the proxies for summer balance. Glacier National Park shows average to high summer ablation rates between ca. the 1720s–1770s and the 1850s–1880s, when low ablation conditions are reconstructed for Peyto Glacier. The most striking difference, however, is the long-term linear trend toward higher summer ablation rates for Peyto Glacier that critically influence its net balance record. These differences between the components of reconstructed net balance may reflect either actual differences in climate (or glacier) regimes between these two areas or differences in the parameters/techniques used to develop the reconstructions. To address this issue, we first investigate the instrumental climate records.

INSTRUMENTAL CLIMATE RECORDS

Most scientists agree that the net balance of glaciers in western North America is controlled primarily by summer temperatures and winter snowpack (e.g., Letréguilly 1988; Bitz and Battisti 1999; Pederson et al. 2004; Watson and Luckman 2004*a*). Therefore, to investigate possible differences in contemporary climate drivers of mass balance between these two regions, we examined instrumental records of snow water equivalent (used to evaluate winter snowpack) and minimum and maximum temperature that could be used to assess the magnitude of summer melting and changes in seasonality in the region.

SNOW WATER EQUIVALENT RECORDS

We assembled the longest, most complete snow water equivalent[2] records from 25 stations located at elevations > 1,000 m along the portion of the Continental Divide located within the study region (Figure 10.1). The majority of these records are from snow courses (Canadian and U.S. data) and snowpack telemetry sensors (most of the U.S. data) located near present-day glaciers, though at lower elevations (mean = 1,570 m, range 1,040–2,030 m). These records probably underestimate actual snowfall amounts at the glaciers, as Peyto Glacier is between 2,140 and 3,180 m and glacier termini in Glacier National Park range from 2,000 to 2,400 m a.s.l. The snow water equivalent data are presented in standardized (i.e., dimensionless) units to permit comparison of records from different elevations and compensate for the considerable spatial variability in snowfall totals even at the same elevation.

The Peyto Glacier winter mass balance reconstruction is positively and significantly ($P < 0.05$) correlated with the record from 22 of the 25 stations over the 1951–87 interval. Moreover, more than 60% of these correlation coefficients exceed 0.50, indicating that these records are suitable for exploring the primary controls of winter mass balance in the instrumental record. They also display coherent decadal-scale variation, with below-normal winter snowpack between 1922 and 1945 and between 1977 and 2003 and above-normal snowpack from 1946 to 1976 except for values slightly below the mean in the early 1960s (Figure 10.3). This pattern and the spatial scale of coherence are consistent with records of snowpack in British Columbia (Moore and McKendry 1996), Glacier National Park (Selkowitz, Fagre, and Reardon 2002), the Pacific Northwest (Mote 2003), and other parts of western North America (Brown and Braaten 1998; Cayan 1996). Similar variation, particularly the sharp decrease in the mid-1970s, has been identified in time series of many other climate-related variables in the western Americas (Ebbesmeyer et al. 1991) and corresponds with interdecadal variations in sea-surface temperatures in the Pacific Ocean (Mantua et al. 1997; McCabe and Dettinger 2002). In the absence of local proxies for winter precipitation totals, these results demonstrate that winter precipitation varies coherently over large areas in the western cordillera, justifying the use of more distant tree-ring-derived proxy records to estimate winter balance in the Canadian Rockies and adjacent Montana.

The mean intercorrelation of each snow water equivalent record with all 24 others over the 1951–87 period exceeds 0.50 except for two of the most northern records, Field and Yellowhead, both in the Rocky Mountain Trench. Interestingly, principal components analysis (with varimax rotation) calculated over the 1951–2003 interval (17 records) identified northern and southern regions (Figure 10.3) that differ primarily in the magnitude rather than the timing of major periods of above- and below-normal snow water equivalent values. In particular, the period of high snowpack centered around 1952 is slightly more pronounced in the southern part of the region. Although snowpack over the past 20 years has been relatively low, mean snow water equivalent values are greater at recording stations in the northern half of the transect. These differences may simply reflect the earlier measurement date of these northern records

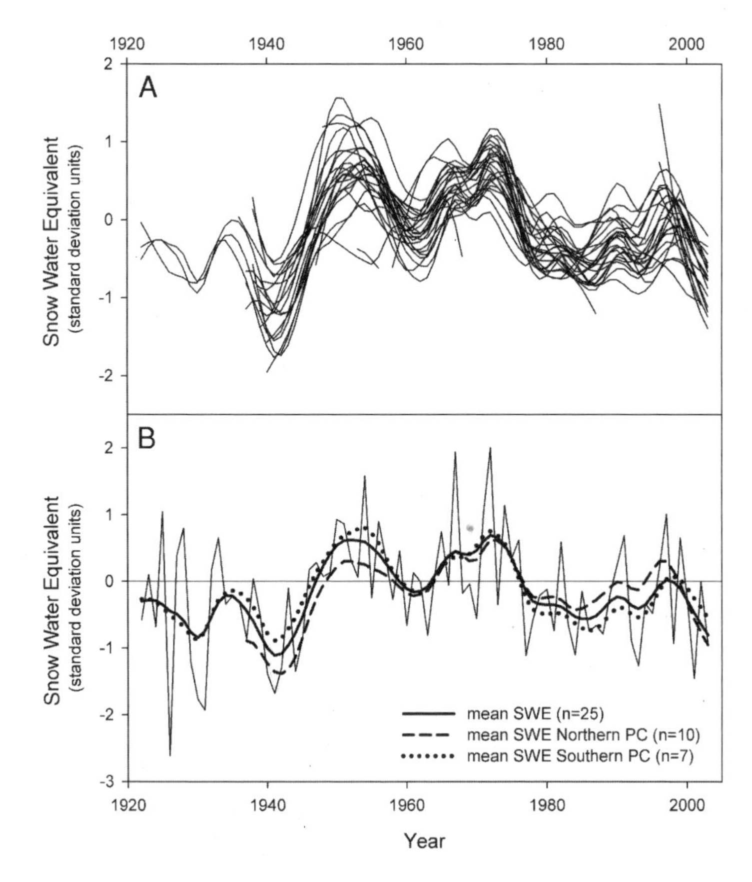

FIGURE 10.3. Standardized snow water equivalent (SWE) records for April 1 (May 1 for the U.S. stations) over the twentieth century. (A) Standardized values at 25 stations. (B) Mean annual values for these stations smoothed with a ten-year spline. The data are presented in standard-deviation units from their 1951–2003 means (several station records end before 2003). Also plotted are smoothed (10-year spline) mean values for the 10 most northerly stations that load on Principal Component 1 (41% variance explained; varimax rotation) and the 7 southern stations that load on Principal Component 2 (35% explained).

(April 1, generally the time of maximum snowpack [Cayan 1996]) compared with the southern records (five of seven are Montana stations that report May 1), but this is unlikely because over the length of the record the northern measurements are not consistently higher than their southern counterparts. The mean elevation of the northern sites (1,611 m) is also slightly higher than that of the southern sites (1,522 m). Mote (2003) and Selkowitz, Fagre, and Reardon (2002) note that the April 1 snow water equivalent is weakly correlated with winter (October–March) temperatures in this part of the Rockies. However, snowpacks are less at the lower-elevation southern stations over the past ~20 years, suggesting that they are perhaps more sensitive to the higher winter (Luckman 1998) and spring (Figure 10.4) temperatures recorded across the region. A more detailed analysis of a larger set of snow water equivalent records and related monthly/seasonal precipitation and temperature records may help identify the cause(s) of these slight north-south differences.

SUMMER TEMPERATURE RECORDS

Temperature records were assembled from the longest and most complete meteorological station records in the region along the Continental Divide ($n = 15$; Figure 10.1). The highest-quality long-term stations are all located in valley bottom sites ranging from 640 to 1,390 m in elevation. Temperature data from these lower-elevation sites probably underestimate warming at higher elevations near glaciers because of nonlinearities in the scale and pace of temperature change with elevation in mountains (Diaz and Bradley 1997; Beniston, Diaz, and Bradley 1997), but the trends documented at valley floor stations may be considered representative, if conservative, estimates of changes at glacier sites.

Figure 10.4 shows summer (June–August) mean maximum and minimum temperatures

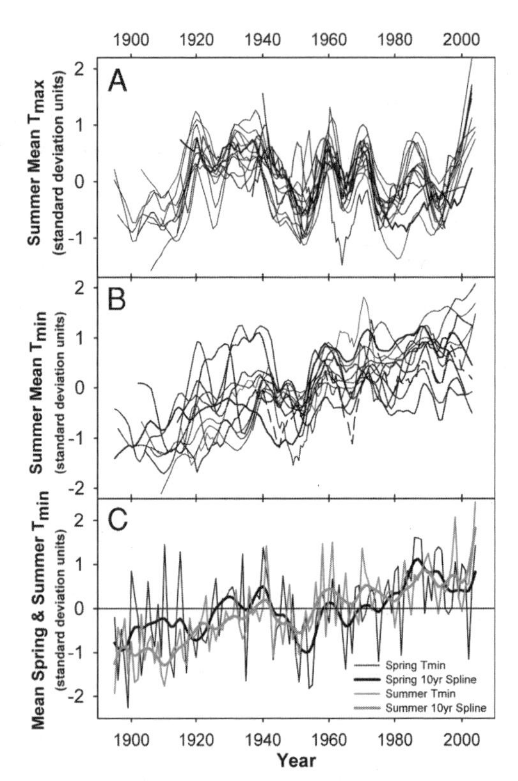

FIGURE 10.4. Standardized seasonal maximum (T_{max}) and minimum (T_{min}) temperature records over the twentieth century. (A) Mean summer T_{max} anomalies for 15 meteorological stations. (B) Mean summer T_{min} anomalies for these 15 stations, smoothed using a ten-year spline. (C) Mean annual and smoothed (ten-year spline) values for spring (March–May) and summer (June–August) T_{min} anomalies using all stations. The majority of the temperature data was obtained from the Web sites of the U.S. Historic Climatology Network (Easterling et al. 1996; ftp://ftp.ncdc.noaa.gov/pub/data/ushcn/) and the Adjusted Historical Canadian Climate Data (Vincent et al. 2002; Environment Canada 2006). Additional unadjusted Canadian station records were obtained from the Canadian National Archive, Meteorological Service of Canada.

from all 15 stations within the study area over the 1895–2003 interval. The records of mean summer maximum temperature display strong decadal-scale coherence but no long-term trend after ca. 1920.[3] Periods of above-average maximum temperature occur from 1920 to 1940 and in the 1960s, 1970s, and 1980s before increasing to the extreme value in 2003. The Peyto Glacier summer balance reconstruction shows significant ($P < 0.05$) correlations (ranging from -0.22 to -0.53) with summer maximum temperature for 11 stations, with the strongest relationships occurring with the more northerly ones. The summer mass balance reconstruction for the Glacier National Park region is significantly ($P < 0.05$) correlated with summer maximum temperature for 13 stations (range -0.36 to -0.65, highest with Kalispell). Thus, the temperature variation exhibited in the instrumental records of summer maximum temperature is representative of a large portion of the variation in summer mass balance for glaciers throughout the region.

Summer minimum temperature records exhibit a different but common mode of variation and change. There is a strong linear increase over the twentieth century, with all stations exhibiting a mean intercorrelation exceeding 0.50. Linear increases in nonstandardized individual station records show changes ranging from 0.5 °C to 4.0 °C (ca. 1895–2003; the length of record varies) that appear to be dependent on elevation and station location in relation to the Continental Divide. Thus the most extreme changes in mean summer minimum temperature exhibited by many of the recording stations ranges from 4–5 °C in the early twentieth century to 7–8 °C in 2003. The strong linear trend in the summer minimum temperature records, combined with the variation in the maximum temperature records, also indicates a decreasing diurnal temperature range that is consistent with studies of other instrumental records (e.g., Skinner and Gullett 1993) and paleoclimatic investigations (Wilson and Luckman 2002). These changes may result in greater summer ablation as temperatures at the glaciers are maintained at higher levels throughout the summer season. Because these trends in minimum temperature tend to be consistent across all seasons, increases in the spring and autumn also extend the period of melting. Averaged spring (March–May) and summer (June–August) minimum temperatures for the standardized records of the 15 stations show strong increases over the twentieth century. Absolute spring minimum temperature values (1895–2003) for these stations ranged from -2 °C to -8 °C in the early twentieth century. By the 1980s, however, average spring

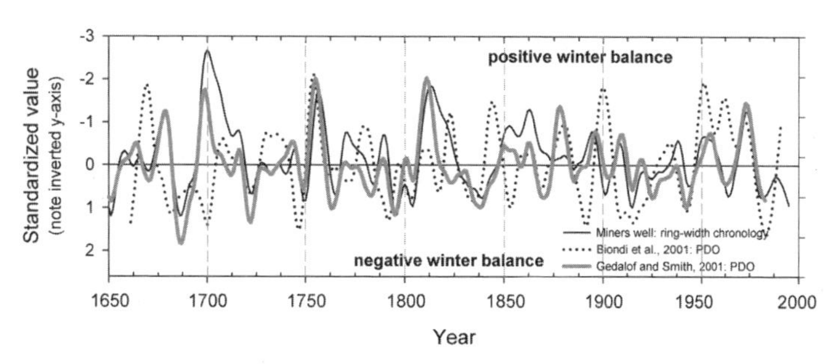

FIGURE 10.5. Tree-ring-derived records related to sea-surface temperatures in the Pacific Ocean and winter mass balance in the study area. The Gedalof and Smith (2001) and Biondi, Gershunov, and Cayan (2001) series are reconstructions of PDO; Miners Well is a *Tsuga mertensiana* (mountain hemlock) tree-ring-width chronology from the Gulf of Alaska developed by G. Wiles, P. E. Calkin, and D. Frank (National Climatic Data Center n.d.). All series have been smoothed with a ten-year spline.

minimum temperature values at the lower or more southerly sites had risen to (or beyond) 0 °C, extending the duration of the annual melt. Similar effects would be expected at higher elevation sites.

Shea, Marshall, and Livingston (2004) find spring, summer, and autumn minimum temperatures to be important predictors for modeling twentieth-century glacier locations in the Canadian Rockies. Summer minimum temperature is not a specific component of the summer balance reconstructions developed for either glacier because of the lack of representative proxy records, though it likely plays an important role in summer ablation rates. Since neither summer balance proxy accounts specifically for the observed increases in minimum temperature, we cannot account for the possible impact on glaciers of changes in the diurnal temperature range. However, estimates of summer ablation for Peyto Glacier show a strong twentieth-century linear increase that is not seen in the instrumental maximum temperature records largely because of the length of the instrumental series. This trend in the summer balance reconstruction results largely from the Athabasca summer maximum temperature reconstruction used as a predictor. Wilson and Luckman (2002) show a strong linear increase in proxy maximum temperature for interior British Columbia beginning ca. 1825 and extending into the early twentieth century, when records begin to exhibit strong decadal variation through the mid-twentieth century. The Athabasca maximum temperature reconstruction exhibits the same behavior (Luckman and Wilson 2005; Figure 10.5). Thus the Peyto Glacier reconstruction mirrors the long-term trend in maximum temperature, but the instrumental records of minimum temperature suggest that estimates of summer ablation rates in both studies might be improved if the long-term trends in minimum temperature could be better quantified back through time.

The regional instrumental temperature records for the twentieth century demonstrate coherence in both pattern and trend, though for certain periods (e.g., 1920s–1930s and 1960s–1990s) the relative magnitude of peaks differs between stations. Given that temperatures are a major control of the magnitude of summer balances, this similarity suggests that the pattern of summer balances should also be similar. A denser network of climate reconstructions will be necessary to explore details of possible differences in summer temperatures and winter snowfall across the region in the preinstrumental period. Nevertheless, given the generally strong regional coherence of the instrumental records of the principal climatic inputs to both summer and winter balances, it seems that significant differences between the reconstructions are unlikely to be related to regional variation in the climatic drivers of mass balance.

TREE-RING BASED SEASONAL MASS BALANCE PROXIES

WINTER MASS BALANCE

Previous studies of late-twentieth-century glacier mass balance in the western cordillera (McCabe, Fountain, and Dyurgerov 2000; Bitz and Battisti 1999; Walters and Meier 1989; Demuth and Keller 2006) have identified connections with conditions in the Pacific Ocean and, in particular, with the documented pattern of decadal-interdecadal sea-surface temperature variation (Mantua et al. 1997). Therefore, given the lack of proximal winter-sensitive tree-ring chronologies, both attempts to reconstruct winter balance targeted predictors related to this variation (Figures 10.1 and 10.2). This strategy assumes that the teleconnected relationships seen in the instrumental record are stable through time and across space (e.g., that the Miners Well chronology is strongly correlated with Peyto Glacier winter balance throughout the last several centuries). It also assumes that the PDO operated in a similar manner (with similar variation) prior to the period of instrumental records. Tree-ring-based studies of past interdecadal variation in the Pacific Ocean (MacDonald and Case 2005; D'Arrigo, Villalba, and Wiles 2001; Gedalof and Smith 2001; Biondi, Gershunov, and Cayan 2001; Evans et al. 2000; Minobe 1997) show little agreement as to conditions prior to ca. 1840 (Gedalof, Mantua, and Peterson 2002). Given potential changes in the behavior of the Pacific and the resultant impact on local climates, the choice of potential predictor variables for winter balance has a strong influence on the resultant reconstructions. This source of variation necessitates verification of these reconstructions through comparisons with other proxy data sources.

In evaluating the winter balance results, we look to independently derived records for additional confirmation of reconstructed trends and variation in mass balance. The dated moraine record compiled from 66 glaciers in the central Canadian Rockies by Luckman (2000) is particularly useful. Terminal moraines form at the downvalley limits of glaciers following the change from positive to negative net mass balance. Major periods of moraine formation in the regional record correspond with or follow the periods of strong positive mass balance (i.e., 1700–1725 and 1825–1850 [Figure 10.2]) identified in the Peyto Glacier reconstruction (Watson and Luckman 2004*a*). The mid-nineteenth-century period is also identified as one of positive mass balance in the Glacier National Park series and corresponds with independent assessments for the Little Ice Age maximum for the region (Carrara 1989; Key, Fagre, and Menicke 2002). However, the interval of high winter balances for Peyto Glacier in the early 1700s is neither as prolonged nor as pronounced in the Glacier National Park reconstruction (Figure 10.2).

Both sets of reconstructions indicate that periods of glacier advance were usually associated with a combination of cool summers and wet winters. The summer balance component of the Glacier National Park reconstruction indicates that summers in the early 1700s were cool and wet (i.e., summer melting was below normal) but winters were relatively dry (Figure 10.2). The PDO reconstruction used to derive the winter balance proxy for the park incorporates temperature-sensitive tree-ring chronologies from Alaska and the Pacific Northwest plus reconstructions of the Palmer Drought Severity Index (PDSI) from northern Mexico (Figure 10.2).[4] If the Gedalof and Smith (2001) PDO reconstruction (Figure 10.5) were used to predict winter balance in Glacier National Park, net balance values in the early 1700s would be strongly positive and consistent with the moraine record for the central Canadian Rockies and the Peyto Glacier reconstruction. This similarity is not surprising given that the Gedalof and Smith PDO and Peyto winter balance reconstructions both use chronologies from Alaska and western British Columbia.

Further differences are noted between the reconstructions over the twentieth century. The

Glacier National Park winter balance series shows more consistently positive values in the latter half of the twentieth century than the Peyto series. Peyto winter balance is estimated as above average for parts of the twentieth century, but high summer ablation (related to warm summer temperatures) results in negative net balances. Given the strong similarity of snowpack (Figure 10.3) and temperatures (Figure 10.4) across the region, the striking differences in net balance over the twentieth century are probably related to the Glacier National Park summer balance reconstruction, which may at times reflect variation in precipitation more strongly than the higher temperatures over this period. Both net balance series do, however, show positive values in the 1960s to 1970s that correspond to the formation at several sites of small readvance moraines in the 1970s and 1980s (Luckman, Harding, and Hamilton 1987; Pederson et al. 2004).

These results demonstrate the strong sensitivity of the winter balance reconstructions to the predictor variables used. Even though many of the predictors are considered to represent the same phenomenon (i.e., variations in North Pacific sea-surface temperatures), they are developed using data from different areas. Several studies (e.g., Gedalof, Mantua, and Peterson 2002; Biondi, Gershunov, and Cayan 2001) have suggested that the PDO itself has varied in intensity over time, and this variation would likely impact the strength and pattern of teleconnections to North American climate. Significant predictors in the D'Arrigo, Villalba, and Wiles (2001) and Biondi, Gershunov, and Cayan (2001) PDO reconstructions are from Mexico and California and are therefore strongly related to conditions in the equatorial Pacific. The similarities between the Peyto mass balance reconstruction, the moraine record for the Canadian Rockies, and the Gedalof and Smith (2001) PDO reconstruction suggest that the stronger weighting of data from Alaska and the Pacific Northwest may better reflect the conditions in the North Pacific that control the strength and movement of winter storms (i.e., the strength and location of the Aleutian Low) and ultimately affect winter mass balance at glaciers in this part of the Rockies.

SUMMER MASS BALANCE

An interesting difference between the Peyto Glacier and Glacier National Park mass balance estimates is the increasingly negative summer and net mass balance values during the twentieth century in the former. The summer balance component for Glacier National Park is estimated from a summer drought reconstruction that is only partially temperature-dependent. Although this drought reconstruction is significantly correlated with 13 of 14 maximum summer temperature records, it is much more strongly correlated with Kalispell precipitation (instrumental drought $r = 0.98$; reconstructed drought $r = 0.60$). This is a significant limitation considering that summer mass balance records collected at Peyto Glacier (1966–present: Demuth and Keller 2006) are not significantly correlated with summer precipitation. Therefore, it would appear that at times the Glacier National Park summer balance estimates contain a mixed temperature and precipitation signal that may not consistently track summer melt.

The instrumental temperature records for this region show a steep positive trend in mean monthly minimum temperatures (Figure 10.4) that is consistent with previous research (e.g., Wilson and Luckman 2002) but not seen in instrumental maximum temperature records. However, estimates of summer balance at Peyto Glacier do not explicitly include or necessarily capture changes in minimum temperature, making it difficult to assess the importance of this temperature variable. The trend toward increasing summer ablation is consistent with independent estimates based on photographic evidence that Peyto Glacier has lost 70% of its volume over the past 100 years (Demuth and Keller 2006; Demuth 1996; Wallace 1995). The Glacier National Park summer balances during the twentieth century are much more variable, and the absence of a trend may be because this drought-driven reconstruction captures little of the long-term trends present in minimum temperature. Although both summer precipitation and temperatures contribute to drought, drought records generally do not exhibit the same amount of low-frequency variability seen in temperature reconstructions. The differences may also be related to the amount of centennial-scale variation that is both preserved and present in the reconstructions. For example, the most heavily weighted predictor in the Peyto summer balance reconstruction (the Athabasca summer maximum temperature reconstruction) was constructed using a standardization procedure (i.e., regional curve standardization [Briffa et al. 1996]) that maximizes the retention of low-frequency climate information. Also, the Athabasca reconstruction spans more than 1,000 years, which is sufficient for preservation of centennial-scale variation, whereas the Glacier National Park summer drought reconstruction extends back only 461 years.

SUMMARY AND DISCUSSION

In this chapter we have compared the first attempts to provide proxy mass balance information for Peyto Glacier in the Canadian Rockies and the glaciers of Glacier National Park, Montana. Though different data and approaches were used, these studies indicate that, with careful selection, tree-ring data can provide effective proxies for the major components of mass balance and thereby assist in the reconstruction of continuous records of past glacier fluctuations that are not available from more traditional glacier studies. These reconstructions also allow a more direct evaluation of the relative importance of temperature and precipitation controls on mass balance variability in the absence of long-term instrumentally derived records.

Evaluation of the reconstructions and longer instrumental climate records from the region indicate that decadal and longer-term trends in both precipitation and temperature have significant influences on mass balance that are not detectable in relatively short measured mass balance series (generally <40 years in length). Although these sites are approximately 500 km apart, examination of appropriate temperature and winter snowpack records confirms that the low-frequency variation in the temperature and snow water equivalent records is very similar during the twentieth century. Slight north-south differences in snow water equivalent were identified, however, and absolute amounts of precipitation and temperature values vary. If this strong coherence across the region has held over the past 300 years, the comparison of the proxy mass balance records and evaluation of the predictor variables suggests that many of the differences between these reconstructions may result from the choice of predictor variables used. Although the strength and intensity of the PDO appear to have varied over time, the scale and mean location of the Aleutian Low in winter and its control on regional snowpack should not have been dramatically different in the past. Therefore, the coherence seen across the region related to the Aleutian Low, through its influence on storm tracks, likely existed as well. However, the slight differences in snow water equivalent values across the region do suggest that there may have been breakdowns in regional coherence in the past (perhaps related to slight deviations in the location of the mean zonal flow). Further investigations of the instrumental climate record and the development of additional reconstructions of climate parameters for sites across the region would help address this issue.

In the cases discussed here, twentieth-century snowfall variability is driven principally by changes in circulation in the Pacific Ocean. Both reconstructions estimate positive winter and net mass balance for the mid-nineteenth century, confirming the dated moraine records in the two regions. Positive net mass balance is also identified in both series for the 1970s, coincident with minor readvances in both regions. There are, however, striking differences in the magnitude of seasonal and net balance estimates for the early 1700s and the twentieth century. It is difficult to distinguish differences in the winter balance reconstructions that may reflect local climate differences within the Rockies from those that may reflect larger-scale differences in teleconnection patterns influencing the proxies used to estimate winter mass balance. Winter balance proxies should ideally be based on more locally developed tree-ring data. In addition, these results indicate that proxies used to estimate summer balance should maximize the temperature signal and minimize the influence of summer precipitation. The development of a summer temperature reconstruction for Glacier National Park may reduce ambiguities related to the mixed precipitation and temperature signal in its summer balance series.

Although these glaciers experience a continental climate regime and have been assumed to be mainly sensitive to variations in solar radiation and hence temperatures, our results indicate that it is important to consider both precipitation- and temperature-related variables in these cases. Studies of the instrumental record and these reconstructions indicate that winter balance values may vary by as much as 30–40% on decadal time scales in response to circulation changes. Therefore, snowfall variation is an important factor regulating net glacier mass balance. Our analysis of the temperature record also suggests that the increase in minimum temperatures over the twentieth century may be a key variable influencing glacier mass loss, producing changes in the duration of the melt season and related changes in the ratio of rain/snow inputs to these glaciers. Neither proxy summer balance series explicitly includes predictors that are related to minimum summer temperatures, and therefore both may underestimate mass balance changes related to these effects. Discovery of tree-ring series or other proxy records that are sensitive to (*a*) minimum in addition to maximum temperatures and (*b*) winter rather than summer precipitation might lead to significant improvement of these mass balance estimates.

The moraine record suggests slight differences in the history of glacier fluctuations between northern sites in Jasper and sites in Banff and Waterton/Glacier Parks. A slight north-south difference in the magnitude of snow water equivalent anomalies was identified in the instrumental record, and it is possible that this difference was greater in the past and caused differences in winter balance between the two regions. The 1700s advance in the more northerly sites is of similar or only slightly larger extent than the mid-1800 advance, and small differences in snowfall or reduced temperatures along this gradient may be sufficient to account for minor differences in the relative extent of eighteenth- and nineteenth-century glaciers.

Reconstruction of mass balance data using proxy climate records derived from tree rings offers the possibility of examining the drivers of glacier fluctuations directly rather than through the filtered and often censored record seen in conventional reconstructions of glacier history. Assessing the degree of similarity between these proxy mass balance series and known glacier fluctuations can help us identify the scale of the forcing factors and therefore possibly identify their causes (in these cases, large-scale Pacific variation versus local controls). This type of analysis may also help us understand the timing and rate of climate changes. These two preliminary attempts indicate the potential for the development of a more comprehensive picture of how glaciers have fluctuated in the past and how future modeled or actual climate changes may influence glaciers in the future.

ACKNOWLEDGMENTS

We thank R. D'Arrigo, F. Biondi, and Z. Gedalof for providing their PDO reconstructions, the scientists who contributed the tree-ring data in the International Tree-Ring Data Bank, and those who have collected and contributed mass balance data. We also thank Steve Gray and the reviewers of this manuscript, whose comments helped improve it. Rob Wilson provided the Athabasca maximum temperature reconstruction used in the Peyto Glacier summer mass balance model. Funding was provided to Pederson and Fagre by the U.S. Geological Survey and to

Watson and Luckman by the Canadian Foundation for Climate and Atmospheric Sciences and the Natural Sciences and Engineering Research Council of Canada. We thank Patricia Connor (Cartographic Section, Geography Department, University of Western Ontario) for preparing part of Figure 10.1.

REFERENCES CITED

Beniston, M.B., H.F. Diaz, and R.S. Bradley. 1997. Climatic change at high elevation sites: An overview. *Climatic Change* 36:233–51.

Biondi, F., A. Gershunov, and D.R. Cayan. 2001. North Pacific decadal climate variability since 1661. *Journal of Climate* 14:5–10.

Bitz, C.M., and D.S. Battisti. 1999. Interannual to decadal variability in climate and the glacier mass balance in Washington, western Canada, and Alaska. *Journal of Climate* 12:3181–96.

Briffa, K.R., P. Jones, F. Schweingruber, W. Karlen, and G. Shiyatov. 1996. Tree ring variables as proxy climate indicators: Problems with low-frequency signals. In *Climatic variations and forcing mechanisms of the last 2000 years,* ed. P. Jones, R.S. Bradley, and J. Jouzel, 9–41. Berlin: Springer-Verlag.

Brown, R.D. and R.O. Braaten. 1998. Spatial and temporal variation of Canadian monthly snow depths, 1946–1955. *Atmosphere-Ocean* 36:37–54.

Carrara, P.E. 1989. *Late Quaternary glacial and vegetative history of the Glacier National Park region, Montana.* U.S. Geological Survey Bulletin 1902. Denver: U.S. Government Printing Office.

Cayan, D.R. 1996. Interannual climate variability and snowpack in the western United States. *Journal of Climate* 9:928–48.

D'Arrigo, R., R. Villalba, and G. Wiles. 2001. Tree-ring estimates of Pacific Decadal variability. *Climate Dynamics* 18:219–24.

Demuth, M.N. 1996. *Effects of short term historical glacier variations on cold stream hydro ecology: A synthesis and case study.* Environment Canada National Hydrology Research Institute Contribution Series CS 96003. Saskatoon: National Hydrology Research Institute.

Demuth, M.N., and R. Keller. 2006. An assessment of the mass balance of Peyto Glacier (1966–1995) and its relation to recent and past century climate variability. In *Peyto Glacier: One century of science,* ed. M.N. Demuth, D.S. Munro, and G.J. Young, 83–133. National Hydrology Research Institute Science Report 8. Saskatoon: National Hydrology Research Institute.

Diaz, H.F., and R.S. Bradley, R.S. 1997. Temperature variations during the last century at high elevation sites. *Climatic Change* 36:253–79.

Easterling, D.R., T.R. Karl, E.H. Mason, P.Y. Hughes, D.P. Bowman, R.C. Daniels, and T.A. Boden, eds. 1996. *United States historical climatology network (U.S. HCN) monthly temperature and precipitation data.* Oak Ridge, TN: Carbon Dioxide Information Analysis Center, Oak Ridge National Laboratory.

Ebbesmeyer, C.R., D.R. Cayan, D.R. McLain, F.H. Nichols, D.H. Peterson, and K.T. Redmond. 1991. 1976 step change in the Pacific climate: Forty environmental changes between 1968–75 and 1977–84. In *Proceedings of the Seventh Annual Pacific Climate (PACLIM) Workshop,* ed. J.L. Betancourt and J.L. Tharp, 129–41. California Department of Water Resources Interagency Ecological Studies Program Technical Report 26. Sacramento: Department of Water Resources.

Environment Canada. 2006. Adjusted historical Canadian climate data. http://www.cccma.bc.ec.gc.ca/hccd (accessed August 21, 2007).

Evans, M.N., A. Kaplan, R. Villalba, and M.A. Cane. 2000. Globality and optimality in climate field reconstructions from proxy data. In *Interhemispheric climate linkages,* ed. V. Markgraf, 53–72. San Diego: Academic Press.

Fritts, H.C. 1976. *Tree rings and climate.* London: Academic Press.

Gedalof, Z., N.J. Mantua, and D.L. Peterson. 2002. A multi-century perspective of variability in the Pacific Decadal Oscillation: New insights from tree rings and coral. *Geophysical Research Letters* 29, doi: 10.1029/2002GL015824.

Gedalof, Z., and D.J. Smith. 2001. Interdecadal climate variability and regime-scale shifts in Pacific North America. *Geophysical Research Letters* 28:1515–18.

Hodge, S.M., D.C. Trabant, R.M. Krimmel, T.A. Heinrichs, R.S. March, and E.G. Josberger. 1998. Climate variations and changes in mass of three glaciers in western North America. *Journal of Climate* 11:2161–79.

Key, C.H., D.B. Fagre, and R.K. Menicke. 2002. Glacier retreat in Glacier National Park, Montana. In *Satellite image atlas of glaciers of the world, Glaciers of North America: Glaciers of the western United States,* ed. R.S. Williams Jr. and J.G. Ferrigno, J365–J381. Washington, DC: United States Government Printing Office.

Letréguilly, A. 1988. Relation between the mass balance of western Canadian mountain glaciers and meteorological data. *Journal of Glaciology* 34:11–18.

Luckman, B. H. 1998. Landscape and climate change in the central Canadian Rockies during the 20th century. *Canadian Geographer* 42:319–36.
———. 2000. The Little Ice Age in the Canadian Rockies. *Geomorphology* 32:357–84.
Luckman, B. H., K. A. Harding, and J. P. Hamilton. 1987. Recent glacier advance in the Premier Range, British Columbia. *Canadian Journal of Earth Sciences* 24:1149–61.
Luckman, B. H., and R. J. S. Wilson. 2005. Summer temperatures in the Canadian Rockies during the last millennium: A revised record. *Climate Dynamics* 24, doi: 10.1007/s00382-004-0511-0.
MacDonald, G. M., and R. A. Case. 2005. Variations in the Pacific Oscillation over the past millennium. *Geophysical Research Letters* 32, doi: 10.1029/2005GL022478.
Mantua, N. J., S. R. Hare, Y. Zhang, J. M. Wallace, and R. C. Francis. 1997. A Pacific decadal oscillation with impacts on salmon production. *Bulletin of the American Meteorological Society* 78:1069–79.
McCabe, G. J., and M. D. Dettinger. 2002. Primary modes and predictability of year-to-year snowpack variations in the western United States from teleconnections with the Pacific Ocean climate. *Journal of Hydrometeorology* 3:13–25.
McCabe, G. J., A. G. Fountain, and M. Dyurgerov. 2000. Variability in winter mass balance of Northern Hemisphere glaciers and relations with atmospheric circulation. *Arctic, Antarctic, and Alpine Research* 32:64–72.
Meteorological Service of Canada. 2000. *Canadian snow data* (CD-ROM). Downsview, Ontario: CRYSYS Project, Climate Processes and Earth Observation Division.
Minobe, S. 1997. A 50–70-year climatic oscillation over the North Pacific and North America. *Geophysical Research Letters* 24:683–86.
Moore, R. D., and I. G. McKendry. 1996. Spring snowpack anomaly patterns and winter climatic variability, British Columbia, Canada. *Water Resources Research* 32:623–32.
Mote, P. W. 2003. Trends in snow water equivalent in the Pacific Northwest and their climate causes. *Geophysical Research Letters* 30, doi: 10.1029/2003GL0171258.
National Climatic Data Center. n.d. International tree-ring data bank. http://www.ncdc.noaa.gov/paleo/treering.html (accessed August 21, 2007).
Pederson, G. T., D. B. Fagre, S. T. Gray, and L. J. Graumlich. 2004. Decadal-scale climate drivers for glacial dynamics in Glacier National Park, Montana, USA. *Geophysical Research Letters* 31: L12203, doi: 10.1029/2004GL019770.
Selkowitz, D. J., D. B. Fagre, and B. A. Reardon. 2002. Interannual variations in snowpack in the Crown of the Continent ecosystem. *Hydrological Processes* 16:3651–65.
Shea, J. M., S. J. Marshall, and J. M. Livingston. 2004. Glacier distributions and climate in the Canadian Rockies. *Arctic, Antarctic, and Alpine Research* 36:272–79.
Skinner, W. R., and D. W. Gullett. 1993. Trends of daily maximum and minimum temperature in Canada during the past century. *Climatological Bulletin* 27:63–77.
Smith, D. J., D. P. McCarthy, and M. E. Colenutt. 1995. Little Ice Age glacial activity in Peter Lougheed and Elk Lakes Provincial Parks, Canadian Rocky Mountains. *Canadian Journal of Earth Sciences* 32:579–89.
Vincent, L. A., X. Zhang, B. R. Bonsal, and W. D. Hogg. 2002. Homogenization of daily temperatures over Canada. *Journal of Climate* 15: 1322–34.
Wallace, A. L. 1995. The volumetric change of Peyto Glacier, Alberta, Canada 1896–1996. M.Sc. thesis, Wilfred Laurier University.
Walters, R. A., and M. F. Meier. 1989. Variability of glacier mass balances in western North America: Aspects of climate variability in the Pacific and western Americas. *Geophysical Monographs* 55:365–74.
Watson, E., and B. H. Luckman. 2004*a*. Tree-ring-based mass-balance estimates for the past 300 years at Peyto Glacier, Alberta, Canada. *Quaternary Research* 62:9–18.
———. 2004*b*. Tree-ring based-reconstructions of precipitation for the southern Canadian Cordillera. *Climatic Change* 65:209–41.
Wilson, R. J. S., and B. H. Luckman. 2002. Tree-ring reconstruction of maximum and minimum temperatures and the diurnal temperature range in British Columbia, Canada. *Dendrochronologia* 20:257–68.
Young, G. J., and A. D. Stanley. 1977. *Canadian glaciers in the International Hydrological Decade Program, 1965–1974: Ram River Glacier, Alberta.* Inland Waters Directorate, Water Resources Branch, Summary of Measurements, Scientific Series 70. Ottowa: Inland Waters Directorate.

11

Glacier Fluctuations in the European Alps, 1850–2000

AN OVERVIEW AND A SPATIOTEMPORAL ANALYSIS OF AVAILABLE DATA

Michael Zemp, Frank Paul, Martin Hoelzle, and Wilfried Haeberli

Fluctuations of mountain glaciers are among the best natural indicators of climate change (Houghton et al. 2001). Changes in precipitation and wind lead to variations in accumulation, while changes in temperature, radiation fluxes, and wind, among other factors, affect the surface energy balance and thus ablation. Disturbances in glacier mass balance, in turn, alter the flow regime and, consequently, after a glacier-specific delay, result in a glacier advance or retreat such that the glacier geometry and altitude range change until accumulation equals ablation (Kuhn et al. 1985). Hence, mass balance is the direct and undelayed signal of annual atmospheric conditions, whereas changes in length are an indirect, delayed, and filtered but enhanced signal (Haeberli 1998).

The modern concept of worldwide glacier observation is an integrated and multilevel one; it aims to combine in-situ observations with remotely sensed data, understanding of process with global coverage, and traditional measurements with new technologies. This concept uses detailed mass and energy balance studies from just a few glaciers, together with length change observations from many sites and inventories covering entire mountain chains. Numerical models link all three components over time and space (Haeberli 2004). The European Union–funded ALP-IMP Project focuses on multicentennial climate variability in the Alps on the basis of instrumental data, model simulations, and proxy data. It represents a unique opportunity to apply this glacier-monitoring concept to the European Alps, where by far the most concentrated amount of information about glacier fluctuations over the past century is available. The World Glacier Monitoring Service (WGMS) has compiled, within the framework of the ALP-IMP Project, an unprecedented data set containing inventory data (i.e., area, length, and altitude range) from approximately 5,150 Alpine[1] glaciers and fluctuation series from more than 670 of them (i.e., more than 25,350 observations of annual front variation and 575 of annual mass balance) dating back to 1850.

In this chapter we offer an overview of the available glacier data sets from the European Alps and analyze glacier fluctuations between

1850 and 2000. To achieve this, we analyze glacier size characteristics from the 1970s, the only time period for which a complete Alpine inventory is available, and extrapolate Alpine glaciation in 1850 and in 2000 from size-dependent area changes from Switzerland. We go on to examine mass balance and front variation series for the insight they provide into glacier fluctuations, the corresponding acceleration trends, and regional distribution patterns at an annual resolution. Finally, we discuss the representativeness of these recorded fluctuation series for all the Alpine glaciers and draw conclusions for glacier monitoring.

BACKGROUND

The worldwide collection of information about ongoing glacier changes was initiated in 1894 with the founding of the International Glacier Commission at the Sixth International Geological Congress in Zurich, Switzerland. At that time, the Swiss limnologist F. A. Forel began publishing the periodical *Rapports sur les variations périodiques des glaciers* on behalf of the commission (Forel 1895). Up until 1961, data compilations constituting the main source of length change data worldwide were published in French, Italian, German, and English. Since 1967, the publications have all been in English. The first reports contain mainly qualitative observations except for the glaciers of the European Alps and Scandinavia, many of which have had extensive documentation and quantitative measurements recorded from the very beginning. After World War I, P. L. Mercanton edited the publications, which began to appear less than annually. From 1933 to 1967 they were published on behalf of the International Commission on Snow and Ice (ICSI), part of the International Association of Hydrological Sciences (IAHS). Since then they have been published at five-year intervals under the title *Fluctuations of Glaciers*, at first by the Permanent Service on the Fluctuations of Glaciers (PSFG [Kasser 1970]) and then, after the merger of the PSFG with the Temporary Technical Secretariat for the World Glacier Inventory (TTS/WGI) in 1986, by the WGMS. An extensive overview of the corresponding literature is given by Hoelzle et al. (2003).

The need for a worldwide inventory of perennial snow and ice masses was first considered during the International Hydrological Decade declared by the United Nations Educational, Scientific, and Cultural Organization (UNESCO) from 1965 until 1974 (UNEP/GEMS 1992). Preliminary results and a thorough discussion of the techniques and standards employed in glacier inventorying were given in IAHS (1980). A status report and the corresponding national literature of all national glacier inventories compiled at that time was published by Haeberli et al. (1989a). More detailed reports on glacier area changes for specific regions or countries, often with special emphasis on developments since 1850, can be found in CGI/CNR (1962) for Italy, Gross (1988a) for Austria, Maisch et al. (2000) for Switzerland, Vivian (1975) for the Western Alps, Maisch (1992) for the Grisons (Switzerland), Böhm (1993) for the Goldberg region (Hohe Tauern, Austria), and Damm (1998) for the Rieserferner group (Tyrol, Austria).

THE DATA

The Alpine glacier information available is of three types: the World Glacier Inventory (WGI), the Swiss Glacier Inventory 2000 (SGI2000), and Fluctuations of Glaciers (FoG). The geographical distribution of the different data sets is shown in Figure 11.1.

THE WORLD GLACIER INVENTORY

The WGI contains attribute data on glacier area, length, orientation, and elevation as well as a classification of morphological types and moraines linked to the glacier coordinates. The inventory entries are based upon specific observation times and can be viewed as snapshots of the spatial glacier distribution. The data are stored in the WGI database (part of the WGMS database) and are published in Haeberli et al. (1989*a*), which summarizes the national inventories for the entire Alps.

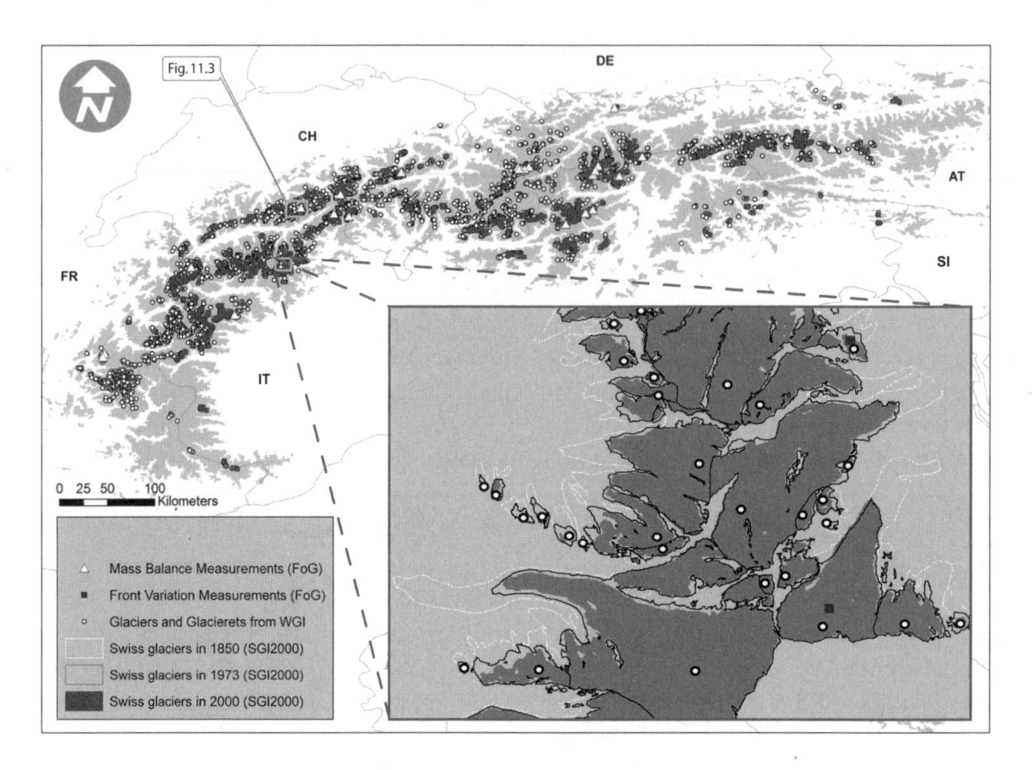

FIGURE 11.1. Geographical distribution of available glacier information in the Alps: WGI data (*white circles*) and mass balance (*white triangles*) and front variation (*dark gray squares*) data from the FoG database. Elevations above 1,500 m a.s.l. are in light gray. *AT*, Austria; *FR*, France; *DE*, Germany; *IT*, Italy; *SI*, Slovenia; and *CH*, Switzerland. The inset shows Swiss glacier polygons for 1850, 1973, and 2000 from the SGI2000.

Complete national inventories for the European Alps are available for Austria (1969), France (1967–71), Switzerland (1973), Germany (1979), and Italy (1975–84). The inventories for Austria, Switzerland, and Germany refer to a single reference year, while the records of France and Italy are compiled over a longer period of time to achieve total coverage (Figure 11.2). However, in every inventory there is a certain percentage of glaciers for which no data from the corresponding reference period/year could be obtained and information from earlier years has been substituted. For example, in the Swiss inventory, data from only 1,550 glaciers date from 1973, while the information for the remaining 274 glaciers refers to earlier years. Glacier identification, assignment, and partitioning (due to glacier shrinkage) are the main challenges for comparisons of inventories overlapping in space or time. Therefore, the total number and areas of glaciers may vary in different studies. Haeberli et al. (1989*a*) sum the area of the 5,154 Alpine glaciers from Austria (542 km^2), France (417 km^2), Switzerland (1,342 km^2), Germany (1 km^2), and Italy (607 km^2) as 2,909 km^2. Because of the inconsistencies just mentioned, the data set used in this study differs slightly from these numbers; the Italian inventory sums up to only 602 km^2 and the number of Alpine glaciers to 5,167. These differences, however, are smaller than 0.3% and therefore negligible.

THE SWISS GLACIER INVENTORY 2000

The SGI2000 has been compiled from multispectral Landsat Thematic Mapper (TM) data acquired in 1998–99 (path-row 194/5-27/8). Glacier information (e.g., area, slope, aspect) was obtained from a combination of glacier outlines with a digital elevation model and the related analysis by a Geographic Information

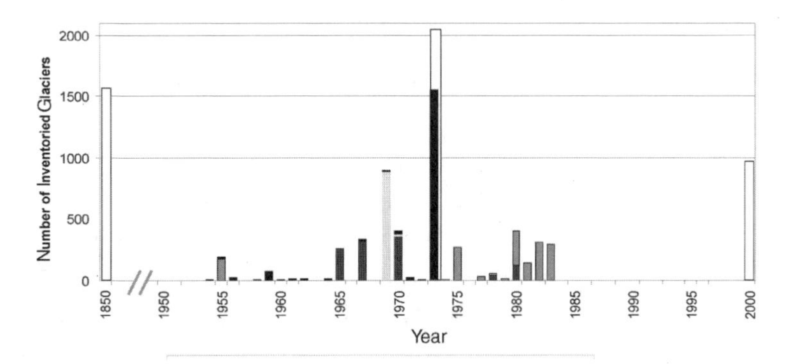

FIGURE 11.2. Numbers of inventoried glaciers in the Alps by year, country, and data source. (For 1973, for example, there are data in the WGI from 6 Italian, 2 Austrian, and 1,550 Swiss glaciers and data in the SGI2000 from 2,057 Swiss glaciers.)

System (Kääb et al. 2002; Paul et al. 2002; Paul 2004). Several glaciers were not properly identified because of cast shadow, snow cover, and debris and were excluded from the statistical analysis. New areas for 938 glaciers were obtained for 2000 and the related topographical information extracted. The glacier inventories from 1850 and 1973 were digitized from the original topographic maps and are now a major part of the SGI2000 (Figure 11.3). The 1973 outlines are also used to define the hydrological basins of individual glaciers in the satellite-derived inventory, in particular the ice-ice divides. However, because different identification codes were used in the inventories of Müller, Caflisch, and Müller (1976), Maisch et al. (2000), and the SGI2000, a direct comparison of glacier areas is not yet possible. Moreover, glacier retreat has caused severe changes in glacier geometry (tongue separation, disintegration, etc.) that prevent direct comparison. For this reason our analysis of glacier changes was based on different samples. The major results of this study have been summarized by Paul et al. (2004).

FLUCTUATIONS OF GLACIERS

The FoG database contains attribute data on glacier changes over time—front variations, mass balance, and changes in area, thickness, and volume—linked to glacier coordinates. The data are stored in the FoG database (part of the WGMS database) and published in the *Fluctuations of Glaciers* series at five-year intervals (latest edition, Haeberli et al. 2005*b*) and biannually in the *Glacier Mass Balance Bulletin* (latest edition, Haeberli et al. 2005*a*).

Regular glacier front variation surveys in the Alps started around 1880. The number of glaciers surveyed and the continuity of series changed over time because of world history and the perceptions of the glaciological community (Haeberli and Zumbühl 2003; Haeberli, this volume). Direct measurements of glacier mass balance in the Alps started at Limmern (Switzerland) and Plattalva (Switzerland) in 1948, followed by Sarennes (France) in 1949, Hintereis (Austria) and Kesselwand (Austria) in 1953, and others. In the last reporting period (1995–2000) 297 glacier front measurements were made, along with measurements of the mass balance of 18 Alpine glaciers (Haeberli et al. 2005*b*). For the analysis here only front variation series with more than nine survey years and mass balance series longer than three years have been considered (Figure 11.4).

There are some reconstructed front variation series for several Alpine glaciers, spanning time periods from centuries to millennia (e.g., Holzhauser and Zumbühl 1996; Holzhauser 1997; Nicolussi and Patzelt 2000; Holzhauser,

FIGURE 11.3. Synthetic oblique-perspective of the Aletsch Glacier region, Switzerland, generated from a digital elevation model (DEM25; reproduced by permission of swisstopo, BA057338) overlaid with a fusion of satellite images from Landsat TM (1999) and IRS-1C (1997) in a grayscale rendition. The Grosser Aletsch Glacier retreated about 2,550 m from 1850 (*white lines*) to 1973 (*black lines*) and another 680 m by 2000.

Magny, and Zumbühl 2005). In addition, there are some studies that estimate secular mass balance trends from cumulative glacier length changes (e.g., Haeberli and Holzhauser 2003; Hoelzle et al. 2003) or from glacier surfaces reconstructed from historical maps (cf. Haeberli 1998; Steiner et al., this volume). These studies, however, have not been prepared within an international framework, and most of the data are not publicly available, so we have not considered them here.

ANALYSIS AND RESULTS

ALPINE GLACIERIZATION IN THE 1970s

The only complete Alpine inventory available is from the 1970s, with 5,154 glaciers and an area of 2,909 km^2 (Haeberli et al. 1989*a*). Paul et al. (2004) have estimated the total ice volume to be about 100 km^3, much lower than the 130 km^3 suggested earlier by Haeberli and Hoelzle (1995). The latter estimated the total ice volume from the total Alpine glacier area and an averaged thickness from all the glaciers (in accordance with semielliptical cross-sectional glacier geometry). Paul et al. (2004) calculated the total volume loss (-25 km^3) for the period 1973–1998/99 from the mean Alpine glacier area (2,753 km^2) and the average cumulative mass balance for eight Alpine glaciers (-9 m water equivalent). Assuming that the relative change in volume is likely to have been larger than the corresponding relative change in area (for geometric reasons), the estimated relative volume loss is roughly -25% and, therefore, the total Alpine ice volume in the 1970s was about 100 km^3.

Eighty-two percent of Alpine glaciers are smaller than 0.5 km^2 and cover 21% of the total glaciated area (Figure 11.5). Glacierets and névés (perennial snowbanks) do not normally show dynamic reactions and therefore are usually excluded from glacier studies. However, neglecting these small glaciers in inventories could introduce significant errors in the assessment of regional glacier change. Only seven glaciers (Grosser Aletsch, Gorner, Fiescher, Unteraar, Unterer Grindelwald, and Oberaletsch in Switzerland and Mer de Glace in France) are larger than 20 km^2 but represent 10% of the total area. Glaciers between 1 and 10 km^2 account for 46% of the Alpine glacier area.

The regional distribution of numbers and areas of Alpine glaciers can be calculated for each Alpine country. Most of the glaciers are located in Switzerland (35%), followed by Italy (27%), France (20%), and Austria (18%). Regarding total glacier area, the majority of European ice is located in Switzerland (46%) and Italy (21%). Austria ranks third, with 19% of the Alpine glacier area, followed by France with

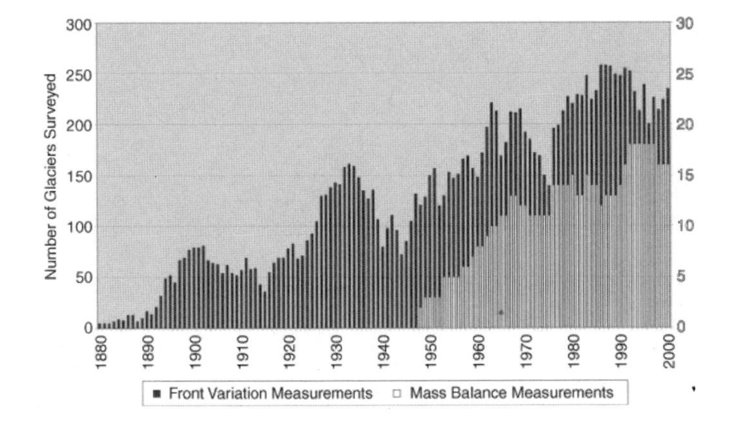

FIGURE 11.4. Frequency of front variation (*black bars, left axis*) and mass balance (*white bars, right axis*) measurements in the Alps, 1880–2000. Only glaciers with more than 18 front variations or three mass balance surveys are considered.

14%. The five German glaciers, with a total area of 1 km², and the two small Slovenian glaciers are not considered in the tables.

Tables 11.1 and 11.2 show the glacier size characteristics in the 1970s. The numbers of glaciers in each area-class are very similar in all countries except for France, where 50% of the glaciers are smaller than 0.1 km². The area distribution in Austria and Italy is dominated equally by small- and middle-sized glaciers. Mer de Glace, with an area of 33 km², corresponds to almost 8% of the French glacierization. In Switzerland the 22 largest glaciers (> 10 km²) account for 37% of the total glacier area.

ALPINE GLACIERIZATION IN 1850 AND 2000

Using the Alpine inventory of the 1970s, the Alpine glacier areas in 1850 and in 2000 can be extrapolated by applying the relative area changes (1850–1973, 1973–2000) of the seven glacier size classes from the SGI2000 to the corresponding Alpine glacier areas in the 1970s (Table 11.3). The estimated Alpine glacier areas amount to 4,474 km² in 1850 and to 2,272 km² in 2000. This corresponds to an overall glacier area loss from 1850 until the 1970s of 35% and almost 50% by 2000—or an area reduction of 22% between the 1970s and 2000. Dividing the total area loss by time provides estimates of area change per decade of 2.9% between 1850 and 1973 and 8.2% between 1973 and 2000. Several methods exist for calculating glacier volume from other variables, based either on statistical relationships (e.g., Müller, Caflisch, and Müller 1976), empirical studies (e.g., Maisch et al. 2000), or physical parameters (e.g., Haeberli and Hoelzle 1995). However, all of them employ glacier size as a scaling factor, and the deviations between individual methods are large. As the individual glacier sizes for the year 2000 are not yet available for all glaciers, we have not attempted to present glacier volume evolution over time. However, a current estimate of Alpine glacier volume in 2000 indicates that approximately 75 km³ remain (Paul et al. 2004).

ALPINE FRONT VARIATIONS

Large valley glaciers have retreated continuously since the Little Ice Age maximum around 1850. Smaller mountain glaciers show marked periods of intermittent advances in the 1890s, the 1920s, and the 1970–80s. The front variations of the smallest glaciers have a high annual variability. In Figure 11.6 front variation series with more than 18 measurement years are plotted and sorted according to glacier size. The advance periods of the 1920s and the 1970–80s and the retreat periods in between and after 1990 show up very clearly. However, on the individual level the climate signal from variations in the front position of glaciers is much more complex. This noise prevails even when the data set is sorted according to

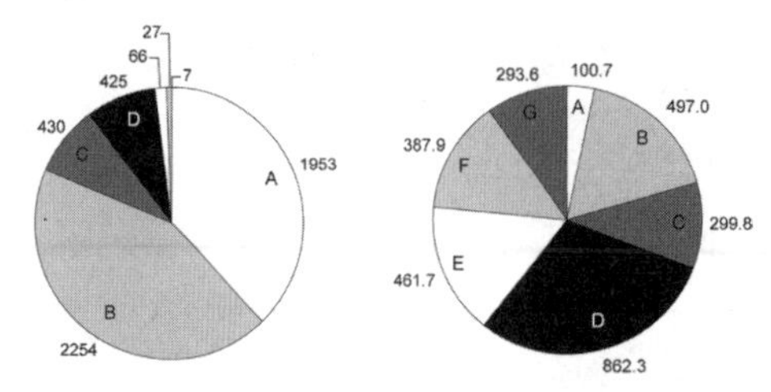

FIGURE 11.5. Distribution of glaciers by number (*left*) and size (*right*) in the Alps for the 1970s. Pie charts give percentages with absolute values indicated. (A) < 0.1 km²; (B) 0.1–0.5 km²; (C) 0.5–1.0 km²; (D) 1.0–5.0 km²; (E) 5.0–10.0 km²; (F) 10.0–20.0 km²; (G) > 20.0 km². The five German and two Slovenian glaciers are not considered in this figure.

response time (see Johannesson, Raymond, and Waddington 1989; Haeberli and Hoelzle 1995) or analyzed in geographical subsamples. Figure 11.6 is dominated by the smaller mountain glaciers, and therefore the signals of the large valley glaciers and the smallest glaciers (including absolute retreat values) are more visible in the graphs of individual cumulative front variation (e.g., Haeberli et al. 1989*b*; Hoelzle et al. 2003).

ALPINE MASS BALANCES

Fifty years of direct mass balance measurements show a clear trend of mass loss. Although some of the glaciers measured gained mass from the 1960s to the 1980s, ice loss has accelerated in the past two decades (Figure 11.7). With respect to the geographical distribution, years with a uniformly positive (e.g., 1965, 1977, 1978) or negative (e.g., 1964, 1973, 1983) Alpine mass balance signal, as well as years with a clear spatial gradient in net balance (e.g., 1963, 1976) or with heterogeneous signals, can be found mainly before 1986. After 1981, uniformly negative mass balance years dominate. Nine Alpine reference glaciers (Careser in Italy, Gries and Silvretta in Switzerland, Hintereis, Kesselwand, Sonnblick, and Vernagt in Austria, and Saint Sorlin and Sarennes in France) with continuous mass balance series over more than 30 years show a mean annual loss of ice thickness close to 37-cm water equivalent per year, resulting in a total thickness reduction of about 13 m water equivalent between 1967 and 2001. The corresponding values for the period 1980–99 are 60 cm water equivalent and 12.3 m water equivalent per year, respectively (Table 11.4).

DISCUSSION

DATA COVERAGE

Glacier studies have a long tradition in the Alps that began with the establishment of systematic observation networks in the 1890s (Haeberli, this volume). In comparison with the rest of the world, the European Alps have the densest and most complete spatial glacier inventory over time (Haeberli et al. 1989*a*). Thus, the inventory data contain information on spatial glacier distribution at certain times, whereas the fluctuation series provides high-resolution temporal information for specific locations. Interestingly, the 1970s is the only period in which an Alpine inventory with total spatial coverage can be compiled, most glaciers being relatively close to steady-state conditions (Figure 11.7; Patzelt 1985). The reconstructed glacier extents at the end of the Little Ice Age (around 1850) and the glacier outlines derived from multispectral satellite data around 2000 from the SGI2000 cover the major parts and the full range of area-classes of Swiss glaciation. Thus, they can be used to extrapolate Alpine glaciation in 1850 and 2000 on the assumption that the relative losses of the different area-classes in Switzerland are representative of other Alpine countries as well. This, of course, is not necessarily the case. The fluctuation series are numerous and well distributed over the Alps, with a minimum number of front variation series in the southwestern part of the Alps. For the fluctuation series, length and completeness of the time series are most relevant.

GLACIER SHRINKAGE

The inventory for the 1970s and the extrapolated area estimates for 1850 and 2000 show dramatic shrinkage of the Alpine glaciers. Despite the high degree of variability in individual glaciers, the European Alps have experienced a 50% decrease

TABLE 11.1

Distribution of Glaciers by Number and Area (Absolute Values) in the Alps in the 1970s

AREA-CLASS (KM^2)		ALPS WGI	AT WGI	CH WGI	FR WGI	IT WGI	ALPS FOG, FV	ALPS FOG, MB
0.0–0.1	Number	1,953	287	636	522	508	16	0
	Area (km^2)	100.7	16.2	29.4	24.5	30.5	1.0	0.0
0.1–0.5	Number	2,254	416	826	361	651	130	3
	Area (km^2)	497.0	92.3	185.5	77.0	142.2	36.2	0.9
0.5–1	Number	430	112	156	73	89	92	4
	Area (km^2)	299.8	77.6	108.4	51.4	62.5	64.0	2.8
1–5	Number	425	95	152	79	99	198	13
	Area (km^2)	862.3	213.2	294.8	153.6	200.7	446.6	35.2
5–10	Number	66	10	32	7	17	56	3
	Area (km^2)	461.7	71.8	223.0	51.0	115.9	392.1	24.6
10–20	Number	27	5	16	2	4	27	2
	Area (km^2)	387.9	71.2	240.1	26.1	50.5	396.0	33.0
>20	Number	7	0	6	1	0	7	0
	Area (km^2)	293.6	0.0	260.5	33.1	0.0	293.5	0.0
Total	Number	5,162	925	1,824	1,045	1,368	526	25
	Area (km^2)	2,902.9	542.2	1,341.7	416.6	602.4	1,629.3	96.5

NOTE: *AT*, Austria, *CH*, Switzerland, *FR*, France, *IT*, Italy; *FV*, front variation surveys (more than nine measurements); *MB*, mass balance surveys (more than three measurements).

in ice coverage over the past 150 years. The area loss over each decade (in percent) between the 1970s and 2000 is almost three times greater than the related loss of ice between 1850 and the 1970s. Variations in glacier front position provide a higher-resolution assessment of the glacier retreat over the past 150 years. Though glaciers have generally been retreating since 1850, there have been several periods of documented readvances—in the 1890s, the 1920s, and the 1970s and 1980s (Patzelt 1985; Müller 1988; Pelfini and Smiraglia 1988; Reynaud 1988; Haeberli et al. 1989b). The area reduction after the 1970s occurred mainly after 1985 (see also Paul et al. 2004), and therefore the acceleration of the glacier retreat in the past two decades was even more pronounced. Mass balance measurements are available only for the past five decades and confirm the general trend of glacier shrinkage. While some glaciers gained mass between 1960 and 1980, ice loss has accelerated in the past two decades. The mean specific (annual) net balance of the 1980s is 18% below the average of 1967–2001, and the value for the 1990s doubles that average ice loss. The most recent mass balance data show a continuation of the acceleration trend after 2000, with a peak in the extraordinary year of 2003, when the ice loss of the nine Alpine reference glaciers was about 2.5-m water equivalent—exceeding the average of 1967–2000 by a factor of nearly seven. Estimated total glacier-volume loss in the Alps in 2003 corresponds to 5–10% of the remaining ice volume (Zemp et al. 2005). The acceleration of glacier shrinkage after 1985 indicates a transition toward rapid down-wasting rather than a dynamic glacier response to a changed climate (cf. Paul et al. 2004).

The general glacier retreat since 1850 corresponds well with the observed warming trend in

TABLE 11.2

Distribution of Glaciers by Number and Area (Percentage) in the Alps in the 1970s

AREA-CLASS (KM²)		ALPS WGI	AT WGI	CH WGI	FR WGI	IT WGI	ALPS FOG, FV	ALPS FOG, MB
0.0–0.1	Number (%)	37.8	31.0	34.9	50.0	37.1	3.0	0.0
	Area (%)	3.5	3.0	2.2	5.9	5.1	0.1	0.0
0.1–0.5	Number (%)	43.7	45.0	45.3	34.5	47.6	24.7	12.0
	Area (%)	17.1	17.0	13.8	18.5	23.6	2.2	0.9
0.5–1	Number (%)	8.3	12.1	8.6	7.0	6.5	17.5	16.0
	Area (%)	10.3	14.3	8.1	12.3	10.4	3.9	2.9
1–5	Number (%)	8.2	10.3	8.3	7.6	7.2	37.6	52.0
	Area (%)	29.7	39.3	22.0	36.9	33.3	27.4	36.5
5–10	Number (%)	1.3	1.1	1.8	0.7	1.2	10.6	12.0
	Area (%)	15.9	13.2	16.6	12.2	19.2	24.1	25.5
10–20	Number (%)	0.5	0.5	0.9	0.2	0.3	5.1	8.0
	Area (%)	13.4	13.1	17.9	6.3	8.4	24.3	34.2
>20	Number (%)	0.1	0.0	0.3	0.1	0.0	1.3	0.0
	Area (%)	10.1	0.0	19.4	7.9	0.0	18.0	0.0
Total	Number (%)	100.0	100.0	100.0	100.0	100.0	100.0	100.0
	Area (%)	100.0	100.0	100.0	100.0	100.0	100.0	100.0

NOTE: *AT*, Austria, *CH*, Switzerland, *FR*, France, *IT*, Italy; *FV*, front variation surveys (more than nine measurements); *MB*, mass balance surveys (more than three measurements).

this period (e.g., Oerlemans 1994, 2001: 110–11; Maisch et al. 2000; Zemp, Hoelzle, and Haeberli 2007). However, the onset of the Alpine glacier retreat after 1850 may have been triggered by a negative winter precipitation anomaly (relative to the mean of 1901–2000) during the second half of the nineteenth century (Wanner et al. 2005). The intermittent periods of glacier advances in the 1890s, the 1920s, and the 1970s and 1980s can be explained by earlier wetter and cooler periods, with reduced sunshine duration and increased winter precipitation (Patzelt 1987; Schöner, Auer, and Böhm 2000; Laternser and Schneebeli 2003). Schöner, Auer, and Böhm (2000) concluded from the study of a homogenized climate data set and mass balance data from the Austrian part of the eastern Alps that the more positive mass balance periods show a high correlation with winter accumulation and a lower correlation with summer temperature, while more negative mass balance periods are closely correlated with summer temperature and show no correlation with winter accumulation. In addition they found that the positive mass balance period between 1960 and 1980 was characterized by negative winter North Atlantic Oscillation index values, which caused an increase of the meridional circulation mode and a more intense northwesterly to northerly precipitation regime (see Wanner et al. 2005). The observed trend of increasingly negative mass balances since 1980 is consistent with accelerated global warming and correspondingly enhanced energy flux toward the earth's surface (Haeberli et al. 2005*b*).

REPRESENTATIVENESS OF THE SGI2000 AND THE FLUCTUATION SERIES

When analyzing national inventories or individual fluctuation series, the question of representativeness often arises. Are the subsample

TABLE 11.3

Alpine Glaciation, 1850, 1970s, and 2000

	SWITZERLAND (SGI2000)								ALPS			
	1850		1973		2000		1850–1973[a]	1973–2000[a]	1970s		1850[b]	2000[b]
AREA-CLASS (KM²)	*Number*	*Area (km²)*	*Number*	*Area (km²)*	*Number*	*Area (km²)*	*Area Change (%)*	*Area Change (%)*	*Number*	*Area (km²)*	*Area (km²)*	*Area (km²)*
< 0.1	297	17.3	1,022	40.1	164	3.6	–55.4	–64.6	1,953	100.7	225.5	35.6
0.1–.5	715	181.3	673	153.9	448	60.3	–52.9	–45.6	2,254	497.0	1,055.0	270.4
0.5–1	249	172.5	151	104.1	131	63.5	–44.3	–29.1	430	299.8	538.0	212.6
1–5	253	524.4	157	296.0	141	217.1	–33.2	–17.9	425	862.3	1,291.1	707.9
5–10	26	195.5	35	249.4	36	232.6	–19.7	–10.8	66	461.7	574.8	412.1
10–20	18	259.9	14	216.3	13	192.8	–14.8	–8.2	27	387.9	455.1	356.1
> 20	9	270.5	5	225.9	5	213.0	–12.3	–5.7	7	293.6	334.8	276.9
Total	1,567	1,621.4	2,057	1,285.7	938	982.9	–27.1	–16.1	5,162	2,902.9	4,474.3	2,271.6

[a] The relative area changes in Switzerland are calculated from the comparable subsamples: 1,567 glaciers for 1850–1973 and 938 glaciers for 1973–2000, respectively.

[b] Alpine glacier area in 1850 and 2000 is extrapolated from the glacier area in the 1970s (WGI) and relative area changes of the seven glacier area-classes in Switzerland (SGI2000).

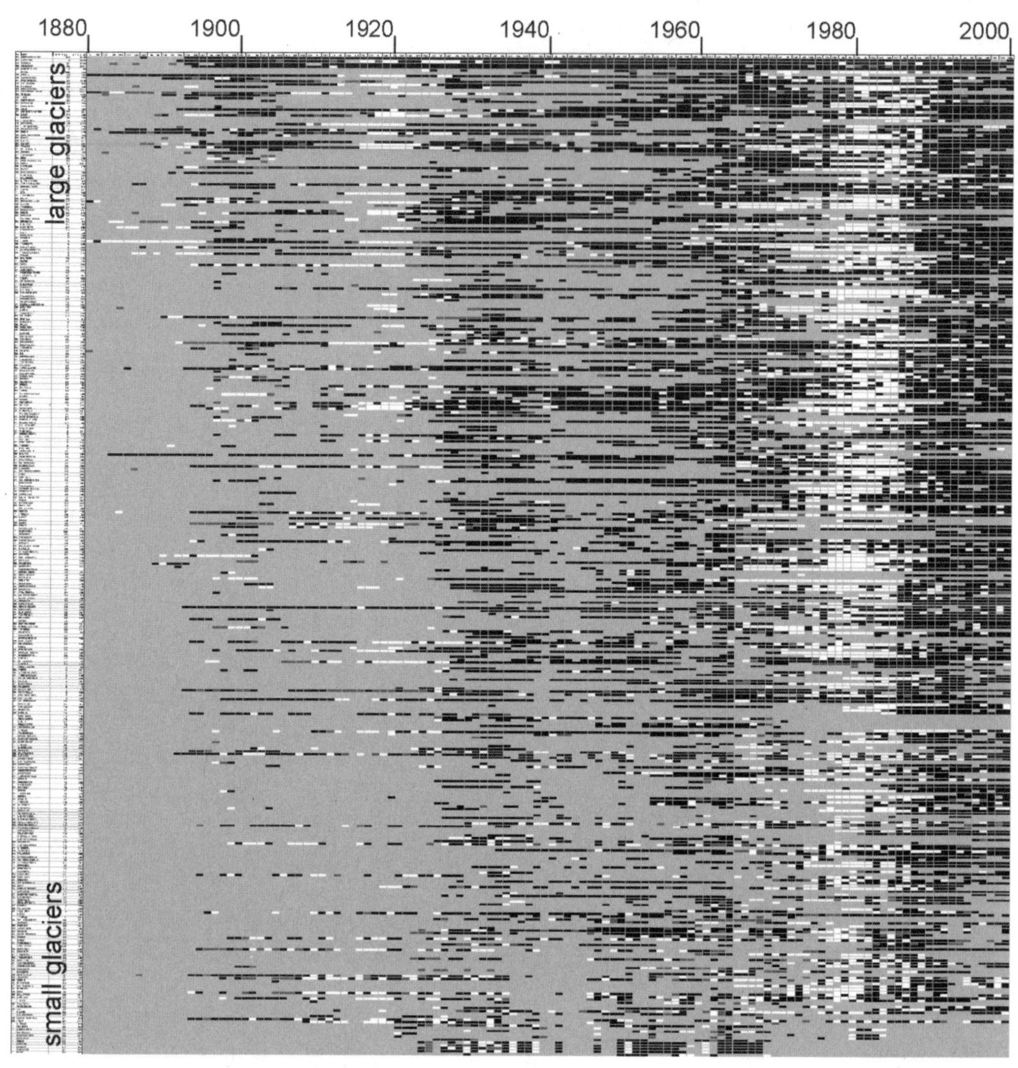

FIGURE 11.6. Alpine front variation series, 1880–2000. Annual front variation values from glaciers with more than 18 measurements are colored white after an advance, black after a retreat; dark gray indicates no apparent variation and light gray no data. Each row represents one glacier. The glaciers are sorted according to length in the 1970s (*y*-axis).

investigated and the glaciers surveyed representative of the entire glacierization? Comparison of the area characteristics of the 1850 and 2000 subsamples of the SGI2000 (on which the extrapolation of the Alpine areas of 1850 and 2000 is based) with the complete Swiss inventory in the WGI shows that the distributions of the area-classes are similar. Nevertheless, small glaciers (<0.1 km^2) are underrepresented, and glaciers in northeastern Switzerland are poorly represented (Paul et al. 2004). However, the SGI2000 subsamples for 1850 and 2000 include 86% of the Swiss glaciers covering 88% of the total area and 51% of the Swiss glaciers covering 87% of the total area, respectively. Thus, the SGI2000 can be considered a representative subsample of Swiss glaciation, which is very similar to the glaciation of the other Alpine countries. The ice coverage of the European countries is equally distributed with respect to the number of glaciers in each area-class, with the largest glaciers being overrepresented in the area distribution. Therefore, the different area-classes were considered when the

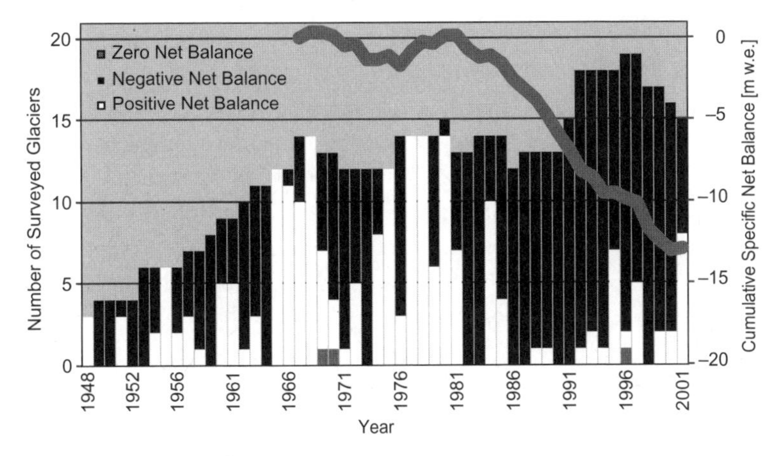

FIGURE 11.7. Alpine mass balance measurements, 1948–2001, showing annual numbers of glaciers (*left axis*) with a zero net balance (*dark gray*), positive net balance (*white*), or negative net balance (*black*) and the mean cumulative specific net balance of the nine Alpine reference glaciers from 1967 to 2001 (*right axis*).

extrapolation was applied to all Alpine glaciers. The large relative area change of the smaller glaciers leads to a more pronounced area change in the entire Alps than in Switzerland (assuming a uniform climate change across the region) because of the greater frequency of large glaciers in the latter.

Front variations are measured mainly on middle-sized and large glaciers, while glaciers smaller than 0.5 km² are underrepresented. This is to be expected because glacierets and névés are often unsuitable for this kind of measurement in terms of their limited accessibility and their low dynamic response. Front variation series with more than nine measurement years exist for about 10% of all Alpine glaciers, which cover more than 50% of the total glacier area. The dynamic response to climatic forcing of glaciers with variable geometry results in striking differences in the recorded curves, reflecting the considerable effects of size-dependent filtering, smoothing, and enhancing of the delayed tongue response with respect to the input (mass balance) signal (Oerlemans 2001). Dynamic response time depends mainly on glacier length, slope, and mass balance gradient (Johanneson, Raymond, and Waddington 1989; Haeberli and Hoelzle 1995). As a consequence, large valley glaciers with a dynamic response time of several decades show the secular climate trend, while smaller mountain glaciers show marked periods of intermittent advances and retreats on a decadal scale. The smallest, somewhat static, low-shear-stress glaciers (cirque glaciers) have altitude ranges that are comparable to or smaller than the interannual variation in equilibrium line altitude and hence, in general, reflect yearly changes in mass balance without any delay (Hoelzle et al. 2003).

Mass balance measurements are labor-intensive and are therefore available from only 25 glaciers, mainly from 0.5 to 10 km² in size, covering only 3% of the glacier area. In spite of their small number, they are geographically well distributed over the entire Alps. Mass balance is the direct and undelayed response signal to annual atmospheric conditions. It documents degrees of imbalance between glaciers and climate due to the delay in dynamic response caused by the characteristics of ice flow (deformation and sliding). Over long time intervals mass balance variations indicate trends of climatic forcing. With constant climatic conditions (no forcing), balances would tend toward zero. Long-term nonzero balances are therefore an expression of ongoing climate change (Haeberli et al. 2005). Summer and winter balance even provide intraannual climate information and should therefore be surveyed on all mass balance glaciers (Dyurgerov and Meier 1999; Vincent 2002). In general, fluctuation series are well distributed across the Alps and represent the range of area-classes quite well. In view of the large contribution of

TABLE 11.4

Mean Specific (Annual) Net Balance of the Alpine Reference Glaciers

TIME PERIOD	NUMBER OF REFERENCE GLACIERS	NET BALANCE (MM W.E.)
1950–59	1–5	–536
1960–69	6–9	–26
1970–79	9	–69
1980–89	9	–437
1990–99	9	–767
1949–2001	1–9	–412
1967–2001	9	–369

glaciers smaller than 1 km² to glacier shrinkage in the past and the prediction of ongoing global warming (e.g., Schär et al. 2004; Beniston 2005), future work should include studies on the influence of atmospheric warming on small glaciers and on current down-wasting processes (see also Paul et al. 2004). However, the climatic sensitivity of glaciers depends not only on glacier size but also on sensitivity to variations in regional climate versus local topographic effects, which potentially complicates the extraction of a regional or global climate signal from glacier fluctuations (Kuhn et al. 1985; Vincent et al. 2004). Mass balance and ice flow models calibrated with available fluctuation data are needed to quantify these effects (Oerlemans et al. 1998; Oerlemans 2001; Paul et al., this volume).

CONCLUSIONS

In the European Alps the growth of the glacier monitoring network over time has resulted in an unprecedented glacier data set with excellent spatial and temporal coverage. The WGMS has compiled information on spatial glacier distribution from approximately 5,150 Alpine glaciers and fluctuation series (front variation and mass balance) from more than 670 of these glaciers. National inventories provide complete Alpine coverage for the 1970s, when the glaciers covered an area of 2,909 km². This inventory, together with the SGI2000, is used to extrapolate Alpine glacier-covered areas in 1850 and 2000 of about 4,470 km² and 2,270 km², respectively. This corresponds to an overall glacier area loss from 1850 of 35% by the 1970s and almost 50% by 2000.

Annual mass balance and front variation series provide a better time resolution of glacier fluctuations over the past 150 years than the inventories. During the general retreat, intermittent periods of glacier advances in the 1890s, the 1920s, and the 1970s and 1980s can still be seen. Increasing mass loss, rapidly shrinking glaciers, and disintegrating and spectacular tongue retreats are clear warnings of the atmospheric warming observed in the Alps during the past 150 years and the acceleration observed over the past two decades.

While inventory data contain information on spatial glacier distribution at certain times, fluctuation series provide temporal information at specific locations. Continuity and representativeness of fluctuation series are thus essential for the planning of glacier monitoring. Furthermore, modeling should be enhanced and integrated into monitoring strategies. It is very important to continue with long-term fluctuation measurements and to extend the series back in time with reconstructions of former glacier geometries. Additionally, it is necessary to integrate glacier monitoring and reconstruction activities into the framework of the Global Land Ice Measurements from Space (GLIMS) project and the WGMS.

ACKNOWLEDGMENTS

We are indebted to the numerous people who have provided the WGMS with data over the years. In particular, we thank all the national correspondents of the WGMS and G. Rossi and W. Schöner for their collaboration in the revision of the Italian and the Austrian data sets. Special thanks go to S. Baumann, M. Dischl, A. Hinterberger, J. Meilwes, W. Peschke, C. Rothenbühler, and A. Stolz for their contribution to the correction and formatting of the data and to I. Woodhatch for editing this report. We gratefully acknowledge the constructive comments of G. Pederson, an anonymous referee, and the scientific editors. Last but not least, we thank R. Frauenfelder and other colleagues from the WGMS for their daily efforts and teamwork. This study was mainly funded by the ALP-IMP Project of the European Community's Environment/Global Change program (BBW. No. 01.0498-2) and the Department of Geography of the University of Zurich.

REFERENCES CITED

Beniston, M. 2005. Warm winter spells in the Swiss Alps: Strong heat waves in a cold season? A study focusing on climate observations at the Saentis high mountain site. *Geophysical Research Letters* 32: L01812, doi: 10.1029/2004GL021478.

Böhm, R. 1993. Kartometrische Daten der Vergletscherung der Goldberggruppe in den Hohen Tauern. *Zeitschrift für Gletscherkunde und Glazialgeologie* 29:133–52.

CGI/CNR (Consiglio Nazionale delle Ricerche/Comitato Glaciologico Italiano). 1962. *Catasto dei ghiacciai italiani*. 4 vols. Varese.

Damm, B. 1998. Der Ablauf des Gletscherrückzuges in der Rieserfernergruppe (Tirol) im Anschluss an den Hochstand um 1850. *Zeitschrift für Gletscherkunde und Glazialgeologie* 34:141–59.

Dyurgerov, M., and M. Meier. 1999. Analysis of winter and summer glacier mass balances. *Geografiska Annaler* 81A:541–54.

Forel, F.A. 1895. Les variations périodiques des glaciers: Discours préliminaire. *Extrait des Archives des Sciences Physiques et Naturelles* 34:209–29.

Gross, G. 1988. Der Flächenverlust der Gletscher Österreichs 1850–1920–1969. *Zeitschrift für Gletscherkunde und Glazialgeologie* 23:131–41.

Haeberli, W. 1998. Historical evolution and operational aspects of worldwide glacier monitoring. In *Into the second century of worldwide glacier monitoring: Prospects and strategies*, ed. W. Haeberli, M. Hoelzle, and S. Suter, 35–51. Paris: UNESCO.

———. 2004. Glaciers and ice caps: Historical background and strategies of world-wide monitoring. In *Mass balance of the cryosphere*, ed. J.L. Bamber and A.J. Payne, 559–78. Cambridge, UK: Cambridge University Press.

Haeberli, W., H. Bosch, K. Scherler, G. Østrem, and C. Wallén, eds. 1989*a*. *World glacier inventory: Status 1988*. Nairobi: IAHS(ICSI)/UNEP/UNESCO/ World Glacier Monitoring Service.

Haeberli, W., and M. Hoelzle. 1995. Application of inventory data for estimating characteristics of and regional climate-change effects on mountain glaciers: A pilot study with the European Alps. *Annals of Glaciology* 21:206–12.

Haeberli, W., and H. Holzhauser. 2003. Alpine glacier mass changes during the past two millennia. *PAGES News* 11(1):13–15.

Haeberli, W., P. Müller, P. Alean, and H. Bösch. 1989*b*. Glacier changes following the Little Ice Age: A survey of the international data basis and its perspectives. In *Glacier fluctuations and climatic change: Proceedings of the Symposium on Glacier Fluctuations and Climatic Change, held in Amsterdam, 1–5 June 1987*, ed. J. Oerlemans, 77–101. Dordrecht, Boston, and London: Kluwer Academic Publishers.

Haeberli, W., J. Noetzli, M. Zemp, S. Baumann, R. Frauenfelder, and M. Hoelzle, eds. 2005*a* *Glacier mass balance bulletin. 8, 2002–2003*. Zurich: World Glacier Monitoring Service/IUGG(CCS)/ UNEP/UNESCO/WMO.

Haeberli, W., M. Zemp, R. Frauenfelder, M. Hoelzle, and A. Kääb, eds. 2005*b*. *Fluctuations of glaciers 1995–2000*. Vol. 8. Zurich: World Glacier Monitoring Service/IUGG (CCS)/UNEP/UNESCO.

Haeberli, W., and H. Zumbühl. 2003. Schwankungen der Alpengletscher im Wandel von Klima und Perzeption. In *Welt der Alpen: Gebirge der Welt*, ed. F. Jeanneret, D. Wastl-Walter, U. Wiesmann, and M. Schwyn, 77–92. Bern: Haupt.

Hoelzle, M., W. Haeberli, M. Dischl, and W. Peschke. 2003. Secular glacier mass balances derived from cumulative glacier length changes. *Global and Planetary Change* 36:295–306.

Holzhauser, H. 1997. Fluctuations of the Grosser Aletsch Glacier and the Gorner Glacier during the last 3200 years: New results. In *Glacier fluctuations during the Holocene*, ed. B. Frenzel, 35–58. Paläoklimaforschung/Palaeoclimate Research 24.

Holzhauser, H., M. Magny, and H.J. Zumbühl. 2005. Glacier and lake-level variations in west-central Europe over the last 3500 years. *The Holocene* 15:789–801.

Holzhauser, H., and H.J. Zumbühl. 1996. To the history of the Unterer Grindelwald Glacier during the last 2800 years: Palaeosols, fossil wood and historical pictorial records, new results.

Zeitschrift für Geomorphologie, n.s., suppl. 104:95–127.
Houghton, J. T., et al. 2001. *Climate change 2001: The scientific basis. Contribution of Working Group 1 to the third assessment report of the Intergovernmental Panel on Climate Change.* Cambridge, UK: Cambridge University Press.
IAHS (International Association of Hydrological Sciences). 1980. *World glacier inventory: Proceedings of the Riederalp Workshop, September 1978.* IAHS publication 126. Paris: International Association of Hydrological Sciences.
Jóhanneson, T., C. Raymond, and E. Waddington. 1989. Time-scale for adjustment of glaciers to changes in mass balance. *Journal of Glaciology* 35:355–69.
Kääb, A., F. Paul, M. Maisch, M. Hoelzle, and W. Haeberli. 2002. The new remote-sensing-derived Swiss glacier inventory. 2. First results. *Annals of Glaciology* 34:362–66.
Kasser, P. 1970. Gründung eines "Permanent Service on the Fluctuations of Glaciers." *Zeitschrift für Gletscherkunde und Glazialgeologie* 6:193–200.
Kuhn, M., G. Markl, G. Kaser, U. Nickus, and F. Obleitner. 1985. Fluctuations of climate and mass balance: Different responses of two adjacent glaciers. *Zeitschrift für Gletscherkunde und Glazialgeologie* 2:409–16.
Laternser, M., and M. Schneebeli. 2003. Long-term snow climate trends of the Swiss Alps (1931–1999). *International Journal of Climatology* 23:733–50.
Maisch, M. 1992. Die Gletscher Graubündens: *Rekonstruktion und Auswertung der Gletscher und deren Veränderungen seit dem Hochstand von 1850 im Gebiet der östlichen Schweizer Alpen (Bündnerland und angrenzende Regionen).* Physische Geographie 33. Zurich: Geographisches Institut der Universität Zürich.
Maisch, M., A. Wipf, B. Denneler, J. Battaglia, and C. Benz. 2000. *Die Gletscher der Schweizer Alpen: Gletscherhochstand 1850, Aktuelle Vergletscherung, Gletscherschwund Szenarien.* 2d ed. Zurich: VdF Hochschulverlag.
Müller, F., T. Caflisch, and G. Müller. 1976. *Firn und Eis der Schweizer Alpen: Gletscherinventar.* Geographisches Institut der ETH Zürich publ. 57. Zürich: Versuchsanstalt für Wasserbrau, Hydrologie und Glaziologie der ETH Zürich.
Müller, P. 1988. *Parametrisierung der Gletscher-Klima-Beziehung für die Praxis: Grundlagen und Beispiele.* Mitteilungen der Versuchsanstalt für Wasserbau, Hydrologie und Glaziologie 95.
Nicolussi, K., and G. Patzelt. 2000. Untersuchungen zur holozänen Gletscherentwicklung von Pasterze und Gepatschferner (Ostalpen). *Zeitschrift für Gletscherkunde und Glazialgeologie* 36:1–87.
Oerlemans, J. 1994. Quantifying global warming from the retreat of glaciers. *Science* 264:243–45.
———. 2001. *Glaciers and climate change.* Lisse, Abingdon, Exton, and Tokyo: A. A. Balkema.
Oerlemans, J., B. Anderson, A. Hubbard, P. Huybrechts, T. Johannesson, W. H. Knap, M. Schmeits, A. P. Stroeven, R. S. W. Van de Wal, J. Wallinga, and Z. Zuo. 1998. Modelling the response of glaciers to climate warming. *Climate Dynamics* 14:267–74.
Patzelt, G. 1985. The period of glacier advances in the Alps, 1965 to 1980. *Zeitschrift für Gletscherkunde und Glazialgeologie* 21:403–7.
———. 1987. Gegenwärtige Veränderungen an Gebirgsgletschern im weltweiten Vergleich. *Verhandlungen des Deutschen Geographentages* 45:259–64.
Paul, F. 2004. The new Swiss glacier inventory 2000: Application of remote sensing and GIS. Ph.D. diss., University of Zurich.
Paul, F., A. Kääb, M. Maisch, T. Kellenberger, and W. Haeberli. 2002. The new remote-sensing-derived Swiss glacier inventory. 1. Methods. *Annals of Glaciology* 34:355–61.
———. 2004. Rapid disintegration of Alpine glaciers observed with satellite data. *Geophysical Research Letters* 31: L21402, doi: 10.1029/2004GL020816.
Pelfini, M., and C. Smiraglia. 1988. L'evoluzione recente del glacialismo sulle Alpi Italiani: Strumenti e temi di ricerca. *Bollettino della Società Geografica Italiana* No. 1–3:127–54.
Reynaud, L. 1988. Alpine glacier fluctuations and climatic changes over the last century. *Mitteilungen der Versuchsanstalt für Wasserbau, Hydrologie und Glaziologie* 94:127–46.
Schär, C., P. L. Vidale, D. Lüthi, C. Frei, C. Häberli, M. Liniger, and C. Appenzeller. 2004. The role of increasing temperature variability in European summer heat waves. *Nature* 427:332–36.
Schöner, W., I. Auer, and R. Böhm. 2000. Climate variability and glacier reaction in the Austrian eastern Alps. *Annals of Glaciology* 31:31–38.
UNEP/GEMS. 1992. *Glaciers and the environment.* UNEP/GEMS Environmental Library 9. Nairobi: UNEP.
Vincent, C. 2002. Influence of climate change over the 20th century on four French glacier mass balances. *Journal of Geophysical Research* 107(D19):4375, doi: 10.1029/2001JD000832.
Vincent, C., G. Kappenberger, F. Valla, A. Bauder, M. Funk, and E. Le Meur. 2004. Ice ablation as evidence of climate change in the Alps over the

20th century. *Journal of Geophysical Research* 109: D10104, doi: 10.1029/2003JD003857.

Vivian, R. 1975. *Les glacier des Alpes occidentales.* Grenoble: Allier.

Wanner, H., C. Casty, J. Luterbacher, and A. Pauling. 2005. 500 Jahre Klimavariabilität im europäischen Alpenraum: Raumzeitliche Strukturen und dynamische Interpretation. *Rundgespräche der Kommission für Ökologie, Klimawandel im 20. und 21. Jahrhundert* 28:33–52.

Zemp, M., R. Frauenfelder, W. Haeberli, and M. Hoelzle. 2005. Worldwide glacier mass balance measurements: General trends and first results of the extraordinary year 2003 in Central Europe. *Materialy Glyatsiologicheskikh Issledovaniy* 99:3–12.

Zemp, M., M. Hoelzle, and W. Haeberli. 2007. Distributed modelling of the regional climatic equilibrium line altitude of glaciers in the European Alps. *Global and Planetary Change* 56:83–100.

12

Tropical Glaciers, Climate Change, and Society

FOCUS ON KILIMANJARO (EAST AFRICA)

Thomas Mölg, Douglas R. Hardy, Nicolas J. Cullen, and Georg Kaser

In recent years glaciers in the tropics have been recognized as particularly valuable indicators (proxy data) of climate change (e.g., Houghton et al. 2001). Not only do they provide insight into regional climate variations from remote high-altitude sites but they also show a greater variety in terms of climate sensitivity than mid- and high-latitude glaciers (Kaser 2001; Kaser et al. 2004*a*). While the mass balance of the latter glaciers is strongly tied to air temperature (Ohmura 2001), tropical glaciers are most sensitive both to atmospheric moisture content (e.g., cloudiness and therefore incoming solar radiation) and to precipitation and thus the reflection of solar radiation on the glacier surface (the so-called albedo) (e.g., Hastenrath 1984; Kaser and Noggler 1991; see Kaser et al. 2004*a* for a review). Studying the behavior of tropical glaciers and the associated climatic controls therefore involves studies on the important link between air temperature and moisture changes (cf. Pielke 2004).

The most characteristic feature of low-latitude climate is that seasonality is due entirely to the annual cycle of moisture (see Hastenrath 1991 and Asnani 1993). In general, this cycle results mainly from the north-south oscillation of the Intertropical Convergence Zone (ITCZ), one of the significant circulation patterns in our climate system. The ITCZ is a zone of strong convection and rainfall that passes through the equatorial regions twice a year and reaches the boundary of its oscillation zone (the outer tropics) once a year. Thus, the oscillation of the ITCZ induces a succession of rainy and dry seasons. The contrast between hygric seasons is small in regions with moisture influx from extensive rain forests, and therefore these regions are more or less humid year-round. The monthly variation in air temperature, however, is small throughout the tropics, even at high altitudes (Hardy et al. 1998). Figure 12.1, A, depicts the annual cycle of rainfall at Moshi, at the foot of Kilimanjaro, in equatorial Tanzania. The so-called long rains occur between March and May, and there is a secondary peak of rainfall in November and December ("the short rains") when the ITCZ moves more rapidly across the Kilimanjaro region. Long-term air temperature data at Moshi (Vose et al. 1992) underline the thermal homogeneity of tropical climate, since monthly means vary less than 5 °C. The vertical

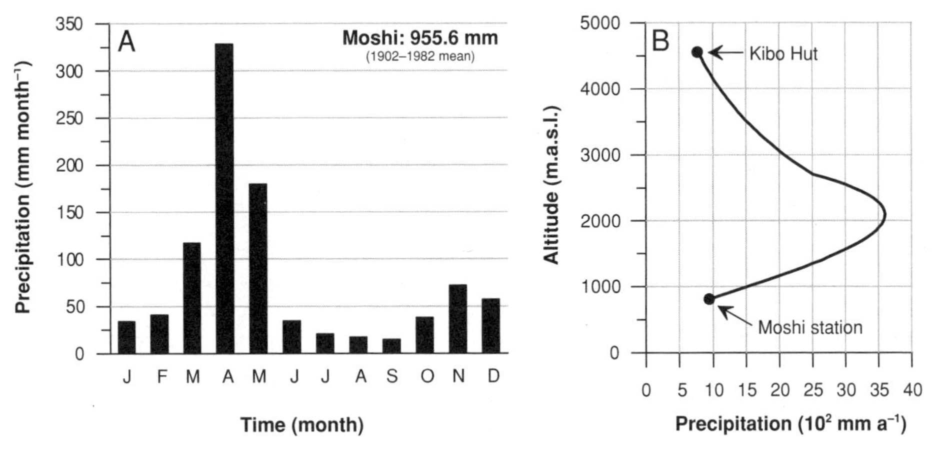

FIGURE 12.1. (A) Mean annual cycle of precipitation at Moshi (813 m a.s.l.), equatorial East Africa, 1902–82, and (B) vertical distribution of annual precipitation on the southern slope of Kilimanjaro. Data from the Global Historical Climatology Network climate data set (Vose et al. 1992); vertical distribution derived from the precipitation-altitude relation after Røhr and Killingtveit (2003), with Moshi as reference station.

distribution of rainfall on the southern slopes of Kilimanjaro (Figure 12.1, B) exhibits the pattern typical of tropical mountains, with a belt of maximum rainfall at mid-altitude locations (Hastenrath 1991). The annual cycle of tropical climate outlined here has a particular impact on energy exchange at the glacier-atmosphere interface and therefore induces the special sensitivity of tropical glaciers to moisture variability (Kaser et al. 2004*a*).

According to the definition introduced by Kaser (1999) (Figure 12.2), tropical glaciers occur in regions where three zones coincide: (1) the astronomic tropics, (2) the area of oscillation of the ITCZ, and (3) any zone in which the diurnal air temperature amplitude exceeds the annual amplitude. This means that almost all tropical glaciers (>99% with respect to glacier surface area [Kaser 1999]) are found in the South American Andes. There are, however, four other glaciated massifs in the tropics, and they are equally important as climate indicators. Three of them, Rwenzori, Mt. Kenya, and Kilimanjaro (summarized to one point in Figure 12.2), lie in equatorial East Africa (Hastenrath 1984). Moreover, a small glaciation remains in the mountains of Irian Jaya in Indonesia (Kincaid and Klein 2004). Apart from brief periods (e.g., Georges 2004), all tropical glaciers have suffered severe mass loss over the past century (Kaser 1999), capturing the attention of scientists and others around the world. Regional studies referring to South America (e.g., Favier, Wagnon, and Ribstein 2004) as well as to East Africa (e.g. Kruss 1983; Mölg, Georges, and Kaser 2003) demonstrate the sensitivity of the glaciers to changes in moisture-related climate parameters.

The remainder of this chapter will concentrate on the retreating glaciers of Kilimanjaro, which have experienced considerable media coverage in recent years. The next section summarizes what we know about the climatic controls of this ongoing retreat, and the following one illustrates the creative stance required for designing future research. The final section shows how the findings from this basic research can be linked to social activities and needs.

KILIMANJARO GLACIERS AND CLIMATE: WHAT WE KNOW

The current glacier recession on Kibo, the glaciated cone of the Kilimanjaro massif, started in about 1880, following at least some decades of stability (Hastenrath 2001). Description and monitoring of this recession began with the early explorers (Meyer 1900; Klute 1920) and resulted in the first detailed map, by Klute (1920), showing the glacier extent in 1912. Hastenrath and

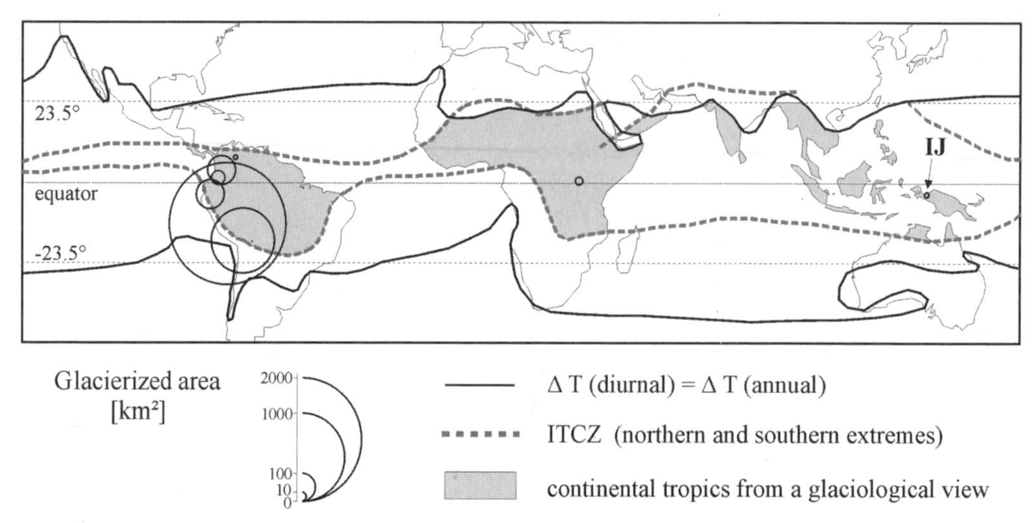

FIGURE 12.2. Global distribution of tropical glaciers at the end of the twentieth century (after Kaser and Osmaston 2002). Circles indicate the glacier surface area in respective countries (IJ indicates Irian Jaya).

Greischar (1997) and Thompson et al. (2002) have derived further glacier extents for different times in the twentieth century. Figure 12.3 summarizes the findings of these studies and illustrates the strong recession in the course of the twentieth century. In the year 2000, only 2.6 km^2 of glacier surface area were left (Thompson et al. 2002) of the ~20 km^2 estimated in 1880 (Osmaston 1989). The glaciers on the summit plateau have a peculiar shape, with vertical ice walls at their margins and a near-horizontal surface. The glaciers on the steep slopes of Kibo show a simpler, sometimes nearly rectangular shape. To understand the connection between glaciers and climate, it is essential to distinguish these three glacier regimes (Kaser et al. 2004*b*): the vertical ice walls on the plateau (regime 1, Figure 12.4, A), the horizontal glacier surfaces on the plateau (regime 2, Figure 12.4, A), and the slope glaciers below the plateau (regime 3, Figure 12.4, B). Basal melting due to geothermal heat (Kibo is a dormant volcano) has probably had a limited effect on modern glacier retreat (Kaser et al. 2004*b*), but this is not entirely clear.

Although the recession of the Kibo glaciers was recognized early, the climatological reasons for it have remained speculative until recently and, in the past few years, have repeatedly been attributed to anthropogenic global warming. Consequently, research on Kilimanjaro's summit climate and the recession of the glaciers was begun in 2000 (e.g., Hardy 2002; Thompson et al. 2002; Kaser et al. 2004*b*) and studies of the physics of climate-glacier interactions in more recent years (e.g., Mölg and Hardy 2004). The studies to date rely primarily on data collected by the University of Massachusetts automatic weather station on the horizontal surface of the Northern Icefield (Figure 12.3) since February 2000. Two new stations were installed in February 2005 by the University of Innsbruck in front of an ice cliff (AWS2) and at the transition into the southern slope glacierization (AWS3) to measure conditions of glacier regimes 2 and 3, respectively. Recorded data show that this high-altitude climate reveals the general features of tropical climate as described above, although precipitation exhibits greater intraannual and interannual variability than expected. Mean annual temperature at the summit is −7.0 °C, with monthly means varying only ~2 °C around the annual mean.

Regarding glacier-climate interactions, two points can be highlighted: (1) The Kibo glaciers seem to be retreating mainly because of the dry climate and abundant solar radiation, and (2) they are extremely sensitive to variation in precipitation.

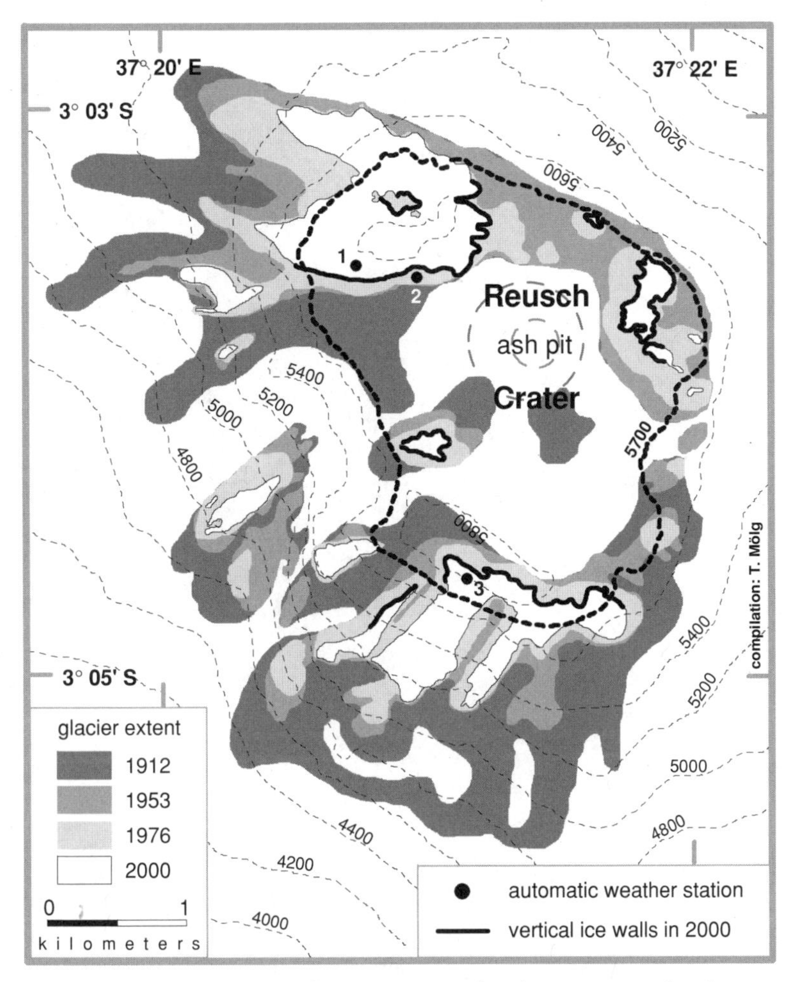

FIGURE 12.3. Glacier extents on Kibo in 1912, 1953, 1976, and 2000 (Hastenrath and Greischar 1997; Thompson et al. 2002); map contours in meters, UTM zone 37 projection. Also shown are the locations of the automatic weather stations (1–3) and the distribution of vertical ice walls (Mölg, Hardy, and Kaser 2003). The highlighted 5,700-m contour delimits Kibo's summit plateau approximately. Dashed circles indicate the conical Reusch Crater. Map compiled by T. Mölg, January 2003/December 2004.

The first point is supported by the model study of Mölg, Hardy, and Kaser (2003), who assumed an idealized ice cap covering the summit plateau in 1880, with a prescribed distribution of vertical ice walls according to early photographs (Meyer 1900). They then modeled mass loss at the ice walls (regime 1) between 1880 and 2000 as a function of absorbed solar radiation. Despite year-round negative air temperatures in the summit plateau area (Mölg and Hardy 2004), the high amounts of solar radiation provided enough energy for melting, which induced severe mass loss at the ice walls. After 120 years of simulation, the broken-up ice cap showed a spatial distribution of ice remnants strikingly similar to the real, observed pattern. This indicates that solar radiation maintains vertical ice wall retreat on the summit plateau in a climate with a lack of mass gain on the glaciers (i.e., a climate that has become too dry to maintain them). Preliminary model experiments on the climate sensitivity of slope glaciers (regime 3) also point to the effects of a drier climate (reduction in precipitation with accompanying reduction in cloudiness and increase in incoming solar radiation), which made slope

FIGURE 12.4. (A) The summit plateau Northern Icefield in July 2005, with a near-horizontal surface (glacier regime 1) and vertical ice cliffs at its margin (glacier regime 2). (B) The southern slope glaciers below the summit plateau in February 2005. (Photos by N. J. Cullen and T. Mölg, respectively.)

glaciers retreat from their 1880 extent. When longer time series from AWS3 become available, these experiments will be more robust.

The extreme sensitivity of the glaciers to variation in precipitation was pointed out by Hardy (2003) and established by Mölg and Hardy (2004) for the horizontal glacier surfaces (regime 2). They applied meteorological data recorded at the University of Massachusetts automatic weather station for 2000–2002 to an energy balance model that calculates all the energy fluxes involved in providing the energy for mass loss at the glacier surface and then verified the results by measured mass loss.

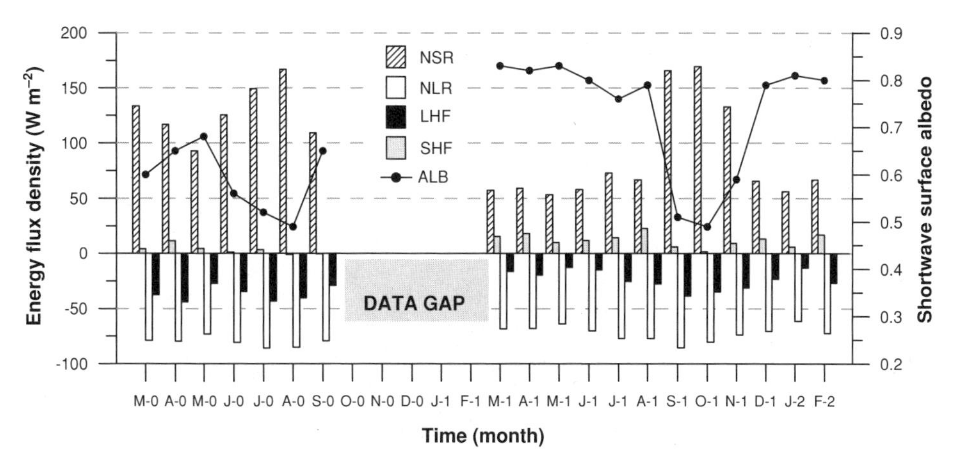

FIGURE 12.5. Energy-balance components at the weather station on the Northern Icefield between March 2000 and February 2002. *NSR*, Net shortwave (solar) radiation; *NLR*, net longwave radiation; *LHF*, turbulent latent heat flux; *SHF*, turbulent sensible heat flux; and *ALB*, shortwave albedo of the glacier surface. Positive values indicate energy gain for the surface, negative ones energy loss. For details of calculation see Mölg and Hardy (2004).

In contrast to the prevailing melting at the ice walls, mass loss on the horizontal surfaces is to a great extent due to sublimation, the direct conversion of water from solid to vapor (Mölg and Hardy 2004). Since sublimation consumes more energy per unit mass than melting, ablation rates on the horizontal surfaces are lower than on the ice walls (Kaser et al. 2004*b*). Figure 12.5 shows the main energy balance components at the weather station. Clearly, the energy balance is governed by variations in net shortwave radiation (i.e., absorbed solar radiation), which, in turn, are controlled by surface albedo. When albedo declines (e.g., between September and November 2001), net shortwave radiation sharply increases, which accelerates mass loss on the glacier surface. Albedo is a function of both precipitation amount and frequency. Local air temperature has hardly any impact at this site, given the small sensible heat flux. The albedo effect illustrated by Figure 12.5 for a 19-month period may certainly also exert a powerful long-term effect on glaciers.

What do other proxy data on East African climate tell us in this context? Several proxy sources support the notion that the climate of East Africa over the past 150 years was characterized by a sudden drop in moisture in the late nineteenth century and drier conditions during the twentieth. Among these proxies are the behavior of glaciers in the two other glaciated massifs of East Africa, Mt. Kenya (Kruss 1983) and Rwenzori (Mölg, Georges, and Kaser 2003), historical and modern accounts of lake levels (Hastenrath 1984; Nicholson and Yin 2001), reconstruction of lake levels from paleolimnological data (Verschuren, Laird, and Cumming 2000), and circulation and rainfall indices that link atmospheric and oceanic currents in the Indian Ocean since ~1850 to East African rainfall (Hastenrath 2001). To get a feeling for the magnitude of this moisture reduction, Nicholson and Yin (2001) applied a physically based water balance model to the lake-level record of Lake Victoria and estimated that mean over-lake precipitation in the 1860s and 1870s (likely coinciding with stable glaciers in East Africa) was about 20% higher than the twentieth-century mean reported by Yin and Nicholson (2002). For a much longer time scale, Thompson et al. (2002) concluded from an analysis of ice cores drilled on the Kibo glaciers in 2000 that glacier bevavior on Kilimanjaro during the Holocene (the past 10,000 years) coincided with well-known phases of African drought and humid conditions. This supports

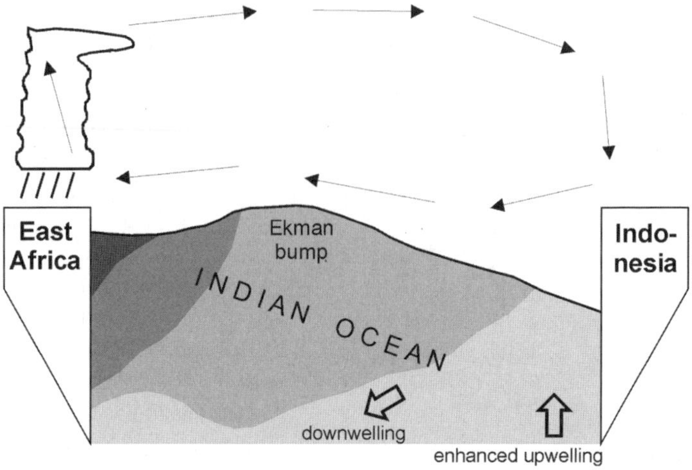

FIGURE 12.6. Vertical cross section at latitude 5°S illustrating ocean-atmosphere mechanisms of the anomalously strong East African short rains of 1997 (after Webster et al. 1999). The color scale from light to dark gray denotes increasing sea temperature; arrows indicate the direction of the atmospheric circulation.

the particular sensitivity of Kibo glaciers to precipitation of recent times.

STRATEGY FOR FUTURE RESEARCH

Studies of the reasons for precipitation variability and changes in (East) Africa have focused on the El Niño Southern Oscillation (ENSO). These studies use statistical methods (e.g., principal components analysis or spectral analysis) to detect relations between the ENSO and rainfall anomalies in Africa (e.g., Nicholson and Entekhabi 1986). Our own future research, however, must particularly consider an aspect arising from a new understanding—that the variation of precipitation in East Africa is tied to the state of the climate of the Indian Ocean and related sea-surface temperature patterns. This has become obvious in both observational studies (e.g., Black, Slingo, and Sperber 2003) and modeling experiments (e.g., Latif et al. 1999). Generally, interannual sea-surface temperature variation in the Indian Ocean is controlled by two modes (An 2004)—the basin-scale mode, which induces a basinwide warming or cooling of the Indian Ocean surface, and the Indian Ocean zonal mode (or dipole mode [Saji et al. 1999]), which alters the east-west sea-surface temperature gradient. There is overall agreement that strong zonal-mode events lead to extreme precipitation over East Africa through unusually strong short rains (Webster et al. 1999; Saji et al. 1999; Black, Slingo, and Sperber 2003) such as those of 1997. The short rains of that year amounted to up to 260% of their mean in the Lake Victoria area and up to 400% in coastal Somalia (Birkett, Murtugudde, and Allan 1999). Figure 12.6 schematically illustrates the causal mechanisms of these extremes (Webster et al. 1999).

The cycle starts with enhanced (reduced) upwelling in the eastern (western) Indian Ocean basin, which completely reverses the climatological sea-surface temperature gradient of a slightly warmer Indian Ocean in the east to an Indian Ocean now warmer in the west. At the same time, anomalous surface easterly winds start to replace the common westerly winds. The persistence of changed upwelling conditions subsequently causes the development of a westward-moving "ridge" in the Indian Ocean, the so-called Ekman bump. The increased moisture influx to East Africa through the easterly winds promotes convection over the landmasses that finally leads to enhanced generation of clouds and rainfall over East Africa. Apparently, this system is maintained by a number of positive feedback loops, in the case of 1997 for three to four months (Black, Slingo, and Sperber 2003).

There is strong evidence that basin-scale-mode events are controlled mainly by the occurrence of ENSO events and follow them with a lag of about a season (An 2004). This supports the statistical view on the ENSO-rainfall correlation stated previously, although the basinwide warming during the ENSO is associated with much weaker precipitation excess than the sea-surface temperature gradient reversal during zonal-mode events (Birkett, Murtugudde, and Allan 1999). The cause of zonal-mode events remains controversial. Some studies link them to the ENSO (Black, Slingo, and Sperber 2003), while others consider them purely internal phenomena (Saji et al. 1999; Webster et al. 1999) initiated and maintained by the positive feedbacks in the Indian Ocean–atmospheric circulation system. An (2004) has recently added a valuable study to this discussion, showing physical evidence for a dynamic link between basin-scale-mode and zonal-mode events.

Whatever the precise cause of these modes is, it appears that glacier recession on Kilimanjaro reflects not just local or regional changes but changes in larger-scale tropical climate. Sea-surface temperature changes in the Indian Ocean, especially changed frequencies of sea-surface temperature anomalies, must therefore be considered as a possible reason for the drier climate in East Africa during the twentieth century. Nicholson (2000) notes that wetter conditions in the recent historical past in Africa may represent mainly the more frequent occurrence of conditions that in the twentieth century typically characterized brief periods (e.g., the conditions of the short rains of 1997).

In the light of the findings presented here, a strategy for future research must operate at several climatological scales. At the microscale, meteorological and glaciological measurements on the Kibo glaciers are expanding. These additional measurements will help to refine the existing climate and mass balance studies (Mölg, Hardy, and Kaser 2003; Mölg and Hardy 2004) and quantify the impact of the local climate on glaciers more precisely. These microscale studies will be combined with model experiments on the connections between local/regional climate and large-scale climate. The employment of a mesoscale model—a numerical circulation model of the atmosphere (or regional climate model in this context)—will enable us to explore a number of aspects of the situation. For example, do sea-surface temperature patterns in the Indian Ocean also govern snowfall at the high altitude of Kibo? This is quite likely, as the greater moisture potential during strong zonal-mode events is visible in vertical atmospheric profiles up to the middle troposphere (5,000–6,000 m) (Black, Slingo, and Sperber 2003). If sea-surface temperature patterns do determine Kibo snow, could a higher frequency of unusually strong short rains prior to 1880 have provided enough snow on Kibo to maintain the glaciers? The results of the microscale studies will provide a clue to the amount of additional local snowfall required to maintain a glacier. Finally, coming back to vertical profiles, how has the stability of the atmosphere, which is critical for convection and therefore the formation of precipitation, changed?

A simple example with the Regional Atmospheric Modeling System (Cotton et al. 2003) will illustrate the approach envisioned. Figure 12.7 shows air flow past an idealized, bell-shaped Kilimanjaro massif that is 5,000 m higher than its surrounding plain and about 50 km in diameter at its base. The model was initialized in a horizontally homogeneous atmosphere with constant stratification (i.e., constant Brunt-Väisälla frequency) and then run for two scenarios (dry, wet) with a constant horizontal velocity at the inflow boundary, surface friction through a constant surface roughness ("no-slip"), nonrotational and three-dimensional. Despite the idealized assumptions, some interesting qualitative details emerge. The absence of pronounced vertical disturbances in the isentropes with only weak gravity wave activity indicates that the flow primarily goes around the mountain and confirms that Kilimanjaro is a good example of the "flow-around regime" discussed by Schär (2002). This implies that abundant precipitation on the summit can form only with convection present, and this creates a direct link to the large-scale scheme (Figure 12.6) in which convection is also one of

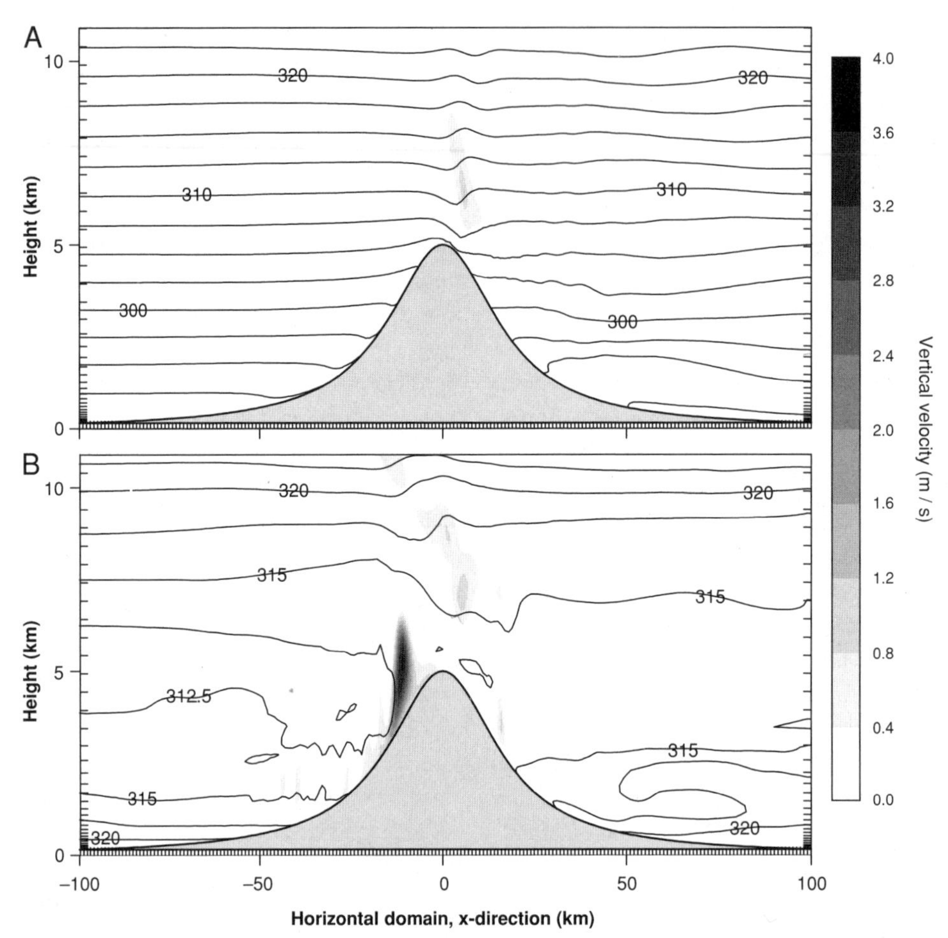

FIGURE 12.7. Vertical cross section of air flow past an idealized Kilimanjaro massif after six hours of integration in a 3-D model atmosphere: (A) dry with 0% relative humidity (RH), and (B) wet (RH = 100% until 5,000 m a.s.l., then linear decrease to 0% at 10,000 m a.s.l.). Shown are isolines of equivalent potential temperature (isentropes) with a 2.5-K spacing and gray-shaded contour surfaces of the upward-directed component of vertical velocity (spacing 0.4 m/s). Air flow follows the isentropes approximately and is from left to right with a constant background flow of 10 m/s. Note the vertical exaggeration of the plots.

the key processes. In fact, the sensitivity experiment (Figure 12.7, B) shows that strong moisture influx in the lower half of the troposphere decreases atmospheric stability and promotes convection near the summit on the windward side that is visible in accelerated upward winds. Plots of horizontal and vertical fields of cloud water content and wind vectors (not shown) confirm that the vertical winds indeed indicate convective cells (and not numerical noise), as these fields show great similarity to characteristic convective patterns (e.g., isolated regions of high cloud water content occur together with pronounced updrafts [Fuhrer and Schär 2005]). Hence, changes in atmospheric stability and related convection, influenced by the local topography (Figure 12.7) as well as the mesoscale transport of moisture to East Africa (Figure 12.6), seem decisive for snowfall on Kibo—which again underlines the need to approach the problem at various scales.

TROPICAL GLACIERS AND SOCIETY

The climate research on retreating tropical glaciers described previously has direct implications for practical measures in response to glacier retreat.

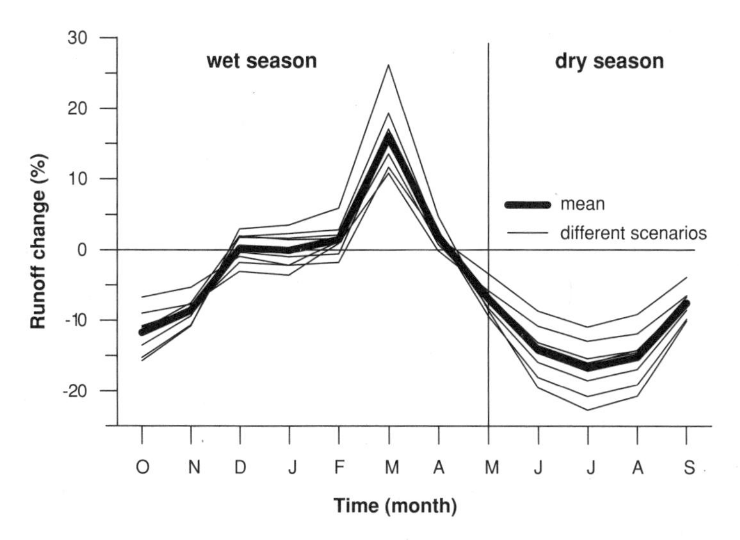

FIGURE 12.8. Changes in future runoff in the Llanganuco catchment area, Peru, with respect to runoff for 1961–90 (after Juen, Georges, and Kaser 2005). Thin lines represent calculations based on different climate scenarios for either 2050 or 2080 (see Juen, Georges, and Kaser 2005 for details), while the bold line shows the cycle of change averaged over all scenarios. The time axis follows the hydrological year (i.e., starts with October).

Most important in this context is that tropical glaciers are essential water resources in the glaciated regions of the Andes, acting as a buffer in the water cycle (Ribstein et al. 1995; Kaser et al. 2003; Juen, Georges, and Kaser 2005). Especially in the dry season, the water available for residents comes almost exclusively from the glaciers (Kaser et al. 2003) and supplies big cities (e.g., Lima) in which water is already scarce. Because of ongoing population growth and the Westernization of lifestyles, water demand is expected to increase. Addressing this issue, Juen, Georges, and Kaser (2005) have modified an existing tropical glacier mass balance model (Kaser 2001) to calculate runoff from glaciated catchment areas and applied their modified model to different scenarios of future climate as suggested by a climate prediction of Hulme and Sheard (1999). Figure 12.8 shows their results for the 33.5%-glaciated catchment area of Llanganuco, Peru. Regardless of which scenario is taken as a reference, dry-season runoff for the years 2050 and 2080 shows a significant decrease. This relationship between water demand and predicted availability in dry seasons points to a serious problem in the future.

Whether the Kibo glaciers contribute significantly to water resources in the Kilimanjaro area is currently under discussion and is vitally important. However, a significant role for them as a water source must be rejected (cf. Kaser et al. 2004*b*). First, as we have seen, a significant part of glacier mass loss is due to sublimation. Our model experiments for two automatic weather stations show that the horizontal glacier surfaces and slope glaciers in the vicinity of the summit plateau lose ~70% of their mass through sublimation. Second, where ice does melt (primarily at the vertical ice walls), it quickly evaporates in the dry atmosphere. During all our field trips, we have observed small meltwater "ponds" at the base of the cliffs but not perennial runoff from them. Third, the Kibo glaciers are very small. If one were to melt all the glacier volume found on the mountain in 2000 (2.6 km^2 surface area × ~25 m mean thickness [Thompson et al. 2002]) and then distribute the melted mass over the entire Kilimanjaro massif (say, 50 km × 80 km), it would amount to only a ~16-mm water column—approximately the amount that can easily be provided by one heavy rain shower. Although certain local communities may have gotten their water from slope glacier runoff in the recent past (cf. Sky Channel News, Johannesburg, August 28, 2002), the majority of the region's population (~1 million) is found in the lowlands of Kilimanjaro (Hemp 2005), whose springs are fed by water from the rain forest belt (see Figure 12.1, B). The maximum zone of annual precipitation is found in the altitude bands around ~2,200 m a.s.l., which coincide exactly with the rain forest belt (Hemp n.d.). Lambrechts et al. (2002) report that 96% of the water flowing from Kilimanjaro originates in the rain forest belt. The maintenance of Kilimanjaro's forests must therefore have highest priority in terms of water supply to the regional population (Agrawala et al. 2003)—an enormous challenge in light of the current land use changes.

In contrast to the limited significance of the Kibo glaciers to hydrology, they show an exceptionally high touristic potential. Thousands of tourists each year are attracted by Kibo and its equatorial ice, and this underscores the need to assess future ice loss. Using a physically based model, Mölg, Hardy, and Kaser (2003) estimate that the plateau glaciers will disappear around mid-twenty-first century if vertical ice wall retreat continues in the present fashion. Thompson et al. (2002) predict the demise of all the glaciers (i.e., both plateau and slope) by 2015–2020. However, from our preliminary model experiments it is not clear yet that the slope glaciers will disappear. Some scenarios indicate that they are likely to stay longer than the plateau glaciers, though probably only in certain areas and to a limited extent. A more precise assessment of this issue will be one of the goals of our project. Agrawala et al. (2003) think that the further retreat or vanishing of glaciers will not reduce Kibo's touristic appeal decisively, as seasonal snow will maintain Kibo's white cap. This characteristic landscape element will be present, however, only during limited periods of the year (i.e., in wet seasons). During dry seasons, when the greatest number of visitors come to Tanzania and Kenya, the glaciers are typically the only natural phenomenon keeping the peak of Kilimanjaro white. Social science research could examine the influence of changing glaciers on tourist appeal in more detail. In any case, the important differentiation between glacier ice and seasonal snow cover must be considered not only in the public discussion of glacier recession but also in consideration of the human perception of the mountain.

A final issue of societal concern is land use change and related shifts in vegetation zones on Kilimanjaro's slopes, which have impacts on water availability and probably on local climate as well. The change is characterized by a decline in forests in the course of the twentieth century. Hemp (n.d.) shows that montane forests and subalpine *Erica* forests lost 9% and 83% of their areas, respectively, within only a few decades (1976–2000). While the former have been destroyed primarily by (illegal) deforestation (but also by agriculture and settlement [Mchallo 1994]), the sharp reduction of subalpine cloud forests is due to the increase in fires as a result of increasing aridity and the subsequent replacement of *Erica* forests with *Erica* bush. Whereas the bush vegetation plays only a marginal role in the local water balance, the forests are essential for filtering and storing water and for intercepting fog (i.e., collecting of cloud water). The strong areal retreat of forests thus significantly affects Kilimanjaro's water supply. The loss in the subalpine belt between 1976 and 2000 alone corresponds to the annual water demand of Kilimanjaro's population if reduced fog interception is counted (Hemp n.d.). Changes in vegetation also have a significant impact on regional and local climate (e.g., Pitman et al. 2004). Generally, the decrease in vegetation cover reduces moist static energy in the atmospheric boundary layer and therefore leads to less convection and rainfall (see examples in Pielke et al. 1992). This makes it likely that a local anthropogenic forcing through deforestation is being superimposed on the large-scale forcing by the Indian Ocean circulation cell. Strict control of the forests is therefore desirable because of their importance for water supply to the lowlands and their possible effect on moisture conditions in the highest reaches of Kibo. The latter will be tested in the mesoscale part of the project through an explicit soil and vegetation submodel coupled with the Regional Atmospheric Modeling System (cf. Cotton et al. 2003).

The societal significance of Kilimanjaro's glaciers nowadays is clearly linked to recent tourism. Many climbers come to see ice and snow under the equatorial sun, while others are drawn by the fact that Kilimanjaro is the highest peak in Africa. When the German geographer Hans Meyer first reached Kibo's summit plateau in 1889 (Meyer 1900) he could not anticipate that 100 years later more than 10,000 people, mainly from Europe and the United States, would ascend the mountain each year (Mchallo 1994). Kibo's attraction was unimaginable even for Europeans in 1848: The missionary Johannes Rebmann's report of the discovery of the mountain's snow cap was dismissed for more than a decade (Meyer 1900). However, after the first efforts to ascend it in the 1860s and 1870s, the mountain's cultural significance for European colonists rose sharply, particularly because of its enormous height. Today's highest point (Uhuru Peak, 5,895 m a.s.l.), for instance, was named "Kaiser-Wilhelm-Spitze" under the German colonial movement (1885–1918) and declared Germany's highest mountain. At the same time, the mountain with its snow and ice cap had—for many centuries—a place in the myths of African tribes living at the base of Kilimanjaro (Gratzl 2000): they viewed it as the shining "house of god," something that one should not approach too closely but adore from a distance. This status has obviously changed, and the long-lasting myth is overshadowed by Kilimanjaro's touristic potential. Many local people today accompany visitors on the mountain (as porters and guides) and rely on that potential. To avoid abuse and overuse of this great natural icon by the stream of tourism (Mchallo 1994), further studies in the cultural, physical, and social sciences are needed, with snow and ice on Kilimanjaro apparently creating an interface between the disciplines.

CONCLUDING REMARKS

The strong retreat of tropical glaciers over the past century provides eminently useful proxy data for the current discussion of climate change. Both field-supported energy balance experiments (e.g., Wagnon et al. 2001; Mölg and Hardy 2004) and pure modeling studies (e.g., Mölg, Georges, and Kaser 2003) indicate the particular sensitivity of tropical glaciers to moisture-related climate parameters and the importance of moisture changes for their behavior. This is true for Kilimanjaro's glaciers, although the peculiar environmental setting of this free-standing mountain requires a special approach to the problem (i.e., the identification of different glacier regimes). The results to date indicate that the glaciers on Kilimanjaro are retreating mainly because of the reduction of moisture in the late nineteenth century. The related local changes in precipitation amount and frequency seem to reflect larger-scale changes in tropical climate, particularly in the Indian Ocean, that merit further investigation. Glacier recession on Kilimanjaro thus demonstrates that global climate change is more complex than local warming, and this is a further argument for studying tropical glaciers as climate proxies.

The relevance of tropical glaciers to society is primarily reflected in their provision of water supplies in parts of South America, mainly during the dry season. As hydroclimatological studies predict a decline in meltwater runoff during the dry season in the coming decades (Juen, Georges, and Kaser 2005), serious future conflicts have to be expected, and this calls for sensitive policy making. Besides the aspect of water supply, tropical glaciers have high touristic potential. Ongoing glacier changes therefore raise interesting questions for combined social and psychological studies. Would a Kilimanjaro that has lost the ice cap on its summit plateau, for instance, continue to be anchored in the minds of people around the world? Climate studies on tropical glaciers represent the first step toward understanding the current changes in these significant symbols of nature and for reconstructing their past and predicting their future.

ACKNOWLEDGMENTS

This research has been funded by the Austrian Science Foundation under Grant No. P17415-N10, the Tyrolean Science Foundation (University of Innsbruck), and the National Science

Foundation under Grant No. ATM-0402557 (University of Massachusetts). Simulations with the mesoscale model were conducted in the RAMS Modeling Group under the supervision of Alexander Gohm and Georg Mayr. Irmgard Juen provided the runoff data for Figure 12.8. Special thanks are due to our local scientific adviser at the Tanzania Meteorological Agency, Tharsis Hyera, the Tanzania Commission for Science and Technology, and Eric Masawe and his team for their assistance on the mountain.

REFERENCES CITED

Agrawala, S., A. Moehner, A. Hemp, M. Van Aalst, S. Hitz, J. Smith, A. Meena, S. M. Mwakifwamba, T. Hyera, and O. U. Mwaipopo. 2003. *Development and climate change in Tanzania: Focus on Mount Kilimanjaro.* Paris: OECD.

An, S. I. 2004. A dynamic link between the basin-scale and zonal modes in the tropical Indian Ocean. *Theoretical and Applied Climatology* 78:203–15.

Asnani, G. C. 1993. *Tropical meteorology.* Pune: Indian Institute of Tropical Meteorology.

Birkett, S., R. Murtugudde, and T. Allan. 1999. Indian Ocean climate event brings floods to East African lakes and the Sudd Marsh. *Geophysical Research Letters* 26:1031–34.

Black, E., J. Slingo, and K. R. Sperber. 2003. An observational study of the relationship between excessively strong short rains in coastal East Africa and Indian Ocean SST. *Monthly Weather Review* 131:74–94.

Cotton, W. R., R. A. Pielke, Sr., R. L. Walko, G. E. Liston, C. J. Tremback, H. Jiang, R. L. McAnelly, J. Y. Harrington, M. E. Nicholls, G. G. Carrio, and J. P. McFadden. 2003. RAMS 2001: Current status and future directions. *Meteorology and Atmospheric Physics* 82:5–30.

Favier, V., P. Wagnon, and P. Ribstein. 2004. Glaciers of the outer and inner tropics: A different behaviour but a common response to climatic forcing. *Geophysical Research Letters* 31:L16403, doi: 10.1029/2004GL020654.

Fuhrer, O., and C. Schär. 2005. Embedded cellular convection in moist flow past topography. *Journal of the Atmospheric Sciences.* 62:2810–28.

Georges, C. 2004. 20th-century glacier fluctuations in the tropical Cordillera Blanca, Perú. *Arctic, Antarctic, and Alpine Research* 36:100–107.

Gratzl, K. 2000. *Mythos Berg.* Vienna and Purkersdorf: Hollinek.

Hardy, D. R. 2002. Eternal ice and snow? In *Kilimanjaro: To the roof of Africa,* ed. A. Salkeld, 224–25. Tampa: National Geographic.

———. 2003. Kilimanjaro snow. In *State of the climate in 2002,* ed. A. M. Waple and J. H. Lawrimore, S48. Bulletin of the American Meteorological Society 84.

Hardy, D. R., M. Vuille, C. Braun, F. Keimig, and R. S. Bradley. 1998. Annual and daily meteorological cycles at high altitude on a tropical mountain. *Bulletin of the American Meteorological Society* 79:1899–1913.

Hastenrath, S. 1984. *The glaciers of equatorial East Africa.* Dordrecht, Boston, and Lancaster: Reidel.

———. 1991. *Climate dynamics of the Tropics.* Dordrecht, Boston, and London: Kluwer.

———. 2001. Variations of East African climate during the past two centuries. *Climatic Change* 50:209–17.

Hastenrath, S., and L. Greischar. 1997. Glacier recession on Kilimanjaro, East Africa, 1912–89. *Journal of Glaciology* 43:455–59.

Hemp, A. 2005. Climate change driven forest fires marginalize the impact of ice cap wasting on Kilimanjaro. *Global Change Biology.* 11:1013–23.

Houghton, J. T., et al. 2001. *Climate change 2001: The scientific basis. Contribution of Working Group 1 to the third assessment report of the Intergovernmental Panel on Climate Change.* Cambridge, UK: Cambridge University Press.

Hulme, M., and N. Sheard. 1999. Escenarios de cambio climático para países de los Andes del Norte. http://www.cru.uea.ac.uk/~mikeh/research/andes.pdf (accessed November 3, 2004).

Juen, I., C. Georges, and G. Kaser. 2005. Modelling observed and future runoff from a glacierized tropical catchment (Cordillera Blanca, Peru). *Global and Planetary Change.* In press.

Kaser, G. 1999. A review of the modern fluctuations of tropical glaciers. *Global and Planetary Change* 22:93–103.

———. 2001. Glacier-climate interaction at low latitudes. *Journal of Glaciology* 47:195–204.

Kaser, G., C. Georges, I. Juen, T. Mölg, P. Wagnon, and B. Francou. 2004*a*. The behavior of modern low-latitude glaciers. *Past Global Changes News* 12:15–17.

Kaser, G., D. R. Hardy, T. Mölg, R. S. Bradley, and T. M. Hyera. 2004*b*. Modern glacier retreat on Kilimanjaro as evidence of climate change: Observations and facts. *International Journal of Climatology* 24:329–39.

Kaser, G., I. Juen, C. Georges, J. Gomez, and W. Tamayo. 2003. The impact of glaciers on the runoff and the reconstruction of mass balance history from hydrological data in the tropical Cordillera Blanca, Peru. *Journal of Hydrology* 282:130–44.

Kaser, G., and B. Noggler. 1991. Observations on Speke Glacier, Rwenzori Range, Uganda. *Journal of Glaciology* 37:313–18.

Kaser, G., and H. Osmaston. 2002. *Tropical glaciers*. Cambridge, UK: Cambridge University Press.

Kincaid, J. L., and A. Klein. 2004. Retreat of the Irian Jaya glaciers from 2000 to 2002 as measured from IKONOS satellite images. Paper presented at the 61st Eastern Snow Conference, Portland, ME, June 9–11.

Klute, F. 1920. *Ergebnisse der Forschungen am Kilimandscharo 1912*. Berlin: Reimer-Vohsen.

Kruss, P. D. 1983. Climate change in East Africa: A numerical simulation from the 100 years of terminus record at Lewis Glacier, Mount Kenya. *Zeitschrift für Gletscherkunde und Glazialgeologie* 19:43–60.

Lambrechts, C., B. Woodley, A. Hemp, C. Hemp, and P. Nnyiti. 2002. *Aerial survey of the threats to Mt. Kilimanjaro forests*. Dar es Salaam: UNDP, GEF Small Grants Programme.

Latif, M., D. Dommenget, M. Dima, and A. Grötzner. 1999. The role of Indian Ocean SST in forcing East African rainfall anomalies during December-January 1997/98. *Journal of Climate* 12:3497–3504.

Mchallo, I. A. J. 1994. The impact of structural adjustment programmes on the natural resources base: The case of tourism development. *UTAFITI News Series* 1(2):88–111

Meyer, H. 1900. *Der Kilimandscharo*. Berlin: Reimer-Vohsen.

Mölg, T., C. Georges, and G. Kaser. 2003. The contribution of increased incoming shortwave radiation to the retreat of the Rwenzori Glaciers, East Africa, during the 20th century. *International Journal of Climatology* 23:291–303.

Mölg, T., and D. R. Hardy. 2004. Ablation and associated energy balance of a horizontal glacier surface on Kilimanjaro. *Journal of Geophysical Research* 109:D16104, doi: 10.1029/2003JD003546.

Mölg, T., D. R. Hardy, and G. Kaser. 2003. Solar radiation-maintained glacier recession on Kilimanjaro drawn from combined ice-radiation geometry modeling. *Journal of Geophysical Research* 108(D23):4731, doi: 10.1029/2003JD003546.

Nicholson, S. E. 2000. The nature of rainfall variability over Africa on time scales of decades to millennia. *Global and Planetary Change* 26:137–58.

Nicholson, S. E., and D. Entekhabi. 1986. The quasi-periodic behavior of rainfall variability in Africa and its relationship to the Southern Oscillations. *Archiv für Meteorologie, Geophysik und Bioklimatologie* A34:311–48.

Nicholson, S. E., and X. Yin. 2001. Rainfall conditions in equatorial East Africa during the nineteenth century as inferred from the record of Lake Victoria. *Climatic Change* 48:387–98.

Ohmura, A. 2001. Physical basis for the temperature-based melt-index method. *Journal of Applied Meteorology* 40:753–60.

Osmaston, H. 1989. Glaciers, glaciations, and equilibrium line altitudes on Kilimanjaro. In *Quaternary and environmental research on East African mountains*, ed. W. C. Mahaney, 7–30. Rotterdam: Balkema.

Pielke, R. A., Sr. 2004. Assessing "global warming" with surface heat content. *Eos* 85:210–11.

Pielke, R. A., Sr., W. R. Cotton, R. L. Walko, C. J. Tremback, W. A. Lyons, L. D. Grasso, M. E. Nicholls, M. D. Moran, D. A. Wesley, T. J. Lee, and J. H. Copeland. 1992. A comprehensive meteorological modeling system: RAMS. *Meteorology and Atmospheric Physics* 49:69–91.

Pitman, A. J., G. T. Narisma, R. A. Pielke Sr., and N. J. Holbrook. 2004. Impact of land cover change on the climate of southwest Western Australia. *Journal of Geophysical Research* 109:D18109, doi: 10.1029/2003JD004347.

Ribstein, R., R. Tiriau, B. Francou, and R. Saravia. 1995. Tropical climate and glacier hydrology: A case study in Bolivia. *Journal of Hydrology* 165:221–34.

Røhr, P. C., and Å. Killingtveit. 2003. Rainfall distribution on the slopes of Mt. Kilimanjaro. *Hydrological Science/Journal des Sciences Hydrologiques* 48:65–78.

Saji, N. H., B. N. Goswami, P. N. Vinayachandran, and T. Yamagata. 1999. A dipole mode in the tropical Indian Ocean. *Nature* 401:360–63.

Schär, C. 2002. Mesoscale mountains and the larger-scale atmospheric dynamics: A review. In *Meteorology at the millennium*, ed. R. Pearce, 29–42. New York: Academic Press.

Thompson, L. G., E. Mosely-Thompson, M. E. Davis, K. A. Henderson, H. H. Beecher, V. S. Zagorodnov, T. A. Mashiotta, P. Lin, V. N. Mikhalenko, D. R. Hardy, and J. Beer. 2002. Kilimanjaro ice core records: Evidence of Holocene climate change in tropical Africa. *Science* 298:589–93.

Verschuren, D., K. R. Laird, and B. F. Cumming. 2000. Rainfall and drought in equatorial East Africa during the past 1,100 years. *Nature* 403:410–14.

Vose, R. S., R.L. Schmoyer, P. M. Steurer, T. C. Peterson, R. Heim, T. R. Karl, and J. Eischeid. 1992. *The Global Historical Climatology Network: Long-term monthly temperature, precipitation, sea level pressure, and station pressure data*. Carbon Dioxide Information Analysis Center NDP-041. Oak Ridge: Carbon Dioxide Information Analysis Center.

Wagnon, P., P. Ribstein, B. Francou, and J. E. Sicart. 2001. Anomalous heat and mass budget of Glaciar Zongo, Bolivia, during the 1997/98 El Niño year. *Journal of Glaciology* 47:21–28.

Webster, P. J., A. M. Moore, J. P. Loschnigg, and R. R. Leben. 1999. Coupled ocean-atmosphere dynamics in the Indian Ocean during 1997–98. *Nature* 401:356–60.

Yin, X., and S. E. Nicholson. 2002. Interpreting annual rainfall from the levels of Lake Victoria. *Journal of Hydrometeorology* 3:406–16.

PART FOUR

Impacts on Human Landscapes

Resources, Hazards, and Cultural Landscapes

13

New Zealand's Glaciers

KEY NATIONAL AND GLOBAL ASSETS FOR SCIENCE AND SOCIETY

John E. Hay and Tui L. Elliott

The New Zealand glacier inventory undertaken in 1978 identified 3,144 glaciers with a surface area greater than 0.01 km² (Chinn 2001). This figure represents a small reduction from 3,155 in the late 1970s. The glaciers extend from latitude 39°15′ to 49°57′S, with elevations ranging between 1,850 and 3,000 m. Two glaciers (the Franz Josef and the Fox) terminate in temperate rain forest at approximately 300 m above sea level (Burrows 2001). The largest glacier (the Tasman) is 29 km in length.

New Zealand's glaciers lie at the wet, warm maritime end of the glacial response spectrum, with dry, cold continental glaciers lying at the opposite end. The glaciers and snow cover reflect New Zealand's humid maritime climate and the transverse (i.e., approximately perpendicular) orientation of its Southern Alps to the prevailing Southern Hemisphere westerlies. Precipitation is evenly distributed through the year, reaching 15,000 mm per annum just west of the Main Divide, with a strong east-west gradient (Chinn and Whitehouse 1980). The mid-latitude, maritime location, coupled with high summer solar radiation inputs, also results in high melt rates and hence a high mass turnover or "activity index." Chinn (1994) estimates that response times to positive mass balance range between 5 and 50 years, while recessional response times are up to 300 years.

The earliest surveys of glacier terminus positions were made in the late 1800s, coinciding generally with the Little Ice Age glacier maximum, and glaciers have been experiencing broad-scale retreat since that time, with minor advances of some superimposed on this trend (Fitzharris, Lawson, and Owens 1999). The inventory estimated the total glacial area to be 1,158 km², or just 0.45% of the New Zealand landmass. In 1978 the ice volume was estimated to be 53.29 km³, having decreased 23–32% during the past century. Since that time there has been no significant change in ice volume overall. Some proglacial lakes have formed.

The mass balance of the Ivory Glacier was monitored between 1969 and 1975 and was consistently negative (Anderton and Chinn 1978). There is no ongoing monitoring of glacier mass balance in New Zealand, but end-of-summer snow lines have been recorded for 50 index glaciers in most years since 1977. The altitude of the annual

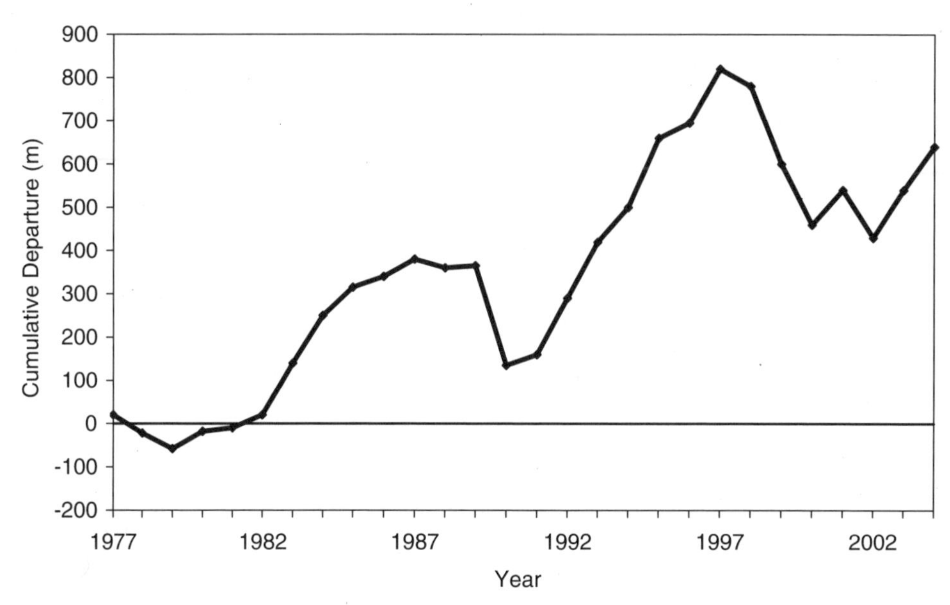

FIGURE 13.1. Cumulative departures in the annual equilibrium line altitude from the steady-state value, 1997 to 2004 (from Chinn, Heydenrych, and Salinger 2005).

snow line at the end of summer is an indicator of mass balance; the higher the snow line, the more negative the mass balance. These annual snow line altitudes are compared with the long-term average snow line that would keep the glacier in equilibrium, or steady state, with its present size. Annual snow-line departures from the altitude of this long-term equilibrium line are used as a surrogate for the change in the annual mass balance (Chinn 1994, 1995; Chinn and Salinger 2001).

Figure 13.1 shows departures of the annual equilibrium snow line altitudes from the steady-state value, plotted as cumulative values, for the period 1997 to 2004. Upward trends represent growth in glacier ice mass and downward trends a loss in ice mass. It is clear that the general regime of positive mass balance that dominated the earlier part of the record has been replaced by more variable mass balance conditions.

The glaciers and snow cover of the Southern Alps provide a resource for hydroelectric power generation, recreation, tourism, and agriculture and horticulture based on irrigation. They also contribute to hazards such as floods and avalanches, and their distributions have critical ecological implications for many protected natural areas such as national parks, the majority of which are located in the mountains (Fitzharris, Lawson, and Owens 1999). This chapter highlights the contributions that New Zealand glacial research is making to international understanding of the earth system and points to the many national and local consequences of the highly dynamic nature of New Zealand's glaciers. Such variations are important because the glaciers have significant social, cultural, and intrinsic value. Finally, it considers the ongoing importance of these glaciers in light of changes in global climate and global tourism, with particular reference to national policy and local resource management practices.

GLOBAL SCIENTIFIC SIGNIFICANCE

New Zealand's glaciers are well placed for the detection and diagnosis of atmospheric circulation changes (Fitzharris, Hay, and Jones 1992; Fitzharris, Lawson, and Owens 1999). They are surrounded by the oceans of the Pacific hemisphere, and the main glaciated ranges are oriented across the prevailing westerly circulation. In addition, the glaciers are located in a nodal area between major centers of atmospheric action located over northern Australia and Tahiti and in the transition area between subtropical and polar air masses and near major ocean fronts (e.g., the Subtropical Front). They are also highly sensitive to climate variations, with high accumulation and ablation rates resulting in rapid advances and retreats. Response time for the glaciers vary between years and decades.

Until recently, conventional wisdom was that the timing and pattern of glaciations were consistent between the Northern and Southern Hemispheres. Broecker (1997) used the apparent synchronicity between ice advances in New Zealand since the Last Glacial Maximum (ca. 21,000 years ago) and the global record as proof of synchronicity over the past 750,000 years. Denton et al. (1999) and others have argued that such synchronous responses to Northern Hemisphere forcing reflect the thermal forcing of glaciation via atmospheric and/or ocean transfers.

Carter and Gammon (2004) address the current debate over whether recent climate fluctuations in mid-latitude New Zealand relate primarily to Northern or Southern Hemisphere polar records. While recent studies suggest that Southern Hemisphere glacial events lead those from the north by approximately 1,500–3,000 years, analyses of data from locations near opposing poles are inadequate to allow the inference of time lags and causality. Carter and Gammon therefore highlight the importance of the 3.9-million-year-long record of glaciation in New Zealand preserved in a sediment core taken east of New Zealand's South Island. It reveals that New Zealand climate cyclicity resembles that of Antarctica at least as far back as about 0.37 million years ago. Carter and Gammon also found that, throughout the Late Pliocene and Pleistocene climate cycling, Southern Hemisphere atmospheric dynamics were tightly correlated and influential across a latitudinal range of at least 45–80°S. There was close agreement between this atmospheric climate signal and the global ocean climate signal contained in oxygen isotope time series at a resolution of approximately 1,000–2,000 years and throughout much of the past 3.9 million years. They use this close matching at short time scales to suggest that nonanthropogenic climate change is primarily atmosphere-driven through interannual oscillations such as the North Atlantic Oscillation and the El Niño Southern Oscillation. These may even have modulated the multimillennial Dansgaard-Oeschger events (Turney et al. 2004).

A similar but independent line of reasoning is advanced by Shulmeister et al. (2004). They refer to studies by Fitzharris and Hay (1989), Fitzharris, Hay, and Jones (1992), Hooker and Fitzharris (1999), and others showing that the strength and trajectory of westerly wind flow across New Zealand and the associated precipitation are critical to modern glacier mass balance in New Zealand. Hay and Fitzharris (1988), Fitzharris, Hay, and Jones (1992), Hooker and Fitzharris (1999), and others have shown the importance of synoptic patterns for glacier mass balance. Persistence of the subtropical high over New Zealand in summer is positively associated with ablation, while an anticyclone over Australia in winter creates a cool, moist southwesterly flow over New Zealand, maximizing snowfall and minimizing melt. Modern steep, fast-reaction-time glaciers respond, with a five-to-seven-year lag, to averaged seasonal synoptic conditions (e.g., Salinger, Heine, and Burrows 1983 and Figure 13.2). The large valley glaciers on both sides of the Southern Alps have longer response times and respond to decadal period forcing. Advances occur after positive southwest-flow-anomaly years, while retreats are associated with northerly anomalies. Changes in westerly flow can be correlated directly with circulation phenomena such as El Niño (Hay et al. 1993), with El Niño years being associated with positive glacier mass balances in New Zealand. The prevalence of positive glacier mass balance since the late 1970s is consistent with the tendency toward more frequent El Niño episodes since the 1970s without intervening La Niña events. The duration of the 1990–95 El Niño was unprecedented in the climate record of the past 124 years (Hay et al. 2003).

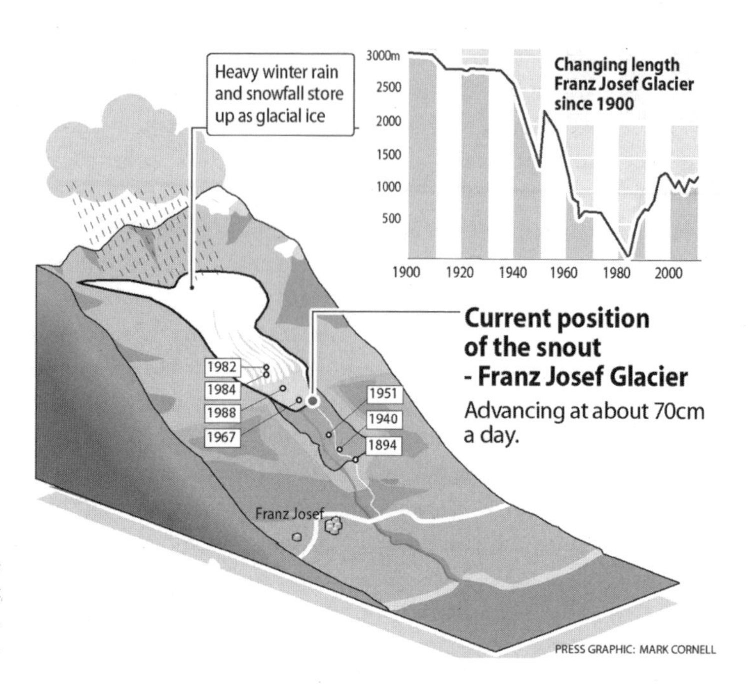

FIGURE 13.2. Variations in the Franz Josef Glacier over the past 100 years (from Madgwick 2005, reprinted by permission).

Shulmeister et al. (2004), relying principally on evidence in terrestrial and marine dust records, oceanic upwelling records, New Zealand vegetation histories, and glacial advances of changes in only the past 20,000 years, conclude that westerly flow has changed significantly through the last glacial cycle. Variations in the latitudinal boundaries in the westerlies may be nearly as large in interannual zonal shifts (about 2° maximum) as they are in glaciation-interglaciation movements (about 3–4°). The changes occur on a variety of time scales from the millennial to the subannual, but the main changes observed in the geological records are millennial and centennial. A westerly maximum occurred at the Last Glacial Maximum, and there is less conclusive evidence for another in the Late Holocene and for a minimum about 11,000 years ago. The strong centennial-scale variation includes, in historical times, strengthening of the southwesterly circulation over southern New Zealand and the westerly circulation over Tasmania during the Little Ice Age (about 1400 to 1850) while the preceding period (800 to 1400) experienced reduced westerly flow. Data from tree rings, ice cores, and sediment records (e.g., Gormez et al. 2004) indicate systematic interannual variability during more recent times.

The preceding studies of systematic variations in the westerly circulation and the modern findings of links between glacial mass balance and westerly wind flow have led Shulmeister et al. (2004) to invoke Milankovich precessional forcing, which drives seasonality, as the most likely source of long-term (millennial) changes in the westerly circulation. This could result in up to 6% more winter (less summer) insolation at low latitudes and 2% more at high latitudes than at present. Similarly, Clement, Cane, and Seager (2001) have proposed that periods of frequent El Niño events are expected during extreme strengths of the seasonal cycle of a semiprecessional cycle, that is, when the perihelion is in either the boreal spring or the autumn. Conversely, El Niño Southern Oscillation shutdown is predicted when perihelion is in either the boreal winter or the summer. Significantly, Turney et al. (2004) found that over the past 45,000 years millennial-scale dry periods indicative of frequent El Niño events are correlated with the Dansgaard-Oeschger millennial-scale warm events in the North Atlantic climate record.

In summary, if enhanced westerly circulation forces glaciation in New Zealand, glacial advances will be synchronous with precessional maxima, brief, and characterized by rapid glacier growth and decay under relatively mild conditions. However, if Northern Hemisphere thermal forcing dominates, glacier growth should occur at precessional minima and be gradual, with overall colder conditions. Shulmeister and colleagues are currently embarked on an ambitious research program designed to identify the dominant source of glacial forcing. Their preliminary results, based on beetle species diversity and abundance, suggest that temperatures during the Last Glacial Maximum were only slightly lower than at present. In addition, loess deposits reveal that, relative to the present, westerly winds were stronger and conditions were drier east of the alpine divide (J. Shulmeister, personal communication, September 15, 2005).

NATIONAL ECONOMIC IMPORTANCE

The economic importance of New Zealand's glaciers arises from the use of meltwater for the generation of electricity and for irrigation of agricultural and horticultural crops and from the glaciers' support of tourism and recreation.

HYDROELECTRICITY

Approximately 60% of New Zealand's electricity production is hydro based. Many of its storage lakes receive runoff from glacierized catchments, with the summer melt helping to ensure close-to-full lake levels by late summer. In New Zealand peak electricity demand occurs in winter, when runoff is low because most precipitation occurs as snow in the upper catchments. Seasonal snow cover in the hydroelectric catchments typically reaches a maximum of 60 km^3 by late spring. Braddock's (1998) estimation of the contribution of meltwater to total runoff for three glaciers in the Southern Alps (Table 13.1) shows that it can be substantial, with, in some instances, more than half the flow into the hydroelectric lakes coming from glacial meltwater. Chinn (2001) reports that over the 72-year period from 1890 to 1962 downwasting of the Tasman Glacier contributed 1.43 cumecs to the discharge of the Tasman River. Recession is now contributing 4.3 cumecs, of which 0.7 cumecs is supplied by direct calving of ice into the proglacial lake.

TABLE 13.1
Contribution of Glaciers to Catchment Runoff

GLACIER	ANNUAL RUNOFF (%)	ABLATION SEASON RUNOFF (%)
Ramsay	34	61
Hooker	14	23
Ivory	9	12

SOURCE: Bradock (1998).

High interannual variability in snow accumulation and runoff has major economic consequences. Braddock showed that the coefficient of variation for annual runoff decreases with increasing glacierization of the catchment, reaching a minimum at around 20% glacierization. However, the relationship is weak and, because of the nature of New Zealand's glaciers, is based on few data points beyond 15%. Fitzharris (1987) found that snow storage accounted for about 40% of the variability in spring runoff, with the remainder attributed to fluctuations in spring precipitation. Coupled with the relatively small storage capacity of the hydro lakes, these interannual variations, including the substantial changes in glacial mass balance, reduce the security of the electricity supply. In the past decade there have been two periods of serious nationwide energy shortage. These can be attributed to the same interannual variations in climate that are associated with glacial advance and retreat.

TOURISM

With European settlement in 1840 came publicity and curiosity about the New Zealand "natural wonderland." Its opportunities for exploration soon drew the adventurous traveler as well as the European migrant. By the 1890s

TABLE 13.2

Annual Numbers of International Visitors, 1998–2004

	1998	1999	2000	2001	2002	2003	2004
Total	1,322,972	1,412,789	1,547,856	1,710,783	1,723,587	1,869,753	2,118,290
Glacier Walk	129,839	141,887	152,796	160,599	145,905	239,298	233,523
Fox and Franz Josef Glaciers	21,129	17,154	28,046	44,911	73,646	113,335	24,493

SOURCE: Tourism Research Council of New Zealand. http://www.trcnz.govt.nz/Surveys/International+Visitor+Survey/Data+and+Analysis/Table-Activities-Attractions-in-NZ.htm.

tourists were being guided on the Tasman Glacier, and climbers were tackling Aoraki/Mt. Cook for the first time in 1882 (Tourism New Zealand 2001). Tourism soon differentiated the natural space of New Zealand, distinguishing urban spaces, the romantically scenic or picturesque alpine and fiord landscapes, and the extraordinary thermal environments of the central North Island's volcanic plateau (Werry 2001). Warren (2002) notes that the "100% Pure New Zealand" campaign, now in its third year, has been successful because it is based on what people say motivates them to come to New Zealand—words such as "fresh," "outdoors," and "invigorating experiences." Warren goes on to report that words such as "geothermal," "Maori," and "glaciers" are now at the top of the list.

Annually there are now more than 2 million short-term visitors to New Zealand each year (Table 13.2). The Tourism Research Council of New Zealand predicts that international visitor arrivals will increase by 6% per year, reaching 3.02 million by 2009. International visitor spending is growing significantly faster than visitor numbers and is expected to increase at 9.7% per annum, reaching $11.7 billion in 2009. New Zealand's glaciers have high tourist appeal due, in particular, to their accessibility, their natural beauty, and a sense of adventure. As the only mountains in Australasia that are snow-covered year-round, they are a source of attraction for nearby Australians, who have never seen ice or snow at home. On an annual basis around 10% of international visitors take a glacier walk, with around 1% of all international visitors visiting the Fox and/or Franz Josef Glaciers. In 2003 more than 300,000 international and domestic visitors visited the Franz Josef Glacier, with more than 80,000 being guided on the glacier.

Maintaining access and ensuring safety have become difficult because of recent rapid changes in glacier behavior. The average annual death toll from avalanches is approximately 1.5 persons, a rate similar to those for Switzerland and Austria when considered on a per-capita basis. The rate has generally increased with recreational use of alpine areas over time. The highest death rate is for mountaineers (Fitzharris, Lawson, and Owens 1999). The potential for a major disaster was revealed when on December 14, 1991, a rock avalanche lowered New Zealand's highest peak (Aorangi/ Mt. Cook, 3,764 m) by about 20 m. Some 14 million m^3 of rock buttress and flanking glaciers traveled 7.4 km and grew to about 55 million m^3 of pulverized rock and snow. The maximum speed was estimated to be 400 to 600 km/hr. There was no known triggering event, though the avalanche created its own magnitude-3.9 earthquake (Chinn, McSaveney, and McSaveney 1992).

Down-wasting of the Tasman Glacier by over 120 m during the twentieth century has resulted in not only the crumbling of the moraine wall but also slumping on the steep valley wall as a result of the release of supporting ice pressure (Chinn 1994). Such changes have caused the loss of lives (e.g., McSaveney, Davies, and Ashby 2002) as well as prime tourist and recreational assets including access roads, alpine trails, and huts. Providing ongoing access to the Franz Josef Glacier has become difficult because of its rapid fluctuations. Not long after completion of a walkway in 1907, the glacier thickened just enough to detach the structure from the valley side. Soon after that the glacier receded and has never again reached such elevations. A large proglacial lake that formed following a rapid decay of the terminus was a popular boating venue until a large block of ice held submerged by debris burst through the lake surface close to a boating party. A hut built in 1981 had to be removed in 1984 because of a rapid advance of the glacier (1 m per day), while a newly constructed swing bridge was destroyed in a glacier outburst flood in December 1965. Ice carpeted the riverbed down to the sea some 16 km away. This outburst flood and the general retreat of the glacier have also released large quantities of sediment, building up the outwash plain to such an extent that an access track and adjacent airstrip have had to be abandoned. Associated flood events and risks have also caused relocation of a township (Warburton 2004).

Chinn (1996) noted increases in the number and size of proglacial lakes and pointed out that they not only accelerate the rate of glacial recession but also create access problems for tourists and recreational visitors. McSaveney, Davies, and Ashby (2002) report that loss of safe access to the upper Ramsay Glacier across the lower glacier tongue now forces climbers to traverse the dangerous foot slopes beside the growing proglacial lake. The length of this dangerous route is increasing as the lake grows. In 2002 a climber using this route was killed by a rapid debris fall caused by the dry raveling (i.e., loosening) of the glacial moraine wall. The collapse was of such a scale, with such large boulders, that it behaved like a large rockfall.

Hazard mapping, forecasting, and active control of avalanches are now common risk-management activities, especially in ski areas and for alpine roads such as the road to Milford Sound.

SOCIAL/CULTURAL IMPORTANCE

New Zealand's glaciers are important to both Maori and the country's more recent inhabitants. The country's best-known glacier, the Franz Josef, is known to Maori as Ka Roimata-o-Hine-hukatere (Tears of the Avalanche Girl) and illustrates the place of glaciers in Maori creation stories. The story is that the mountain-dwelling maiden Hine-hukatere was to marry Wawe, a man of the coastal plains. Wawe was determined to overcome his fear of the mountains, which his future wife loved. One day he went climbing with her, only to slip and fall to his death. Hine-hukatere's tears filled the valley where he lay, and the gods transformed them into the river of ice that today is known as the Franz Josef Glacier (Horne 1986).

While initially Maori stayed close to the coast, their subsequent quests for resources and knowledge took them into alpine areas. The glaciers and surrounding mountains were obstacles to movement but contributed to a sense of place—a symbol of "home"—and helped define tribal boundaries. Every *hapu* (subtribe) had a special association with a hill or mountain representing its point of contact with Rangi, the sky father (Orbell 1995). In most cases Maori gave glaciers names that reflected the danger they represented, but popular use of these names did not survive colonization. Today, however, the alternative names are often used together. Generally, glaciers were feared by Maori in part because the noise they made resembled that of a human in anguish or possessed but also because Maori had no knowledge of safe movement on a glacier. While glaciers themselves had no value for Maori, the wider alpine areas were of special importance, being a source of *pounamu* (greenstone or jade). Lacking metal, they found *pounamu* an ideal material for weapons, ornaments, and tools despite the difficulty of locating and retrieving it. Through trade and gift giving the prized stone was distributed throughout New Zealand.

Early European explorers were guided by Maori, who avoided areas occupied by glaciers and chose to traverse the mountains using glacier-free passes. The early British explorers also felt uncomfortable on such unfamiliar terrain, but explorers and gold prospectors from alpine Europe were attracted to such areas. Many of the names given to the glaciers recognize these latter explorers. From the time of the earliest European settlement, similarities with the Alps of Europe were recognized, and the glaciers and their surroundings have held special direct or vicarious significance for most New Zealanders. Initially through writings and paintings and subsequently through photography, they became an important part of the New Zealand identity, both locally and internationally. The heyday of New Zealand mountaineering coincides with the heroic age of its literary nationalism—the conquest of a hostile landscape in the service of inventing a national identity. Schooled on the icy terrain of the Southern Alps, New Zealand climbers readily adapted to the Himalayas, participating in the first ascent of Mt. Everest (Newton 1999).

Besides having aesthetic and intrinsic value, New Zealand's glaciers and alpine areas in general have become increasingly important for recreation, including skiing, hiking, climbing, rafting, and aerial sports. Again, safety and access are ongoing issues. In the years between 1981 and 1998 there were 539 reported avalanche events involving people or property, with between 13 and 56 and an average of around 30 events per year. In the same period there were 37 fatalities directly attributed to avalanches (Irwin and MacQueen 1999).

IMPLICATIONS OF GLOBAL CHANGES

SCIENTIFIC RESEARCH

As we have seen, recent and ongoing research into past and modern glacial cycles and fluctuations is placing increasing emphasis on spatial coupling and interannual variability. Such findings are reinforcing the importance of cooperative and interdisciplinary research, both nationally and internationally.

GLOBAL WARMING

"Best-guess" projections suggest that one-third of New Zealand's glaciers will cease to exist by the time atmospheric greenhouse-gas concentrations have doubled (Chinn 1989), with a loss of 79–88% of the present ice volume and an increase in the equilibrium line altitude to 580 m (Ruddell 1995). Fitzharris (1989) determined that increases of 3 °C for temperature and 15% for precipitation could result in the raising of the snow line by 300–400 m, a decrease in snow accumulation below 2,300 m, and a reduction in winter snow-covered area for South Island hydro catchments from 45% to 28%. These changes will also affect seasonal river flows in the central and southern South Island, resulting in increased inflow to storage lakes of 40% in winter and a decrease of 13% in summer. Chinn (2001) also reports meltwater runoff scenarios. Since large glaciers such as the Tasman respond very slowly to climate change, in part because of their large ice mass, they are expected to persist for at least several centuries under all warming scenarios currently considered (Ministry for the Environment 2001). Some glaciers would likely experience increased average snowfall that could balance the increased melting, even leading to a temporary advance. All these changes favor increased hydroelectric power generation and reduce the demand for winter storage (Fitzharris and Garr 1996). In the present climate regime, river flows in hydro catchments tend to be lowest in winter, when the demand for electricity is greatest, and rise in spring and summer because of snow and glacier melt. Under warmer conditions, the seasonal asymmetry of runoff is reduced because of increased winter rainfall over the Main Divide, less seasonal snow storage, and a reduced contribution from shrinking glaciers. The seasonal change in supply will coincide with a reduced electricity demand during winter brought about by warmer conditions.

Electricity demand models predict that with a warming of about 2 °C by 2100, annual average electricity demand would decrease by about 6%. Future climate change may therefore be expected to bring net benefits to electricity supply through reduced summer storage needs and increased generation potential during winter peak demand. Summer demand for electricity could increase through a growing use of air conditioning in buildings, but in the current climate this is not very sensitive to summer temperatures (Ministry of the Environment 2001).

TOURISM AND RECREATION

Glacier walks are an important component of the activity package for visitors to New Zealand (see Table 13.2). Projected reductions in glacier numbers, extent, and aesthetics as a consequence of global warming may well decrease the competitiveness of New Zealand as an international tourism destination and increase pressures on remaining tourist focal points, including glaciers. Barringer (1989) has suggested that the projected changes in seasonal snow cover will have a significant adverse impact on New Zealand's ski industry.

RESOURCE MANAGEMENT

Global changes will also have an impact on resource management at both national and local levels. The Resource Management Act of 1991 guides the use and development of New Zealand's land, water, and air and is therefore the principal instrument for natural hazard management. Significantly, it does not include snow and ice in an otherwise comprehensive list of natural hazards.

Until recently an abundance of freshwater in New Zealand has been taken for granted. Recent events have made it clear that management of the resource has not always been sustainable and has not kept pace with economic, cultural, social, and environmental changes. Despite record low inflows to the storage lakes, the two recent serious nationwide energy shortages were managed successfully through voluntary power savings, temporary relaxation of transmission security, and greater use of thermal generation. As a result of recent reforms, the market now sets the price of electricity, producing rapid and large price hikes when drought is anticipated or imminent and reducing the incentive to invest in additional or alternative generation capacity. During the most recent power shortage, electricity spot prices peaked at NZ$0.41/kWh, compared with average spot prices of NZ$0.05–0.10/kWh.

While the Resource Management Act provides the legal framework and mechanisms for water resources management, it does not provide a basis for apportioning water rights between, for example, hydroelectric power generation and irrigation. Such shortcomings were recently highlighted by the large number of competing demands for water from the Waitaki River. The river includes a series of storage lakes that receive runoff from glacierized catchments. In addition to being a strategic national asset for electricity production, the river is also important for irrigation and for its natural, recreational, community, and fishery values and is significant to Ngai Tahu as a defining feature of who they are as a people and as individuals. In September 2004 Parliament passed the Resource Management (Waitaki Catchment) Amendment Act to create an improved decision-making process for allocating water from the Waitaki Catchment. A new water allocation board is charged with developing a regional water allocation plan outlining the objectives, policies, methods, and rules for allocating the catchment's water. At the national level a water program is to incorporate the best ways of managing the country's freshwater resources and thereby ensure that its lakes, rivers, wetlands, and other freshwater resources are fairly used and protected.

Implications for resource management at the local level include enhancing hazard characterization and risk-management processes for both current and emerging risk events. This calls for greater cooperation between local government, civil society, and the private sector. Finally, reduced reliance on glacier- and mountain-based tourism will require diversification of local economies.

SUMMARY AND CONCLUSIONS

New Zealand's glaciers play a critical role in our understanding of global and regional climate variation and change over both geological and historic time scales and for future projections. Recent research is resulting in growing recognition of the importance of internannual variations in atmospheric circulation patterns at both geologic and historic time scales and beginning to reveal the relative importance of dynamic and physical drivers of glacier behavior.

The glaciers are also of national and local importance as resources and sources of risk. National and local policies and practices need to be strengthened to address changing pressures on resources and the changing nature of the associated risks.

REFERENCES CITED

Anderton, P.W., and T.J.H. Chinn. 1978. Ivory Glacier, New Zealand: An IHD basin study. *Journal of Glaciology* 20:67–84.

Barringer, R.J.F. 1989. Changes in snowline altitude and snowfalls on the Remarkables (1930–1985) and their possible significance for the ski industry on Central Otago. In *Proceedings, Fifteenth New Zealand Geography Conference*, ed. R. Welch, 271–77. Hamilton: New Zealand Geographical Society.

Braddock, D.H. 1998. The influence of glaciers on runoff, Southern Alps, New Zealand. M.Sc. thesis, University of Otago.

Broecker, W.S. 1997. Future directions of paleoclimatic research. *Quaternary Science Reviews* 16:821–25.

Burrows, C.J. 2001. *The Franz Josef Glacier, Westland, in the 19th and 20th centuries AD*. Christchurch: Rebus Publications.

Carter, R.M., and P. Gammon. 2004. New Zealand maritime glaciation: Millennial-scale southern climate change since 3.9 Ma. *Science* 304:1659–62.

Chinn, T.J.H. 1989. Glaciers and snowlines. In *Climate change: The New Zealand response*, 238–40. Wellington: New Zealand Hydrological Society.

———. 1994. What's happening to our glaciers? *New Zealand Alpine Journal* 47:96–100.

———. 1995. Glacier fluctuations in the Southern Alps of New Zealand determined from snowline elevations. *Arctic and Alpine Research* 27:187–98.

———. 1996. How much ice has been lost? *New Zealand Alpine Journal* 49:88–95.

———. 2001. Distribution of perennial snow and ice water resources in New Zealand. *Journal of Hydrology (NZ)* 40:139–87.

Chinn, T.J.H., C. Heydenrych, and M.J. Salinger. 2005. *New Zealand glacier snowline survey 2004*. National Institute of Water and Atmosphere Client Report AKL2004-0XX. Auckland: National Institute for Water and Atmosphere.

Chinn, T.J.H., M.J. McSaveney, and E.R. McSaveney. 1992. *The Mount Cook rock avalanche of 14 December, 1991*. Wellington: Geology and Geophysics, Department of Scientific and Industrial Research.

Chinn, T.J.H., and M.J. Salinger. 2001. *New Zealand glacier snowline survey, 2000*. Technical Report 98. Wellington: National Institute for Water and Atmospheric Research.

Chinn, T.J.H., and I.E. Whitehouse. 1980. Glacier snow line variations in the Southern Alps, New Zealand. In *World glacier inventory*, 219–28. Wallingford: International Association of Hydrological Sciences.

Clement, A.C., M.A. Cane, and R. Seager. 2001. An orbitally driven tropical source for abrupt climate change. *Journal of Climate* 14:2369–75.

Denton, G.H., C.J. Heusser, T.V. Lowell, P.L. Moreno, B.G. Andersen, L.E. Heusser, C. Schluchter, and D.R. Marchant. 1999. Interhemispheric linkage of paleoclimate during the last glaciation. *Geografiska Annaler* 81A:107–53.

Fitzharris, B.B. 1987. A method for indexing the variability of alpine snow cover over large areas. In *Proceedings of the Vancouver Symposium*, 139–50. Wallingford: International Association of Hydrological Sciences.

Fitzharris, B.B. 1989. Climate change and the future of New Zealand's snow cover and snowline. In *Climate change: The New Zealand response*, 238–40. Wellington: New Zealand Hydrological Society.

Fitzharris, B.B., and C. Garr. 1996. Climate, water resources, and electricity. In *Greenhouse: Coping with climate change*, ed. W.J. Bonma, G.I. Pearman, and M.R. Manning, 263–80. Collingwood: CSIRO.

Fitzharris, B.B., and J.E. Hay. 1989. Glaciers: Can they weather the storm of climate change? In *Proceedings, Fifteenth New Zealand Geography Conference*, ed. R. Welch, 284–91. Hamilton: New Zealand Geographical Society.

Fitzharris, B.B., J.E. Hay, and P.D. Jones. 1992. Behaviour of New Zealand glaciers and atmospheric circulation changes over the past 130 years. *The Holocene* 2:97–106.

Fitzharris, B.B., W. Lawson, and I. Owens. 1999. Research on glaciers and snow in New Zealand. *Progress in Physical Geography* 23:469–500.

Gormez, B., L. Carter, N. Trustrum, A. S. Palmer, and A. P. Roberts. 2004. El Niño-Southern Oscillation signal associated with Middle Holocene climate change in intercorrelated terrestrial and marine sediment cores, North Island, New Zealand. *Geology* 32:653–56.

Hay, J. E., and B. B. Fitzharris. 1988. The synoptic climatology of ablation on a New Zealand glacier. *Journal of Climatology* 8:201–15.

Hay, J.E., N. Mimura, J. Campbell, S. Fifita, K. Koshy, R.F. McLean, T. Nakalevu, P. Nunn, and N. de Wet. 2003. *Climate variability and change and sea-level rise in the Pacific Islands region: A resource book for policy and decision makers, educators, and other stakeholders.* Apia: South Pacific Regional Environment Programme.

Hay, J. E., M. J. Salinger, B. B. Fitzharris, and R. Basher. 1993. Climatological seesaws in the Southwest Pacific. *Weather and Climate* 13:9–21.

Hooker, B. L., and B. B. Fitzharris. 1999. The correlation between climatic parameters and the retreat and advance of the Franz Josef Glacier. *Global and Planetary Change* 22:39–48.

Horne, A. 1986. *New Zealand: A special place.* Auckland: Lansdowne Press.

Irwin, D., and W. MacQueen. 1999. *Report on avalanche incidents and accidents, 1981–1998.* Wellington: New Zealand Mountain Safety Council.

Madgwick, P. 2005. Coast glaciers surge. *The Press* (Christchurch), August 31.

McSaveney, M. J., T. R. Davies, and G. L. Ashby. 2002. *The fatal Ramsay Glacier rockfall of 9 November 2002.* Institute of Geological and Nuclear Sciences Science Report 2003/2. Wellington: Institute of Geological and Nuclear Sciences.

Ministry for the Environment. 2001. *Climate change impacts on New Zealand.* Wellington.

Newton, J. 1999. Colonialism above the snowline: Baughan, Ruskin, and the South Island myth. *Journal of Commonwealth Literature* 34:85–96.

Orbell, M. 1995. *The illustrated encyclopaedia of Maori myth and legend.* Christchurch: Canterbury University Press.

Ruddell, A. 1995. Recent glacier and climate change in the NZ Alps. Ph.D. diss., University of Melbourne.

Salinger, M. J., M. J. Heine, and C. J. Burrows. 1983. Variations of the Stocking (Te Wae Wae) Glacier, Mount Cook, and climatic relationships. *New Zealand Journal of Science* 26:321–38.

Shulmeister, J., I. Goodwin, J. Renwick, K. Harle, L. Armand, M. S. McGlone, E. Cook, J. Dodson, P. P. Hesse, P. Mayewski, and M. Curran. 2004. The Southern Hemisphere westerlies in the Australasian sector over the last glacial cycle: A synthesis. *Quaternary International* 118–19:23–53.

Tourism New Zealand. 2001. *100 years of pure progress.* Wellington.

Turney, C.S. M., A.P. Kershaw, S.C. Clemens, N. Branch, P.T. Moss, and L.K. Fifield. 2004. Millennial and orbital variations of El Niño/Southern Oscillation and high-latitude climate in the last glacial period. *Nature* 428:306–10.

Warburton, R. 2004. *Glacier country: My years at Franz Josef.* Dunedin: Longacre Press.

Warren, S. 2002. Branding New Zealand: Competing in the global attention economy. *Locum Destination,* 54–56.

Werry, M. L. 2001. Tourism, ethnicity, and the performance of New Zealand nationalism, 1889–1914. Ph.D. diss., Northwestern University.

14

The Impact of Mining on Rock Glaciers and Glaciers

EXAMPLES FROM CENTRAL CHILE

Alexander Brenning

Glaciers and rock glaciers in the semiarid Andes constitute natural stores of water that control the runoff of mountain rivers, especially in the dry summer months. They are responsible for the water supply to the agglomerations of Santiago, Chile (5.3 million inhabitants), and Mendoza, Argentina (1.1 million inhabitants), and the irrigated land in the surrounding lowlands (Corte 1976; Barsch 1988; Schrott 1996; Brenning 2003, 2005*a*, *b*). Rock glaciers are of minor hydrological and geomorphological importance in the European Alps. However, the amount of water stored in them per unit area in the Andes between Santiago and Mendoza is one magnitude higher than in the Alps, and they constitute the only ice bodies in many catchments with summit altitudes of up to 4,500–5,000 m a.s.l. (Brenning 2005*b*). The importance of rock glacier water storage increases toward the arid north (Schrott 1996; Brenning 2005*b*). Their significance in the dry Andes is, however, very little known in Chile.

Rock glaciers have been characterized as the geomorphological expression of creeping bodies of ground ice in areas of mountain permafrost and have been attributed to periglacial and glacigenic conditions (Barsch 1988, 1996; Whalley et al. 1994; Figure 14.1). Since the formation of rock glaciers can in most cases be assumed to have begun in earlier periods of the Holocene or Pleistocene (Haeberli et al. 1999, 2003), most of their ice content (assumed to be 40–60% by volume) may be considered fossil frozen groundwater (Barsch 1977, 1996; Hoelzle et al. 1998; Arenson, Hoelzle, and Springman 2002). Only the ice contained within the seasonally frozen active layer of a rock glacier is exchanged on a yearly basis.

In the semiarid Andes, glaciers *sensu stricto* are most abundant in the Andean main range around the latitude of Santiago and Mendoza (latitude 33–34°S), where summit elevations frequently exceed 5,500 m a.s.l. and reach a pan-American maximum of 6,959 m at Cerro Aconcagua (Lliboutry 1956). Their importance is very limited outside these central parts and rapidly decreases northward.

The main human activities in the semiarid Andes are summer extensive pasturage of cattle, horses, and goats and mining; tourism is very limited. Major mining projects exploit mainly copper and gold reserves of low grade in remote

FIGURE 14.1. Active rock glaciers in a glacier-free cirque above 3,500 m a.s.l. on the west side of Cerro Catedral (4,765 m), Andes of Santiago, February 2002. (Photo by A. Brenning.)

areas of the Andes. Their development since the 1980s is characterized by enormous expansion projects that have made Chile the world's largest copper-producing country and one of the largest gold producers (CCAEC 1996). For example, from 1980 to 2000 Chilean gold and copper production showed sevenfold and fourfold increases, respectively (Lagos et al. 2002). This study focuses on the impact on rock glaciers and glaciers of mining activities at CODELCO División Andina, Los Bronces, and Pascua-Lama (Figure 14.2). Trends in public and governmental awareness and action are observed and related to recent political and legislative developments in Chile.

Although explicit environmental legislation goes back only to 1994, several earlier laws address environmental issues in Chile. In particular, the use of surface and underground water is regulated by the water code and administered by the Dirección General de Aguas (Dourojeanni and Jouravlev 1999). According to this law, any use of both glaciers and rock glaciers would require government approval. The Chilean Environmental Impact Assessment System was established in 1994 by Law No. 19.300 (Fundamental Environmental Law) and supplementary supreme decrees. Environmental impact studies and declarations have been obligatory since 1997 for most industrial and mining activities, and the Comisión Nacional del Medio Ambiente (CONAMA) and its regional agencies (COREMA) are responsible for their implementation. According to this law, the environmental agencies cooperate with the corresponding ministerial services in the evaluation process (Pizarro and Vasconi 2004), which is public and includes hearings designed to foster public participation (Padilla 1996; Sabatini and Sepúlveda 1996). Environmental impact studies and declarations are publicly accessible, in some cases even online (http://www.seia.cl; http://www.conama.cl).

DIVISIÓN ANDINA AND LOS BRONCES: LARGE-SCALE MINING IN A PERIGLACIAL ENVIRONMENT

Rock glaciers were first studied in Chile by Lliboutry (1961, 1986) in the area of the current Los Bronces and División Andina mines. This area was later also visited by the rock glacier researcher D. Barsch in January 1982 (cf. Barsch 1988 and Barsch 1996: 26), a time when local morphology still had not been much altered by mining. Since then, the overwhelming growth of both mines in the area has produced a strong geomorphological impact (Figure 14.3).

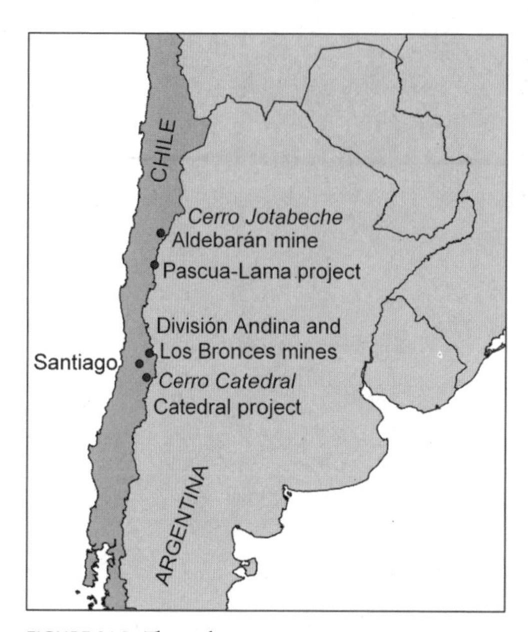

FIGURE 14.2. The study area.

The development of the Los Bronces mine started in the 1830s at several small copper and silver extraction sites situated around 3,500 m a.s.l. Since the exploitation of high-grade minerals was limited by the basic technology, the construction of an underground mine in 1915 and the fusion of several small companies increased production. As a consequence of the conflicts arising during this concentration, the company was named Disputada de Las Condes (*Minería Chilena* 1993).

Disputada was nationalized in 1972 and later sold to Exxon Minerals during the military regime in 1978. The new owner initiated a process of modernization, which increased copper production from 8,400 to 37,000 tons per year by around the year 2000. Today the possibility of an increase to 200,000 tons per year is being evaluated (Editec 2000). The Los Bronces rock mill at 3,500 m a.s.l. is connected by a 57-km-long ore pipeline with the concentrator plant of Las Tórtolas, situated in the forelands of the Andes north of Santiago (*Minería Chilena* 1993). Since 2001 Disputada has been owned by the South African company Angloamerican.

Just north of Los Bronces at División Andina, the state-owned Corporación Nacional del Cobre de Chile (CODELCO) exploits the same copper reserve as Disputada. Extraction began here in 1864 in the western part of the current mine (Holmgren and Vela 1991), and an underground mine started to operate in 1970. After nationalization in 1971, the mine was incorporated into CODELCO in 1976 under the name of División Andina. In 1980, open-pit mining of high-grade minerals began at the Sur-Sur pit (Arcadis Geotécnica 2001). In 1998 an expansion project almost doubled copper production at División Andina, reaching 249,000 tons of refined copper in 1999 (Editec 2000). Current expansion projects are intended to reach an annual production of 400,000 tons of copper in 2006 and of 650,000 tons in 2012 (Arcadis Geotécnica 2001; *Minería Chilena* 2005c). Figure 14.3 is a cartographic representation of the development of mining activities in the División Andina and Disputada de Las Condes areas.

The vast rock glacier areas in the Upper Blanco catchment have been affected by mining activities at least since the Sur-Sur pit of División Andina began to operate in 1980. Since then, two rock glaciers identified by Lliboutry (1961) have disappeared almost completely (Table 14.1). The current expansion project at División Andina provides for an enlargement of the Sur-Sur mine to a total area of 375 ha and the construction of two new waste-rock disposal areas with a total surface area of 497 ha, according to the approved environmental impact study (Arcadis Geotécnica 2001; Figure 14.3, Table 14.1). Thus, more than 8 km^2 of high mountain area will be strongly impacted. The operations imply the destruction or degradation of about 1.4 km^2 of rock glaciers, according to the same study. Furthermore, my inquiries at the Dirección General de Aguas (Departamento de Administración de Recursos Hídricos and Centro de Información de Recursos Hídricos, Santiago) indicated that the agency was unaware of any of these past or proposal removals and alterations of rock glaciers at División Andina.

In the area of the Los Bronces mine of Disputada de Las Condes, a comparison of aerial photographs (Hycon 1955 and Geotec 1997) shows that rock glaciers have also been removed

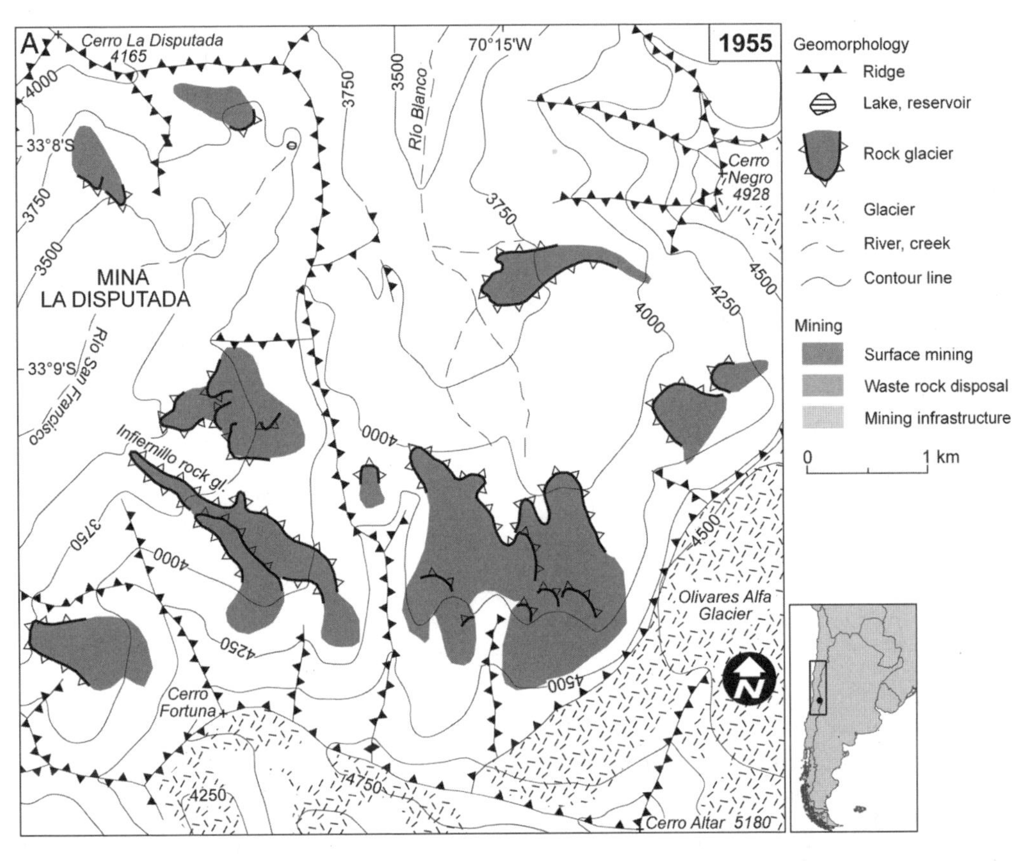

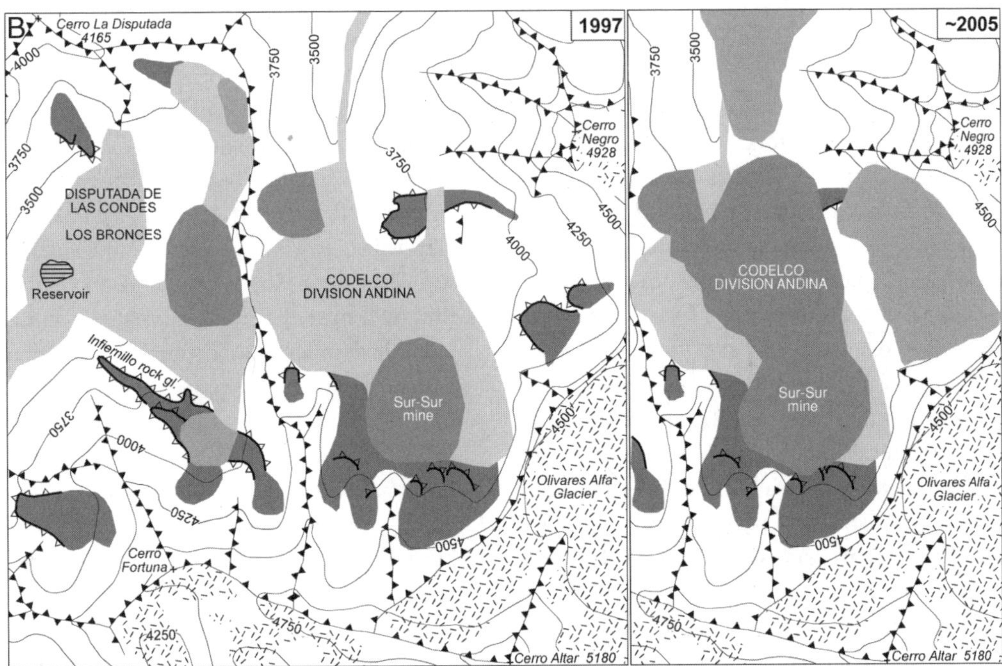

FIGURE 14.3. Geomorphological impact of open-pit mining on rock glaciers at División Andina and Los Bronces. (A) 1955, (B) 1987 and ~2005. Cartography based on Lliboutry (1961), Arcadis Geotécnica (2001), and aerial photography of 1955 and 1997 (Hycon, no. 4300, and Geotec, flight Juncal, no. 5585).

TABLE 14.1

Rock Glacier Area and Ice Volumes Affected by División Andina and Los Bronces Mines

	DIVISIÓN ANDINA	LOS BRONCES
Original rock-glacier area[a] (km^2)	2.6	1.9
Alteration until 1997[a]		
Removed by open-pit mining (km^2)	0.5	0.2
Covered by waste rock deposits (km^2)	—	0.2
Affected by mining infrastructure (km^2)	0.2	0.4
Water equivalent affected until 1997[b] (10^6 m^3)	>5	>6
Alteration 1997–2005[c]		
Removed by open-pit mining (km^2)	0.82	n.a.
Degraded (waste rock, infrastructure) (km^2)	0.58	n.a.
Water equivalent affected 1997–2005[b] (10^6 m^3)	10	n.a.

[a]Calculated from aerial photographs (Hycon, no. 4300, and Geotec, flight Juncal, no. 5585) and the environmental impact studies of Geotécnica Consultores (1996) and Arcadis Geotécnica (2001).
[b]Calculated assuming a minimum permafrost thickness of 20 m, an ice content of at least 40%, and an ice density of 0.9 g/cm^3 (Brenning 2005*b*).
[c]Expansion project of División Andina, data from Arcadis Geotécnica (2001).

and altered (Figure 14.3, Table 14.1). Most information available from Los Bronces concerns the deposition of mine waste on the Infiernillo rock glacier beginning in August 1990. The Infiernillo is an active tongue-shaped rock glacier 2.5 km long. It extends from 3,600 to 4,300 m a.s.l. and covers 1.0 km^2. The upper central part (~0.2 km^2) of it, at ~4,000 m a.s.l., has been covered with waste rock from Los Bronces since August 1990. Its displacement has been monitored at several topographic points, and boreholes from prior to the deposition and some of the results have been published. Contreras and Illanes (1992) reported superficial displacement rates between 0.3 and 1.2 cm per day under natural (predepositional) conditions with strong seasonal variation and highest velocities in spring and especially autumn. The initial deposition of 14 million tons of waste rock led to an immediate increase in rock glacier surface velocity to a peak of about 20 cm per day. This value is extremely high compared with measurements of rock glacier velocities under natural conditions (Barsch 1996; Grebenets, Kerimov, and Bakcheev 1997; Arenson, Hoelzle, and Springman 2002; Roer 2003). Velocities stabilized after this initial period but at higher levels than before the deposition (Contreras and Illanes 1992). Further data on waste-rock deposition and creep rates have not been published, but the addition of 30 million tons of waste between 1992 and 1997 was planned.

The deposition of debris on a rock glacier may have various long-term effects, some of which are rather speculative in the absence of observational data. First, geochemical weathering of the waste rock is likely to produce acid rock drainage that may affect water discharge from the rock glacier even after mine closure (Ripley, Redman, and Crowder 1995; Andía, Lagos, and Danielson 1999; EPA 2001). Second, natural geothermal heating will raise the lower permafrost boundary within decades after deposition and affect rock glacier stability, which is partially temperature driven (Burger, Degenhardt, and Giardino 1999). Third, new permafrost may develop in the waste rock within years to decades (Grebenets, Kerimov, and Bakcheev 1997). This permafrost is, however, unlikely to be ice-rich because of its distance from the rock glacier's rooting zone and the artificial compaction of the deposited material, and it will probably be patchy and vary in ice content depending on local material properties.

The issues discussed above need further investigation to determine the potential hazards arising from the Infiernillo rock glacier. A destabilization of the waste-laden rock glacier was tentatively considered by Contreras and Illanes (1992); possibly triggered, for example, by seismic activity, such an event might produce a catastrophic mass movement affecting the industrial areas of the mines (especially Los Bronces at the Infiernillo rock glacier) and extend downstream, perhaps to the lower parts of the river as far as Santiago. The 1965 tailings dam failure at Disputada that killed 200 people (Aliste, Moraga, and Alvarez 1966) and a (natural) landslide in the Colorado catchment (Andes of Santiago) in 1987 (Casassa and Marangunic 1993; González-Ferrán 1994) demonstrate the effects of such hazards. Therefore, the warning of Burger, Degenhardt, and Giardino (1999) to avoid rock glaciers in the siting of essentially all structures must be kept in mind in the context of waste-rock deposits.

THE PASCUA-LAMA PROJECT

Pascua-Lama is the name of a mining project of the Canadian Barrick Gold Company aiming at the exploitation of a binational gold (14.1 million ounces), silver (461 million ounces), and copper (180,000 tons) reserve in the arid III Region of Chile and the Argentine Province of San Juan. The extraction of minerals is planned to start in 2008. The mining activities will affect, according to the project's environmental impact study, a total area of 17.5 km^3 situated between 4,400 and 5,300 m a.s.l. (Geotécnica Consultores 2001). Analysis of 1996 aerial photographs of 1996 reveals the presence of glaciers and rock glaciers and hence of mountain permafrost.

As a consequence of the superficial extraction of minerals, which will mainly take place on Chilean territory, the removal of 10 ha of glacier is planned. The release of the environmental impact study for the project led to requests and petitions by local nongovernmental organizations and individuals concerned about its possible impact on the water supply for irrigation (*AreaMinera* 2004; *Minería Chilena* 2005*a*, *b*; *MiningWatch Canada* 2005).

In April 2001 the regional environmental agency COREMA (Atacama) approved the Pascua-Lama mining project with the precondition of complying with a glacier management plan. This plan permitted the proposed removal of up to 10 ha of glacier with blastings and excavators and stipulated their redeposition at some nearby location of similar geomorphological characteristics and at a similar altitude. From a glaciological point of view it was obvious that these conditions would in no way guarantee the conservation of the ice removed or lead to the reconstitution of a glacier at the place of deposition. On the contrary, it was likely that the redeposited pieces of glacier ice would disappear within years or decades. Only a comprehensive study of the energy and mass balance of the glacier to be removed and of the artificial ice body to be built would constitute a sound scientific base for such a relocation. Current glaciological and climatological knowledge in the remote and arid Andes of northern Chile is far from sufficient for this kind of research (cf. Kull, Grosjean, and Veit 2002; Corripio and Purves 2003).

After a period of low gold prices had delayed the development of the project, it was resubmitted in December 2004, facing the firm opposition of environmental groups and irrigation farmers (Arcadis Geotécnica 2004; *MiningWatch Canada* 2005). The environmental impact study was finally approved by COREMA Atacama in February 2006 subject to several limitations, including the prohibition of destruction or alteration of the glaciers in the project area (COREMA Atacama 2006).

DISCUSSION

Further examples of potential future impacts of mining on rock glaciers in the Chilean Andes may be added to those presented above. For instance, road construction for the limestone mining project at Cerro Catedral in the Andes of Santiago (Compañía Minera Catedral, a subsidiary of the South American Gold Company), which has been explored and is now awaiting funding, and at the gold mine of Nevado Jotabeche in the

Atacama region (the Aldebarán Project of Compañía Minera Casale, a subsidiary of Arizona Star Resources of Canada and Angloamerican of South Africa), which is ready to operate, has already affected rock glaciers.

In the context of the tremendous growth of the Chilean mining industry (Moussa 1999; Lagos et al. 2002) and the existence of huge ore reserves in the high Andes, further cases of future degradation and destruction of rock glaciers and glaciers have to be expected in Chile. Therefore the question arises how these water resources can be effectively protected. It must be emphasized, however, that current legislation already formally provides the instruments that are necessary for impeding the destruction or degradation of glaciers and rock glaciers. Current politics lead contrarily to administrative decisions such as those described previously, which prioritize economic development and produce a strong discrepancy between what is written on paper and what is practiced on the ground (Carruthers 2001).

The conclusions to be drawn are, however, not entirely pessimistic: Recent developments in the case of the construction of the GasAndes pipeline in 1996 and the Pascua-Lama project show that the local population has started to make use of the participatory instruments provided by the new environmental legislation (Padilla 1996; Sabatini and Sepúlveda 1996; *AreaMinera* 2004; *MiningWatch Canada* 2005; COREMA Atacama 2006). This process has to be seen in the context of 17 years of a military regime (1973–90) and the postdictatorship collapse of grassroots movements that made way for a deeply embedded neoliberalism (Carruthers 2001). If the trend toward increasing awareness and participation persists, then a better trade-off between mining and the environment may become possible. A prerequisite for this awareness is, however, comprehensive scientific knowledge of the natural environment and the diffusion of this knowledge into the society and governmental institutions. In the case of the Chilean Andes, much remains to be done in this respect.

Turning to a global perspective, several examples of direct human impact on glaciers and rock glaciers in other parts of the world and by activities other than mining have been reported (Fisch, Fisch, and Haeberli 1977; Haeberli and Keusen 1983; Giardino and Vick 1985, 1987; Haeberli 1992; Burger, Degenhardt, and Giardino 1999; Diolaiuti et al. 2001; Jurt 2004; Schwegler 2004). For example, the use of glacier crevasses as an unauthorized waste disposal site in South Tyrol (Italian Alps) has been observed (Jurt 2004). The operation of ski lifts on glaciers in the European Alps is well known, even though the use of hazardous substances such as fuel on these unprotected ice bodies and the alteration of physical surface properties affecting the glacier's energy balance are both problematic (Haimayer 1989; Diolaiuti et al. 2001). Rock glaciers have been partly removed in the construction of ski runs (Haeberli 1992). The Davidov glacier in the Tien-Shan Mountains (Kyrgyz Republic) was partly covered with gold mining waste in the late twentieth century (Aizen and Chugunov 1988; Aizen and Zakharov 1988; Homeniuk 2000).

These observations show that the anthropogenic destruction of glaciers and rock glaciers is neither a local Chilean phenomenon nor restricted to developing countries; it is in fact a global problem that raises the question of whether more effective protection of the mountain cryosphere in general is necessary on a global scale.

ACKNOWLEDGMENTS

This work was financed by research scholarships from the German Academic Exchange Service in 2002 and 2004, which are gratefully acknowledged. I also thank the Centro de Información de Recursos Hídricos of the Chilean Dirección General de Aguas (Santiago), Raquel Yamal, and Gonzalo Barcaza for providing valuable information.

REFERENCES CITED

Aizen, V. B., and A. P. Chugunov. 1988. Estimation of the Davidov Glacier response to artificial destruction of its terminus [in Russian]. *Data of Glaciological Studies (MGI)* 67:202–6.

Aizen, V. B., and V. G. Zakharov. 1988. Mass balance and flow velocity of the Davidov Glacier from researches in 1984–1985 [in Russian]. *Data of Glaciological Studies (MGI)* 67:67–82.

Aliste, J., A. Moraga, and L. Alvarez. 1966. Efectos del sismo de marzo de 1965. *Boletín del Instituto de Investigaciones Geológicas de Chile* 20. Santiago: Instituto de Investigaciones Geológicas.

Andía, M. E., G. E. Lagos, and L. J. Danielson. 1999. The challenges posed by mine closure in Chile. In *Proceedings of the Copper 99 International Conference, October 10–13, 1999, Phoenix, Arizona.* Warrendale, PA: Minerals, Metals, and Materials Society.

Arcadis Geotécnica. 2001. *Estudio de impacto ambiental Proyecto de Expansión División Andina.* Santiago.

———. 2004. *Estudio de impacto ambiental modificiones Proyecto Pascua-Lama.* Santiago/Copiapó.

AreaMinera. 2004. Pascua-Lama y sus posibles impactos ambientales y sociales: Interview with César Padilla, August 16. http://www.areaminera.com/ (accessed November 23, 2004).

Arenson, L., M. Hoelzle, and S. Springman. 2002. Borehole deformation measurements and internal structure of some rock glaciers in Switzerland. *Permafrost and Periglacial Processes* 13:117–35.

Barsch, D. 1977. Alpiner Permafrost: Ein Beitrag zur Verbreitung, zum Charakter und zur Ökologie am Beispiel der Schweizer Alpen. *Akademie der Wissenschaften zu Göttingen, Mathematisch-Physikalische Klasse* 3(31):118–41.

———. 1988. Rockglaciers. In *Advances in periglacial geomorphology*, ed. M. J. Clark, 69–90. Chichester: Wiley.

———. 1996. *Rockglaciers.* Berlin: Springer.

Brenning, A. 2003. La importancia de los glaciares de escombros en los sistemas geomorfológico e hidrológico de la Cordillera de Santiago: Fundamentos y primeros resultados. *Revista de Geografía Norte Grande* 30:7–22.

———. 2005*a*. Geomorphological, hydrological, and climatic significance of rock glaciers in the Andes of Central Chile (33–35°S). *Permafrost and Periglacial Processes* 16:231–40.

———. 2005*b*. Climatic and geomorphological controls of rock glaciers in the Andes of Central Chile: Combining statistical modelling and field mapping. Ph.D. diss., Humboldt-Universität zu Berlin.

Burger, K. C., J. J. Degenhardt, and J. R. Giardino. 1999. Engineering geomorphology of rock glaciers. *Geomorphology* 31:93–132.

Carruthers, D. 2001. Environmental politics in Chile: Legacies of dictatorship and democracy. *Third World Quarterly* 22:343–58.

Casassa, G., and C. Marangunic. 1993. The 1987 Rio Colorado rockslide and debris flow, Central Andes, Chile. *Bulletin of the Association of Engineering Geologists* 30:321–30.

CCAEC (Canada-Chile Agreement on Environmental Cooperation). 1996. *Environmental management in Chile.* http://canchil.gc.ca/English/Resource/Reports/ChileEM/ ChileEM_Index.cfm (accessed November 11, 2004).

Contreras, A., and J. L. Illanes. 1992. Depósito de lastre glaciar Infiernillo Sur Mina Los Bronces. In *43a Convención del Instituto de Ingenieros de Minas de Chile, La Serena, October 1992.* Santiago: Instituto de Ingenieros de Minas de Chile.

COREMA Atacama (Comisión Regional del Medio Ambiente, Región de Atacama). 2006. Resolución de calificación ambiental, studio de impacto ambiental modificaciones Proyecto Pascua-Lama. Resolución Exenta 024/2006, February 15. Copiapó, Chile.

Corripio, J. G., and R. S. Purves. 2003. The influence of penitentes on the energy balance of high-altitude glaciers in the dry Central Andes. *Geophysical Research Abstracts* 5:12738.

Corte, A. E. 1976. The hydrological significance of rock glaciers. *Journal of Glaciology* 17:157–58.

Diolaiuti, G., C. D'Agata, M. Pavan, G. Vassena, C. Lanzi, M. Pinoli, M. Pelfini, M. Pecci, and C. Smiraglia. 2001. The physical evolution of and the anthropic impact on a glacier subjected to a high influx of tourists: Vedretta Piana Glacier (Italian Alps). *Geografia Fisica e Dinamica Quaternaria* 24:199–201.

Dourojeanni, A., and A. Jouravlev. 1999. *El Código de Aguas de Chile: Entre la ideología y la realidad.* Serie Recursos Naturales e Infraestructura 3. Santiago: United Nations.

Editec. 2000. *Compendio de la minería chilena.* Santiago.

EPA (Environmental Protection Agency). 2001. *Abandoned mine site characterization and cleanup handbook.* EPA 530-C-01-001. Seattle.

Fisch, W., Sr., W. Fisch, Jr., and W. Haeberli. 1977. Electrical D.C. resistivity soundings with long profiles on rock glaciers and moraines in the Alps of Switzerland. *Zeitschrift für Gletscherkunde und Glazialgeologie* 13:239–60.

Geotécnica Consultores. 1996. *Estudio de impacto ambiental Compañía Minera Disputada de Las Condes, Proyecto de Epansión–2, mina Los Bronces.* Santiago.

———. 2001. *Estudio de impacto ambiental Proyecto Pascua-Lama.* Santiago.

Giardino, J.R., and S.G. Vick. 1985. Engineering hazards of rock glaciers. *Bulletin of the Association of Engineering Geologists* 22:201–16.

———. 1987. Geologic engineering aspects of rock glaciers. In *Rock glaciers,* ed. J.R. Giardino, J.F. Shroder, and J.J.D. Vitek, 265–87. Boston: Allen and Unwin.

González-Ferrán, O. 1994. *Volcanes de Chile.* Santiago: Instituto Geográfico Militar.

Grebenets, V.I., A.G. Kerimov, and D.S. Bakcheev. 1997. Dangerous movements of technogenic rock glaciers, Norilsk, Russia. In *Engineering geology and the environment,* ed. P.G. Marinos, G.C. Koukis, G.C. Tsiambaos, and G.C. Stournaras, 689–92. Rotterdam: Balkema.

Haeberli, W. 1992. Construction, environmental problems, and natural hazards in periglacial mountain belts. *Permafrost and Periglacial Processes* 3:111–24.

Haeberli, W., D. Brandova, C. Burga, M. Egli, R. Frauenfelder, A. Kääb, M. Maisch, B. Mauz, and R. Dikau. 2003. Methods for absolute and relative age dating of rockglacier surfaces in alpine permafrost. In *Permafrost: Proceedings of the Eighth International Conference on Permafrost, July 2003, Zürich, Switzerland,* ed. M. Phillips, S. Springman, and L. Arenson, 343–48. Lisse: Balkema.

Haeberli, W., A. Kääb, S. Wagner, D. Vonder Mühll, P. Geissler, J.N. Haas, H. Glatzel-Mattheier, and D. Wagenbach. 1999. Pollen analysis and ^{14}C age of moss remains in a permafrost core recovered from the active rock glacier Murtèl-Corvatsch, Swiss Alps: Geomorphological and glaciological implications. *Journal of Glaciology* 45:1–8.

Haeberli, W., and H.R. Keusen. 1983. Site investigation and foundation design aspects of cable car construction in alpine permafrost at the "Chli Matterhorn," Wallis, Swiss Alps. In *Proceedings of the Fourth International Conference on Permafrost, Fairbanks, Alaska,* 601–5. Washington, DC: National Academic Press.

Haimayer, P. 1989. Glacier-skiing areas in Austria: A socio-political perspective. *Mountain Research and Development* 9:51–58.

Hoelzle, M., S. Wagner, A. Kääb, and D. Vonder Mühll. 1998. Surface movement and internal deformation of ice-rock mixtures within rock glaciers at Pontresina-Schafberg, Upper Engadin, Switzerland. In *Proceedings of the Seventh International Conference on Permafrost, June 1998, Yellowknife, Canada,* ed. A.G. Lewkowicz and M. Allard, 465–71. Nordicana 57. Quebec City: Centre d'Études Nordiques.

Holmgren, C., and I. Vela. 1991. Visita a la mina Los Bronces. In *Actas Congreso Geológico Chileno 6, Guía de Excursión* IC–5.

Homeniuk, L. 2000. Kumtor Gold Project, Kyrgyz Republic. *CIM Bulletin* 93 (1038):67–73.

Jurt, C. 2004. Perceptions of glacier retreat within an alpine municipality in South Tyrol. *Abstracts of Oral Presentations of the Wengen Workshop "Mountain Glaciers and Society," Wengen, Switzerland, October 6–8, 2004.* Fribourg: University of Fribourg.

Kull, C., M. Grosjean, and H. Veit. 2002. Modeling modern and Late Pleistocene glacio-climatological conditions in the North Chilean Andes (29–30°S). *Climatic Change* 52:359–81.

Lagos, G., H. Blanco, V. Torres, and B. Bustos. 2002. Minería, minerales y desarrollo sustentable en Chile. In *Minería y minerales de América del Sur en la transición hacia el desarrollo sustentable,* ed. Mining, Minerals and Sustainable Development Project, South America Work Team. Santiago: CIPMA/IDRC/IIPM.

Lliboutry, L. 1956. *Nieves y glaciares de Chile: Fundamentos de glaciología.* Santiago: Editorial Universitaria.

———. 1961. Phénomènes cryoniveaux dans les Andes de Santiago (Chili). *Biuletyn Peryglacjalny* 10:209–24.

———. 1986. Rock glaciers in the dry Andes. *Data of Glaciological Studies (MGI)* 58:18–24; 139–44.

Minería Chilena. 1993. Disputada de Las Condes: De los capachos a una faena que se proyecta al nuevo siglo. 145:43–7.

———. 2005*a*. Pascua-Lama: Chile movilizado por el traslado de 3 glaciares. http://www.mch.cl/(accessed April 19, 2005).

———. 2005*b*. Proyecto de Oro: Glaciares y contaminación. http://www.mch.cl/(accessed April 28, 2005).

———. 2005*c*. Puesta en marcha para el año 2012: División Andina desarrolla primeras fases de proyecto a 230.000 tpd. http://www.mch.cl/(accessed August 31, 2005).

MiningWatch Canada. 2005. Barrick Gold faces determined opposition at Pascual-Lama and Veladero. *MiningWatch Canada Newsletter* 20. http://www.miningwatch.ca/ (accessed December 23, 2005).

Moussa, N. 1999. *El desarrollo de la minería del cobre en la segunda mitad del siglo XX.* Serie Recursos Naturales e Infraestructura 4. Santiago: United Nations.

Padilla, A. 1996. Participación ciudadana en el Proyecto Gasoducto GasAndes: Poniendo a prueba el desarrollo sustentable. *Ambiente y Desarrollo* 12:7–16.

Pizarro, R., and P. Vasconi. 2004. *Una nueva institucionalidad ambiental para Chile.* Análisis de Políticas Públicas 26.

Ripley, E.A., R.E. Redman, and A.A. Crowder. 1995. *Environmental effects of mining.* Delray Beach, FL: St. Lucie Press.

Roer, I. 2003. Rock glacier kinematics in the Turtmanntal, Valais, Switzerland: Observational concept, first results, and research perspectives. In *Permafrost: Proceedings of the Eighth International Conference on Permafrost, July 2003, Zürich, Switzerland,* ed. M. Phillips, S. Springman, and L. Arenson, 971–75. Lisse: Balkema.

Sabatini, F., and C. Sepúlveda. 1996. Lecciones del conflicto del gasoducto en el Cajón del Maipo: Negociación ambiental, participación y sustentabilidad. *Ambiente y Desarrollo* 12:19–24.

Schrott, L. 1996. Some geomorphological–hydrological aspects of rock glaciers in the Andes (San Juan, Argentina). *Zeitschrift für Geomorphologie* 104, suppl.:161–73.

Schwegler, D. 2004. Weltnaturerbe unter Beschuss. *Der Schweizerische Beobachter* 21/2004:33–35.

Whalley, W.B., C.F. Palmer, S.J. Hamilton, and J.E. Gordon. 1994. Ice exposures in rock glaciers. *Journal of Glaciology* 40:427.

15

Glacier Changes and Their Impacts on Mountain Tourism

TWO CASE STUDIES FROM THE ITALIAN ALPS

Claudio Smiraglia, Guglielmina Diolaiuti, Manuela Pelfini, Marco Belò, Michele Citterio, Teresa Carnielli, and Carlo D'Agata

High Alpine areas have been seeing more tourists in the past few decades, in part because of the increased infrastructure and transport now available at higher elevations. Glacial and periglacial areas are becoming foci of attraction for tourists during the summer months, and climbing, mountaineering, and trekking are becoming more and more popular, attracting masses of sports-minded visitors. At the same time, accelerated glacier retreat linked to changing climate (Haeberli and Beniston 1998) is presenting more severe and more frequent hazards to mountain tourism.

The main types of hazards in mountains have been discussed (Beniston 2000; Huggel, Kääb, and Salzmann 2004; Haeberli 2005; Glaciorisk EU 2001–3), but less attention has been paid to the hazards associated with the use of Alpine glaciers for summer skiing and with the formation of glacier lakes. The importance of these problems lies in the large number of people attracted to these sites and the intense effects on such sites of ongoing glacier retreat. The conservation of the landscape is also of fundamental importance, making it necessary to evaluate risks (for management and mitigation) and adopt strategies to reduce the pollution and environmental damage caused by tourism.

To address these topics, we have selected two case studies in the Italian Alps: a glacier used for summer skiing since the first half of the last century and an ice-contact lake visited by sightseers for over two centuries. In both cases, our selection was based on a tradition of visitor use, as this criterion made it possible to analyze the types of risk and evaluate the tourist impacts over time. The first study site is the Vedretta Piana Glacier (Ortles-Cevedale Group, Italian Alps), the glacier most used for summer skiing in Italy. The glacier is highly affected by ongoing retreat and is used by thousands of skiers every year (a yearly average of about 170,000 trips on the main Stelvio cable-car system was registered in the period 1981–2001). The second is Miage Lake (Mont Blanc Massif, Italian Alps), an ice-contact lake characterized by active calving and memorable emptying episodes that has been described in the literature for over 200 years.

TOURIST USE OF AN ALPINE GLACIER

Although Alpine glaciers have been popular sites for recreational activities in the past, the use of glaciers for summer skiing is a relatively recent phenomenon. The Stelvio glaciers (Stelvio Pass, 2,757 m a.s.l.) and particularly Vedretta Piana are among the best-known sites for summer skiing in Europe. However, the current phase of glacier reduction has considerably reduced the skiable area over time, while economic and social changes have influenced usage patterns. When the Imperial Road of the Stelvio was built in 1825, new trade routes opened up between the Tyrol and Lombardy, and a series of exploratory ascents were made. This led to greater human use of the area, the beginning of ski tourism, and the construction of the first Alpine huts, interrupted only by World War I. Skiing resumed in the 1930s with a gradual development of infrastructure and changes in attitude toward the glacier. The whole Alpine area witnessed a change from the respect for mountains and their hazards typical of the early period of mountaineering to a more superficial approach. Visitors could easily reach an altitude of 3,000 m, where they often found themselves in an unfamiliar environment that was full of risks. Even now, many summer skiers on glaciers treat the experience as if it were the same as winter skiing, overlooking the fact that the glacier mass is dynamic and has crevasses in summer.

Vedretta Piana experienced fluctuations similar to those of other Alpine glaciers in the second half of the twentieth century. Images from the early twentieth century show that the glacier front had a bulging appearance and was visible from the road leading to the pass. Along with most Alpine glaciers, it had experienced a marked advance as a result of the brief but widespread reduction in temperatures in that period. Subsequently, the glacier receded throughout the twentieth century except for a brief period of advance between 1970 and the end of the 1980s associated with a series of cool summers and winters with abundant snowfall. Accelerated recession resumed in the early 1990s. The area of the glacier has changed from 1.7 km^2 at the end of the nineteenth century (Richter 1888) to 0.60 km^2 in 2001 (Diolaiuti et al. 2006*a*), a loss of 64% in the past hundred years.

TABLE 15.1
Vedretta Piana Skiable Area, 1965–2002

YEARS	SKIABLE SURFACE AREA (KM^2)
1965	0.90
1972	0.75
1985	0.68
1998	0.41
2002	0.38

LONG-TERM VARIATIONS OF THE SKIABLE AREA

The skiable surface area from the construction of the earliest lift facilities to the present was determined from historical maps. The skiable areas were mapped on the basis of the positions of lift facilities and the respective downhill runs in 1965, 1972, and 1985 and direct observation in 1998 and 2002 (Table 15.1). The situation in 1965, when the first real lift facilities were set up, suggests that the skiable area was very extensive. Historical photographs show ample zones for skiing and gentle slopes, especially in the upper part of the glacier. However, the number of lifts was limited because of the novelty of the technology. By 1998, because of the considerable reduction in the skiable surface area, two separate areas were clearly distinguishable. The boundaries of the two zones were marked by crevasses that became more evident during the ablation period. During this period, the snow cover (particularly at the end of the summer season) was too limited to prevent skiers from falling into the crevasses, and the crevasses had to be artificially filled by transporting snow from the glacier accumulation basin to the ablation zone. In addition, to get from one ski run to another skiers had to follow trails in the ablation zone, trails also made by transporting snow from the accumulation basin. The skiable surface area was restricted to the upper part of the glacier because several lift facilities had been declared unfit for use and shut down. The skiable surface calculated on the basis of the existing lift systems must take into account seasonal variations: facilities that are open at the beginning of the season may be closed later because of weather conditions and then reopened when the snow cover provides safer skiing conditions.

SKIABLE AREA: SEASONAL VARIABILITY

Field investigations and surveys during the summers of 2000 and 2001 evaluated the short-term relations between glacier dynamics, variations in the skiable surface area, and the presence of tourists. The results show a strong correlation between the variation of these parameters and seasonal meteorological evolution. These two summer seasons were markedly different: The 2000–2001 winter season was exceptional, with abundant snowfall (57% more than in the winter of 1999–2000) that remained through the summer because of cool daytime temperatures and overcast skies for most of the ablation season. The average 2001 summer temperature (June–October) was +3.3 °C. In contrast, the summer of 2000 was characterized by higher air temperatures (average +5.7 °C) that immediately reduced the scanty residual winter blanket of snow. The minimal area for summer skiing was 393,000 m^2 at the end of August 2000 and 502,000 m^2 in mid-August 2001 (Diolaiuti et al. 2006*a*). During the summer of 2000, there was marked reworking of the pistes in the ski district, and the melting of the snow led to continuous redesigning of the downhill ski runs. The number of trips on the Stelvio lifts was 835,955 in 2000 and 1,013,879 in 2001, a 21% increase. This is particularly interesting considering that over the previous five years the number of trips had steadily decreased. The reversal in trend is thus attributable to the favorable conditions for skiing in 2001, surely influenced by the snowfall in the previous winter and spring.

THE DEVELOPMENT OF TOURISM ON THE GLACIER

At present, the accommodation facilities, which experienced a peak of economic development in the 1970s, have excess capacity. The influx of tourists to the glacier has seen a marked reduction in recent summers for interrelated social, economic, and climatic-environmental reasons. The current glacier shrinkage phase, with decreased amount and duration of snow cover, has limited options for skiing, and changes in popular tastes, together with economic factors (adding the cost of ski equipment and lift passes to the cost of hotels, etc.), have come into play. Moreover, changes in the type of tourist interests have also taken place over time. In the early decades of the twentieth century, mountain enthusiasts and climbers would come for seven-day "skiing vacations" at Stelvio Pass, and the summer skiing schools at the pass alternated downhill skiing with glacier traverses and high-altitude trekking. Today the average tourist is exclusively a downhill skier and is often uninformed about the glacier environment and unprepared for or simply uninterested in mountaineering. These tourists (mainly from Italy but recently also from other European countries and from the United States and Japan) are often willing to devote no more than two or three days to summer skiing.

RISK AND SAFETY

Safety and risk are important issues for Vedretta Piana and, generally speaking, for all the glacier surfaces in the Stelvio area used for summer skiing. In fact, because all the lifts are installed on the glacier surface there is a greater risk for skiers than in winter. According to statistics on accidents on Stelvio glacier ski runs provided by the Moena Police Department's Skiers and Mountain Emergency Detachment Unit, there

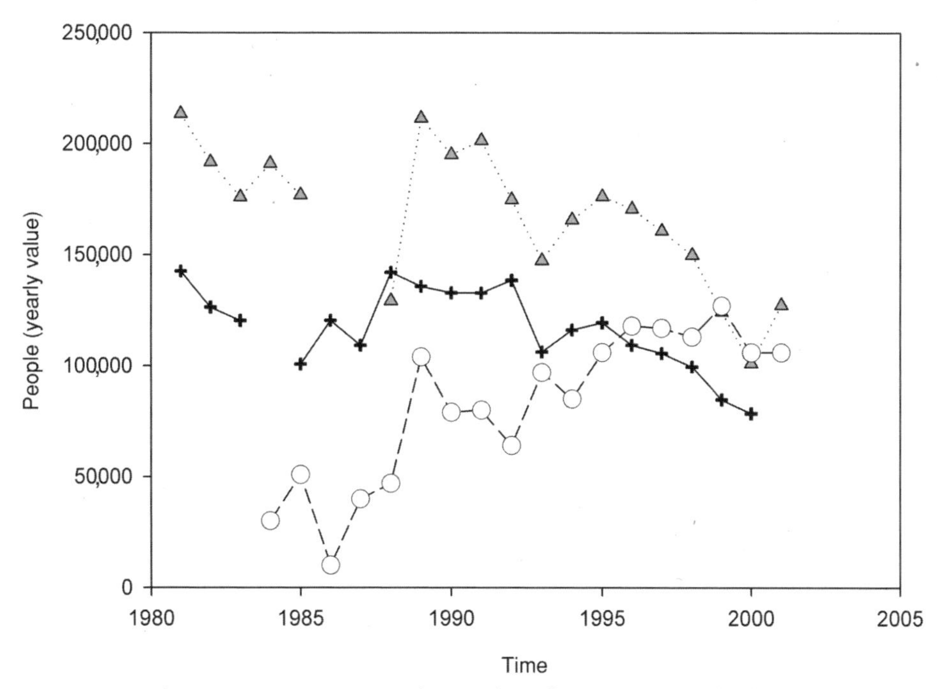

FIGURE 15.1. Vedretta Piana rescue operations (*dots*), numbers of visitors (*crosses*), and cable-car trips (*triangles*) from 1983 to 2001.

was a clear rise in the number of rescue operations from 1983 to 2000 (Figure 15. 1). Data provided by the Bormio Tourist Promotion Office and the Tourist Association of Gomagoi show a negative trend in the number of visitors to Stelvio Pass in the past 19 years, reaching a minimum of 78,504 in 2000. The number of trips on the main Stelvio cable-car system from 1981 to 2001 generally decreased, reaching a minimum of 101,074 in 2000. What is immediately noticeable is the great difference in the number of trips compared with the number of hotel guests. This difference may reflect the changing trends mentioned previously, highlighting the appearance of new sports-minded visitors. A comparison of the trip data with the data on emergency operations revealed that a decrease in cable-car trips corresponded with an increase in the number of emergency operations, which were never fewer than 100 per year. In 17 years, the number of such operations increased fourfold while the number of trips decreased by 50%. This trend is underlined by the data on the risk of injury per 1,000 trips (number of emergency operations/number of trips), pointing to a clear increase from 1984 to 2000 (Figure 15.2).

There may be a number of reasons for the increase in persons requiring emergency assistance on the Vedretta Piana and Stelvio glaciers. Although data on the types of emergencies are not available, it should be considered that skiing on a glacier involves exposure to risks in addition to the usual types of skiing accidents, especially accidental falls into crevasses and moulins. The opening of crevasses depends upon the snow cover and the glacier's flow rate and is particularly evident during the ablation season. The risk of falling into a crevasse thus increases in August and early September, when the snow cover is scanty. Moreover, this risk is strongly interrelated with the maintenance of the facilities, the reporting and marking of crevasse locations, and weather conditions (e.g., fog, storms, overcast skies, snowfall, and rain).

In recent years Vedretta Piana has been characterized by large portions of exposed ice, revealing multiple crevasses and debris-covered

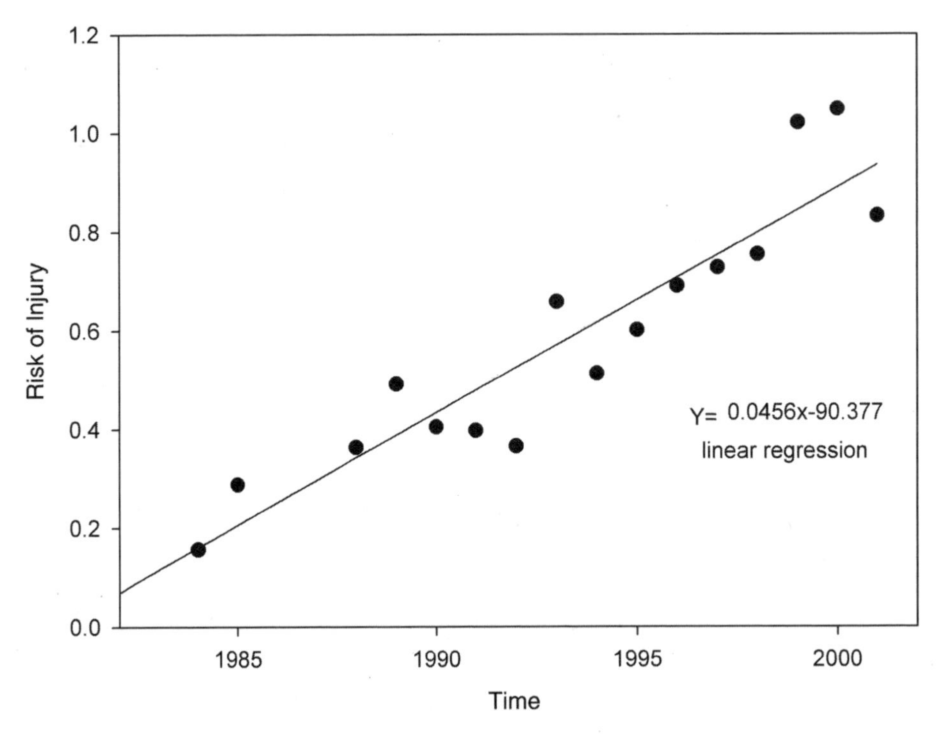

FIGURE 15.2. Risk of injury per 1,000 visits on Vedretta Piana Glacier over time.

zones, at the end of the summer season. During our survey work, observations and measurements were also made on the conditions and surface morphology of the glacier. The 2000 fieldwork, focused largely on the presence of crevasses, revealed a dense presence of crevasses in the summer, when 55 were observed. Skiing is permitted under such conditions only in delimited areas where snow is brought in artificially from the accumulation zone. In periods of intense melting and a higher snow line (particularly in August and early September), the scanty snow cover forces skiers to follow marked trails to bypass the most hazardous points along the ski run, where crevasses have opened up. Changes in the density of the snow, which is frozen in the morning and slushy in the afternoon, also increase the risk for those who are not expert skiers. This reduction of the skiable area forces skiers to maneuver within a very limited space, requiring a good command of the ski equipment and good skiing skills. Moreover, the reduction of the skiable surface area leads to overcrowding, which sometimes results in collisions involving lift structures and between skiers themselves.

Several types of maintenance work on the pistes mitigate the synergic effect of all these risk factors, in particular the transfer of firn from the upper accumulation basin to the lower ablation zone. The aim of these repairs is to make the skiable surface larger and longer by filling in the crevasses and covering the debris-covered ice that are the main causes of skier accidents. It must be emphasized that, while pushing and bulldozing firn and snow downslope to increase skiable surface allows for more skiing in the short term, it hastens glacier shrinkage in the medium term by reducing the supply of the accumulation basin.

Legal title to the areas that the tourists visit and ownership of the main hotel, the lifts, and the other infrastructure are held by small local and regional private firms, banks in the immediate region, the local alpine club, and the local cable-car company. It devolves on this group to set

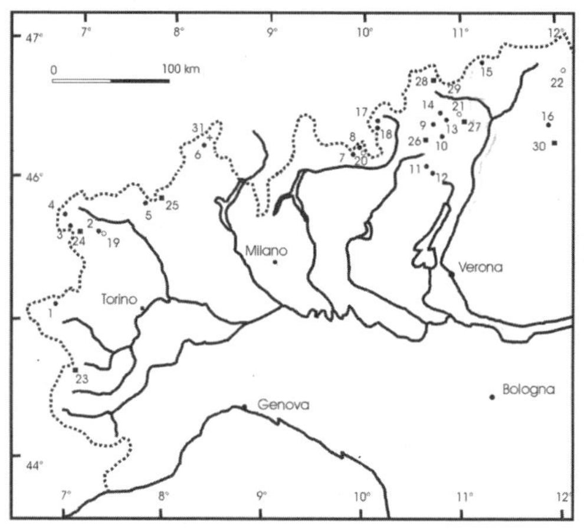

FIGURE 15.3. Lake Miage (*right*) in summer 2003 and location map of the calving glaciers surveyed in Italy. The symbols indicate active (*solid circles*), inactive (*solid triangles*), uncertain activity (*outlined circles*), and artificial calving (*crosses*). Calving site 4 corresponds to Miage Lake. (Photo by G. Diolaiuti.)

the regulations for access to these areas, but the technical decisions about glacier management and security are left to the cable-car company.

ALPINE ICE-CONTACT LAKES

Calving glaciers, a consequence of ongoing glacier shrinkage, are becoming more widespread in the Alps. The retreat of Alpine glacier termini, which has accelerated in recent decades (Haeberli et al. 1999), frees new deglaciated areas, and the presence of moraine ridges (in particular the recessional moraines formed during the last small advance phase of the second half of the twentieth century) produces favorable conditions for the formation of water ponds or small ice-contact lakes in which calving takes place. In the World Glacier Inventory database compiled from data sources at the end of the 1980s, only four Italian glaciers are listed as calving glaciers, but a current list of Italian Alpine glaciers where calving processes have been surveyed includes at least 20 (Figure 15.3). This increase reveals the increasingly wide distribution of the phenomenon due to the formation of more and more ice-contact lakes. Most of these lakes disappear after a few years or become proglacial (the ice front no longer terminates at the lake), and therefore the life span of active calving is generally limited. Calving persists longer where the ice is covered by thick debris that reduces ice ablation (Østrem 1959; Fujii and Higuchi 1977; Mattson, Gardner, and Young 1993; Rana et al. 1997) and stabilizes the position of the glacier terminus (Deline 1999; Thomson, Kirkbride, and Brock 2000; Diolaiuti, D'Agata, and Smiraglia 2003; D'Agata et al. 2005).

The impact of calving on Alpine glacier mass balances and volume reduction is usually less important than that of surface melting (Diolaiuti et al. 2006*b*). Nonetheless, in some cases the temperate ice (i.e., ice at melting pressure) of Alpine glaciers has increased the calving rate and triggered accelerated deglaciation or even catastrophic retreat (e.g., Trift Glacier in the Gadmental Valley in the Bernese Alps of Switzerland [Funk 2004]). Calving sites and icebergs are significant attractions for tourists and the small glaciological community, and until very recently they were observable only in polar regions. Alpine glaciers that are actively calving are visited by increasing numbers, particularly if they are easily accessible and if iceberg production is abundant and frequent. However, calving can create a serious hazard through the propagation of large waves if people are allowed to get too close to them. Eleven sightseers were injured by a wave generated by the calving of Miage Glacier (Mont Blanc Massif, Italy) in the summer of 1996.

MIAGE LAKE

To examine the relationships between calving glaciers and potential hazards to tourists and to suggest viable approaches to risk mitigation, we selected a study site with fully developed calving phenomena and a long history of tourist visitation. The oldest known Italian calving site is Miage Lake (Mont Blanc Massif, Italy), located at an altitude of 2,020 m a.s.l. (Figure 15.3). Historical observations, iconographic documents, and field surveys permit a reconstruction of the lake's history and a description of the active calving processes and their relations with visitors over the past 200 years (Diolaiuti et al. 2005).

Miage Lake is located on the south side of the debris-covered Miage Glacier, the third-largest glacier in Italy (11 km^2). Many tourists, trekkers, and climbers visit the glacier every year because it is located on the standard Italian trail to the Mont Blanc summit and because of its striking natural and glaciological features. Its scientific and historical prominence is the basis for a pending application to UNESCO to place the Miage Glacier on its Geosites list of international environmental assets. Every summer hundreds of tourists stand on the little beach close to the ice-contact zone to admire the breaking ice and iceberg production, sometimes with unfortunate consequences. The lake is also famous for its periodic emptying episodes, which have occurred several times in the past century for reasons that are not entirely clear. This phenomenon creates an intriguing landscape that attracts visitors but presents them with additional hazards.

CALVING AND EMPTYING EPISODES

Surveys performed at Miage Lake during the summers of 2002–05 and winter 2004 allowed identification of the main calving features and analysis of the processes driving iceberg production. Observations on calving and on the influence of the debris cover on the ice-contact zone were also carried out (Diolaiuti et al. 2005). The Miage ice cliff exhibits annual cycles of summer retreat and winter advance. Calving occurs only during the summer melt, and daily activity peaks in the early afternoon, coinciding with the daily melt maximum, with a discharge of large quantities of meltwater into crevasses and into the subglacial drainage system.

At Miage Lake, the peak hours of the day for maximum iceberg production are also the peak hours for the maximum influx of tourists. During the summer, dozens of sightseers flock daily to the lakeshore and stand as close as possible to the ice cliff to admire the breaking away of the ice, which occurs several times a day during the hottest periods. The mean daily ice loss through calving in the summer of 2004 was about 1,900 m^3, making the seasonal volume discharge into the lake about 141,000 m^3. This discharge corresponds to approximately 40% of the lake's total water volume (Diolaiuti et al. 2006*b*). In July 1996, about 10,000 m^3 of ice broke away in a single block; falling into the lake, it created a 5-m-high wave.

Miage Lake is also famous for the emptying episodes that have occurred with varying frequency in its history (Giardino, Mortara, and Bonetto 2001; Deline et al. 2004). The last emptying occurred in the summer of 2004, permitting direct observation of the lake bed and revealing an ice floor, debris-covered and deeply crevassed, that was contiguous with the glacier ice at the ice cliff contact zone. It extended 100 m out from the cliff, almost reaching the two well-preserved moraine ridges that separate the lake into two parts, one with and one without direct contact with the glacier ice. Mapping of the lake bed following the 2004 drainage event indicated that ca. 287.5×10^3 m^3 of water, approximately 80% of the normal lake volume (340×10^3 m^3), was discharged from the lake between September 4 and September 8 (Diolaiuti et al. 2006*b*).

The fieldwork we performed between 2002 and 2005 revealed that calving was more frequent and intense during the summer of 2003, the warmest summer of the past few decades. That summer the quantity of ice discharged by the glacier through calving amounted to about 51% of the lake's total volume.

HAZARDS

The presence and evolution of an active calving cliff are closely interconnected with a number of hazardous phenomena such as hanging cliff collapses, supraglacial debris falling into the lake, and waves created by this debris. These processes can be expected to intensify and spread in the coming years because of the general tendency of Alpine glaciers to become more and more debris-covered and to create new ice-contact lakes in which calving takes place. Analysis of the risk issue must acknowledge the highly variable intensity and unpredictability of these three main sources of danger. The observed magnitude of ice and rock collapses ranges from negligible releases of ice and pebble-sized rockfalls to major cliff sector collapses and boulders tumbling from the top of the cliff. Given the often shallow, irregular nature of these lakes and shorelines, wave heights should also be expected to be highly variable. These hazards have a higher seasonal probability of occurrence when the calving rate reaches its maximum and during the warmest hours of the day. Rising water level in the lake can also promote calving, while lowering of the water level has been observed to reduce it.

Visitors to the site may be divided into three general classes: uninformed occasional visitors (tourists), informed occasional visitors with educational or other experience (researchers, locals, and skilled alpinists), and specialists with detailed knowledge of this environment. Experienced visitors are generally aware of the hazards, but inexperienced visitors and overconfident researchers may unwittingly expose themselves to them. Unfortunately, reports of accidents in the Alps have become common in recent years, and experience has shown how difficult it can be to dissuade people from approaching the calving cliff and even sunbathing there. In the event of a serious accident involving visitors, there may be impacts on economic activities at the mountain resorts, whose efforts to attract summer tourism make them the more exposed to risk situations. Their vulnerability will depend on the effort they make to safeguard their customers by providing them accurate scientific information.

The risk associated with all these elements is likely to increase steadily, given the present increase in the number of calving glaciers in the Alps (Diolaiuti et al. 2006*b*), the growing interest in the natural mountain environment on the part of the public, and the exotic images that the very word "iceberg" can evoke in people living at temperate latitudes. Additional risk factors come into play in the event of emptying episodes involving ice-contact lakes, because emptying makes the most hazardous area at the bottom of the ice cliff more accessible and increases the risk from ice and rockfalls accordingly.

The risk related to calving hazards in the Alps is real, and it will increase with increasing exposure to the hazards. This risk is characterized by a highly variable probability of occurrence in time and space conditioned by the calving rate, the water level, meteorological conditions, and lakeshore and lake-bed morphology. It is therefore not readily quantifiable without detailed on-site study tailored to the particular glacier examined. Any direct control of the sources of danger would require undesirable intervention aimed at eliminating otherwise precious natural resources and/or complete closure of the area to visitors. Excluding these extremes, we believe that the risk can be managed by reducing the vulnerability of visitors to the area. This may be achieved in two ways:

First, better education for uninformed occasional visitors can be provided by guides and by signs posted along the route that identify and encourage a safer approach to the hazardous areas of this environment. Second, access to dangerous areas should be prohibited. We are now in the process of developing remote-controlled platforms to perform instrumental investigations in these areas.

Clearly, there is a need for cooperation with the local authorities: the Commune of Courmayeur, which has the responsibility for safety in these areas, and the Aosta Valley District, which is capable of supporting research aimed at detecting environmental hazards and risks. The aim of this cooperation (which is currently ongoing) is to increase their awareness of the problem and to get more information about the ice cliff and the lake evolution and the effects of lake-bed morphology, shore profiles, and expected wave heights. The final products will be a risk map defining the off-limits areas and the creation of a path to guide visitors away from those areas. These measures will be updated as the glacier and the lake evolve.

CONCLUSIONS

These two case studies demonstrate the changing nature of glacier-related hazards in the Italian Alps and some of the implications for tourists and tourism. The increase in the number of ice-contact lakes and sites affected by calving phenomena increases the risks to visitors to these sites. Calving and lake-emptying episodes attract visitor interest, with potentially positive impacts for the local economy in the mountain areas where they occur, but they lead to new types of hazards and risks. The safety measures to be implemented by the local and regional authorities have to encompass all of the effects of these events. They should be aimed at limiting the influx of tourists and marking some sightseeing points from which it is possible to enjoy the view with the least risk (e.g., with trail maps indicating hazards and risk levels).

The ongoing climate change is also causing a strong reduction in skiable glacier areas and increasing the risks for summer skiers. Visits to the Vedretta Piana Glacier have decreased since the 1980s for socioeconomic and environmental reasons. After a brief cool and snowy period, with snow lines at lower altitudes and positive glacier balances, intense deglaciation has resumed (Patzelt 1985; Wood 1988: Haeberli et al. 1999). There is increased risk to skiers despite an almost steady reduction in the number of trips on lift facilities.

Finally, it can be predicted from the current thickness of the glacier (about 70 m) and recent mass balance data of about −2 m per year (Diolaiuti et al. 2001, 2006*b*) that the Vedretta Piana Glacier may disappear altogether in about 35 years. This crude estimate neglects any drastic changes in climatic conditions, and obviously, a sharp reduction of the accumulation area, an increase in the supraglacial debris cover, and a widening of the crevassed areas will come before the actual extinction of the glacier, making summer skiing more and more difficult. This ominous forecast will apply if climate conditions follow the trends reported by the Intergovernmental Panel on Climate Change (Houghton et al. 2001), although a brief cold phase such as that of the period 1965–85, which contributed to a replenishment of glacier accumulation basins (Wood 1988), would make summer skiing on Vedretta Piana possible longer.

ACKNOWLEDGMENTS

This research was supported by the Italian Ministry of University and Research in the framework of the 2003 MIUR-COFIN Project (under national director P. R. Federici and local director C. Smiraglia) "The Ongoing Transformation of Alpine Italian Glaciers" and of the 2004 MIUR-COFIN Project (under national director M. Panizza and local director M. Pelfini) "The Physical Landscape in the Alpine Environment: A Resource for Sustainable Tourism, Risks and Impacts Consequent to Its Use." We thank the Regione Valle d'Aosta, the Commune of Courmayeur, and the Fondazione Montagna Sicura in Courmayeur for logistic assistance.

REFERENCES CITED

Beniston, M. 2000. *Environmental change in mountains and uplands*. London: Arnold.

D'Agata, C., A. Zanutta, F. Mancini, and C. Smiraglia. 2005. The recent variations of a debris covered glacier (Brenva Glacier) in the Italian Alps monitored with the comparisons of maps and digital orthophotos. *Journal of Glaciology* 52:183–85.

Deline, P. 1999. Le variations Holocènes récentes du Glacier du Miage (Val Veny, Val d'Aoste). *Quaternaire* 10:5–13.

Deline, P., G. Diolaiuti, M. P. Kirkbride, G. Mortara, M. Pavan, C. Smiraglia, and A. Tamburini. 2004. Drainage of ice-contact Miage Lake (Mont Blanc Massif, Italy) in September 2004. *Geografia Fisica e Dinamica Quaternaria* 27:113–20.

Diolaiuti, G., M. Citterio, T. Carnielli, C. D'Agata, M. Kirkbride, and C. Smiraglia. 2006*b*. Rates, processes, and morphology of fresh-water calving at Miage Glacier (Italian Alps). *Hydrological Processes* 20:2233–44.

Diolaiuti, G., C. D'Agata, M. Pavan, G. Vassena, C. Lanzi, M. Pinoli, M. Pelfini, M. Pecci, and C. Smiraglia. 2001. The physical evolution and the anthropic impact on a glacier subjected to a high influx of tourists: Vedretta Piana Glacier (Italian Alps). *Geografia Fisica e Dinamica Quaternaria* 24:199–201.

Diolaiuti, G., C. D'Agata, and C. Smiraglia. 2003. Variations in Belvedere Glacier (Monte Rosa, Italian Alps) tongue thickness and volume in the second half of the 20th century. *Arctic, Antarctic, and Alpine Research* 35:255–63.

Diolaiuti, G., M.P. Kirkbride, C. Smiraglia, D.I. Benn, and L. Nicholson. 2005. Calving processes and lake evolution at Miage Glacier (Mont Blanc, Italian Alps). *Annals of Glaciology* 41:207–14.

Diolaiuti, C., C. Smiraglia, M. Pelfini, M. Belò, M. Pavan, and G. Vassena. 2006*a*. The recent evolution of an Alpine glacier used for summer skiing (Vedretta Piana, Stelvio Pass, Italy). *Cold Regions Science and Technology* 44:206–16.

Fujii, Y., and K. Higuchi. 1977. Statistical analysis of the forms of the glaciers in Khumbu Himal. *Journal Japan Society Snow Ice (Seppyo)* 39:7–14.

Funk, M. 2004. Glaciological and hydraulic investigations in connection with a recently formed proglacial lake. In *Abstracts of Oral Presentations of "Mountain Glaciers and Society: Perception, Science, Impacts and Policy," Wengen, Switzerland, October 6–8, 2004*. Fribourg: University of Fribourg.

Giardino, M., G. Mortara, and F. Bonetto. 2001. Proposta per la realizzazione di un catasto aerofotografico dei ghiacciai italiani. *Geografia Fisica e Dinamica Quaternaria* suppl. 5:89–98.

Glaciorisk EU Project. 2001–2003. Didier Richard, CEMA GREF. http://glaciorisk.grenoble.cemagref.fr/projectglaciorisk.htm (accessed August 2007).

Haeberli, W. 2005. Climate change and glacial/periglacial geomorphodynamics in the Alps: A challenge of historical dimensions. *Geografia Fisica e Dinamica Quaternaria* suppl. 7:9–14.

Haeberli, W., and M. Beniston. 1998. Climate change and its impact on glaciers and permafrost in the Alps. *Ambio* 27:258–65.

Haeberli, W., R. Frauenfelder, M. Hoelzle, and M. Maisch. 1999. On rates and acceleration trends of global glacier mass balances. *Geografiska Annaler* 81A:585–91.

Houghton, J.T., et al. 2001. *Climate change 2001: The scientific basis. Contribution of Working Group 1 to the third assessment report of the Intergovernmental Panel on Climate Change*. Cambridge, UK: Cambridge University Press.

Huggel, C., A. Kääb, and N. Salzmann. 2004. GIS-based modelling of glacial hazards and their interactions using Landsat-TM and IKONOS imagery. *Norwegian Journal of Geography* 58(2):61–73.

Mattson, L. E., J. S. Gardner, and G. J. Young. 1993. Ablation on debris-covered glaciers: An example from the Rakhiot Glacier, Punjab, Himalaya. In *Snow and glacier hydrology*, ed. G. J. Young, 289–96. IAHS publ. 218. Wallingford, UK: International Association of Hydrological Sciences.

Østrem, G. 1959. Ice melting under a thin layer of moraine and the existence of ice in moraine ridges. *Geografiska Annaler* 41A:228–30.

Patzelt, G. 1985. The period of glacier advances in the Alps 1965 to 1980. *Zeitschrift für Gletscherkunde und Glazialgeologie* 23:173–89.

Rana, B., M. Nakawo, Y. Fukushima, and Y. Ageta. 1997. Application of a conceptual precipitation–runoff model in the debris-covered glacierized basin of Langtang Valley, Nepal Himalaya. *Annals of Glaciology* 25:226–31.

Richter, E. 1888. *Die Gletscher der Ostalpen*. Stuttgart: Handbucher zur deutschen Landes- und Volkskunde.

Thomson, M.H., M.P. Kirkbride, and B.W. Brock. 2000. Twentieth-century surface elevation change of the Miage Glacier, Italian Alps. In *Debris-covered glaciers*, ed. M. Nakawo, C.F. Raymond, and A. Fountain. 219–25. IAHS publ. 264. Wallingford, UK: International Association of Hydrological Sciences.

Wood, F. 1988. Global alpine glacier trends 1960s to 1980s. *Arctic and Alpine Research* 20:404–13.

16

Mama Cotacachi

HISTORY, LOCAL PERCEPTIONS, AND SOCIAL IMPACTS OF CLIMATE CHANGE AND GLACIER RETREAT IN THE ECUADORIAN ANDES

Robert E. Rhoades, Xavier Zapata Ríos, and Jenny Aragundy Ochoa

Environmental journalists, activists, and scientists alike are voicing urgent concerns that tropical mountains are losing their glaciers at an alarming rate. Downstream impacts of melting snow and ice provide, they warn, an early glimpse of what might be in store for earth's densely populated lowlands unless action is taken against global warming. Among the negative effects for human society are devastating glacier lake outbreaks (Mool, Bajracharya, and Joshi 2001), loss of alpine biodiversity (EPA 2000), the demise of mountain farming (Price and Barry 1997), declining sources of freshwater for cities (FAO 2002), destruction of sacred mountain sites, and loss of tourism (BBC World News 2003). If only a few of these future scenarios come true, the economic, ecological, and social costs to humanity will be huge.

Despite convincing arguments in call-to-action Web postings, Internet bulletins, and popular articles, little systematic and empirically grounded information is available about the human responses to or societal impacts of diminishing glaciers in mountain regions. While advances have been made in the physical monitoring of mountain glaciers, limited research has been undertaken on the effects of climate change and glacier retreat on human populations. This gap in knowledge needs attention before action or policy aimed at minimizing human costs can be formulated. As a step in building informative case studies, we present research findings on the history, local perceptions, and social impacts of the recent loss of the glacier on Cotacachi, a volcanic peak 4,039 m in elevation situated 35 km north of the equator in Ecuador. Cotacachi is among the first Andean mountains in the past half-century to completely lose its glacier as a result of recent accelerated global warming. As an early example of what might ultimately be in store for mountain glaciers and communities around the world, it deserves careful scrutiny.

Cotacachi, the highest of the northernmost cluster of Ecuadorian volcanoes, exhibits many of the ecological and socioeconomic characteristics typical of the western cordillera of the northern Andes. Covering only 15 km, the mountain rises rapidly from an inter-Andean valley floor at 2,080 m through sharp altitudinal gradients to almost 5,000 m and a zone of what was until recently permanent snow. Agroecological zones of maize, hardy grains, tubers, and high pasture reflect these gradients. Just to the south of Cotacachi is Cuicocha (3,063 m), a crater lake. The lake and the surrounding high-altitude grassland within the Cotacachi-Cayapas Ecological Reserve are the region's main tourist attractions. Indigenous communities are interspersed among remnants of former haciendas at higher and middle elevations while mestizo farms and towns are located in the lower zones. Agriculture is the main activity in indigenous communities, although many young adults work in nearby towns or the capital city of Quito. In the lower zones, the economy is based on commerce, tourism, and an emerging floriculture and agro-industrial greenhouse industry. All social groups depend on drinking and irrigation water derived from Cotacachi and Lake Cuicocha.

To document the demise of the glacier on Cotacachi and the human response to this change, we utilized several different research methods. As is typical of Andean countries, little systematic socioeconomic information is readily available in Ecuador on climate change or glacier retreat. To overcome this gap, we gathered information from historical documents, photographic archives, historical paintings, meteorological station records, and government publications. In addition, we conducted ethnographic interviews and collected oral histories with local people. We also held workshops in which we used photographs and three-dimensional physical models as technical aids to assist in the recall of knowledge that could delineate changes and impacts related to the glacier. Finally, we conducted field observations and monitoring of water availability in selected streams and springs and in Lake Cuicocha. While the information gathered from any single source was thin, we were able to construct a broader vision of the demise of Cotacachi's glacier and its meaning to the local population.

HISTORICAL SOURCES

Chroniclers, travelers, mountaineers, and scientists fascinated with Ecuador's mountains have provided detailed accounts of their glaciers from the sixteenth century to the present. While typically based on brief visits by outsiders and reflecting a Eurocentric natural-history viewpoint, these accounts provide our only long-term data on the Cotacachi glacier (Table 16.1).

The earlier observations place the terminus of the glacier between 4,400 and 4,500 m, while in recent years it was only 200 m below the peak. Many chroniclers included paintings and sketches of Cotacachi that provide visual verification of the glacier's extent around the turn of the twentieth century (Figure 16.1).

Repeat photography, which relies on a time series of photographs across several decades, is a useful method for documenting glacier change (Byers 1987). The first photographs of Cotacachi were likely taken in the last two decades of the nineteenth century (Figure 16.2). An anonymous photograph taken from Cotacachi's town plaza with a view of the eastern slope verifies Whymper's 1880 observation by showing a large glaciated area between the higher and lower peaks of Cotacachi. A 1978 photograph by B. Wuth (reproduced in Vacas 1978, 161) taken from the south shows Cotacachi as a snow-capped mountain; occasionally even today the mountain will receive a thin layer of seasonal snow. By 2002, a photograph by Rhoades taken of the east face shows no evidence of a glacier. Finally, our analysis of aerial photographs as part of a study of land use change from 1963 to 2000 verified what the historical sources and photographs revealed about glacier retreat. In 1963 the east glacier found in Whymper's 1880 sketch was still in place, while by 1993 only a small remnant remained and by 2000

TABLE 16.1

Selected Accounts of Cotacachi Glacier, Nineteenth and Twentieth Centuries

SOURCE	REPORT
von Humboldt (1853, 21), 1802	"Mount Pichincha is located in the same direction and axis as the snow capped mountains Illiniza, Corazón, and Cotacachi."
Wagner (1870, 627–28), 1858–59	"The permanent snow line of Cotacachi in May is 14,814 feet (4,515 m) above sea level."
Orton (1870), 1867	"Twenty-two summits are covered with permanent snow. . . . The snow limit at the equator is 15,800 feet (4,815 m). Cotacachi is always snow-clad."
Dressell (quoted in Hastenrath 1981, 99), 1877	Cotacachi summit is "covered with perpetual and compacted snow."
Whymper (2001 [1892], 260), 1880	"permanent snow, in large beds, as low as 14,500 feet (4,419 m). . . . It is not likely that a crater lies buried beneath the glacier which at present occupies the depression between its peaks."
Wolf (1892, 98)	"Cotacachi is the only snow-capped mountain that is found between the Guayllabamba Valley and the Mira River. It has a large glacier on the east side."
Bermeo (1987, 39–41, 126–29), Luis Bolanos, personal communication, 2004	"There has been an impressive decrease of glaciers on the [Cotacachi] mountain."
von Hillenbrandt (1989, 33)	"Cotacachi is the only mountain in the northern part that has glaciers. There are moraine deposits at the limit as low as 3,900 m."
Sustainable Agriculture and Natural Resource Management Project, 1997	Remnants of the glacier observed and photographed during a reconnaissance trek on the mountain.
Ekkehard and Hastenrath (1998)	Glacier has an area of 0.06 km^2, with the lowest glacier terminus at 4,750 m.
Eric Cadier and Bernard Francou, personal communication, 2004	Glacier nearing its ultimate demise.

permanent ice cover was no longer discernible (Zapata et al. 2006).

METEOROLOGICAL RECORDS

Climate data from private and state meteorological stations collected since the 1960s are limited but suggestive of temperature and precipitation trends (Ontaneda, García, and Arteaga 2002; Gutiérrez 2004, 26–27). Ecuador's (National Institute of Meteorology and Hydrology, or Instituto Nacional de Meteorología e Hidrología [INAMHI]) provides empirical data on changes in temperature and precipitation patterns throughout the country. Data from 15 INAMHI meteorological stations show an increase in the mean, maximum, and minimum annual temperatures and decreasing precipitation over the past century (Cáceres, Mejía, and Ontaneda 2001). INAMHI stations around Cotacachi show mean annual precipitation and mean annual temperature variation for a period of 40 years. While a mean annual temperature increase is clear, the results for

FIGURE 16.1. Images of Cotacachi from Wolf (1892), Whymper (2001 [1892]), and Troya (1913).

precipitation do not always show systematic trends. However, two stations near Cotacachi—San Pablo and Hacienda Maria—recorded a clear decrease in precipitation. Although the few climate stations around Cotacachi are dispersed and generalizations are difficult, the available data point strongly to a climate that is growing warmer and drier—changes that underlie the glacier's retreat.

ORAL HISTORIES AND LOCAL PERCEPTIONS

Although outside visitors and secondary sources provide useful insights and measurements of the physical nature of the glaciers on Cotacachi, these observations are devoid of cultural or social context. Therefore, we elicited local people's observations and responses to climate change and glacier retreat through ethnographic research focused on agriculture, plants, folklore, and communities' histories (Rhoades 2006).

The volcano plays a key mythological and religious role for indigenous Cotacacheños. Following an Andean pattern, they personify and describe Cotacachi with metaphors drawn from the human body and social relations. The volcano is feminine, a *mama* or "mother," and is described as having a body and a head. Indigenous people say that they live on her broad "skirt." (The metaphor of "skirts" of a mountain is widely used in Spain as well.) Local people described Cotacachi as constantly interacting with other features in the landscape and told of her intimate sexual relations with Tayta [father] Imbabura, a volcano located directly across the valley to the east. She has relations with her human inhabitants as well, expressing anger and disappointment with them and at times interfering in the life of the community, especially when social conflict and behavior meet with her disapproval (for similar indigenous North American accounts, see Cruikshank 2005).

FIGURE 16.2. Cotacachi's east (*left and below*) and south faces in 1890, 1978, and 2002. (Photos anonymous, B. Wuth, and R. E. Rhoades, respectively.)

Etched in the collective memories of the people and handed down through their oral histories is an awareness of the volcano's enormous destructive force (Nazarea et al. 2004). Stories and folktales tell of mayhem and death from earthquakes, landslides, rockfalls, volcanic eruptions, and other calamities. The volcano figures strongly in indigenous oral traditions (Nazarea et al. 2004). One local folktale relates how Tayta Imbabura and Mama Cotacachi fell in love. Cotacachi was a beautiful young girl who owned a large hacienda. Tayta Imbabura, an aging womanizer, tired of chasing after other females, Cayambe and Tungurahua (also glaciated peaks), decided to remain with Mama Cotacachi. From their union came a child, a smaller volcano beneath Cotacachi called Yana Urcu. Imbabura became sick and wrapped his head with a white cloth. This, it is said, is why Imbabura is covered with snow only in winter while Mama Cotacachi has permanent snow. Another folktale tells that when Mama Cotacachi has new snow on her peak in the morning, it is a sign that her lover, Tayta Imbabura, has visited her during the night. In yet another, Tayta Imbabura falls in love with the volcano Cayambe, higher and more glaciated than Cotacachi, and Cotacachi stares darkly at him. We were unable to determine the age of this folktale, which seems to take Cotacachi's recent fate into consideration.

To gain insight into present-day perceptions of climate change and the glacier, we organized workshops with farmers representing various Cotacachi communities. To facilitate discussion, we used a three-dimensional physical model with a scale of 1:10,000 to encourage people to discuss the locations of rivers, flows of springs, and the extent of the glacier

FIGURE 16.3. Researchers listen to indigenous farmers explaining climate change using the model. (Photo Jenny Aragundy.)

through time (Figure 16.3). We also used aerial and historical photographs from various years to encourage informants to recall the past.

Previous surveys conducted in the area showed that local farmers listed climate change more frequently than other factors as the most important cause of agricultural change (Campbell 2006). The comments of workshop participants supported these findings, though their accounts of the causes and consequences of this change varied considerably. Elderly people and those from more remote villages believed that it was due to Mama Cotacachi's punishment, while younger people with formal education and contacts with local nongovernmental organizations pinpointed global climate change. Workshop comments focused on three themes: seasons and rain patterns, the glacier and snow on Cotacachi, and water availability.

ORAL ACCOUNTS OF ENVIRONMENTAL CHANGE

Though the demise of the glacier is the most visible impact of climate change at Cotacachi, seasonal rain patterns were more frequently mentioned as a day-to-day concern by participants. They stressed that rainfall had decreased and become more irregular in its timing. Farmers noted greater difficulty in planning field preparation and planting. They reported that lower and midzone crops (e.g., maize) are today being produced in the higher zones. Many referred to the rain as "playing" with people. A female farmer's response was typical: "It doesn't rain as much anymore. It seems the weather is changing so that there is only a drizzle today. It used to be more abundant. The climate seems to be playing. The rains were harder and longer. Today the clouds are polluted and there are only strong winds and everything is dry." Participants noted that 30 years ago Cotacachi was permanently snowcapped and ice was harvested and transported to cities such as Ibarra, where it was used in food preparation and for medicinal purposes. In a representative comment, a farmer said, "There was more snow on Cotacachi; it was whiter when it rained. They say that when there is snow you can see animal figures like there are on Imbabura. There are figures of a rooster and of cattle lying down." Another woman said,

"What I remember of Cotacachi is that it had more snow that came down to near the road to Intag." Another recalled that "snow was used to make ice cream with crushed ice. The mestizos went to fetch it with donkeys." Finally, one farmer from Morales Chupa noted that Cotacachi today is "black" compared with the "white" of Cotapaxi and Chimborazo: "I remember that Cotacachi had snow in the past, but in contrast it has hardly any. These days what we see is black. We used to climb to the *páramo* and observe that there was snow, but now there is none. What I remember of the mountain is that it had snow and looked white. The whole corridor of Cotacachi now looks black."

While there is a universal local awareness that Mama Cotacachi is losing her snow and ice, the community seems unwilling to remove the historic image from its memory. As an exercise, we left the model in the indigenous headquarters for locals to complete by painting it with their own conceptions of landscape features (e.g., communities, sacred places, water bodies). We later learned that they had painted Mama Cotacachi's peak bright white. The mountain continues to be portrayed as snow-capped on billboards, brochures, and tapestries. In the minds and memories of local people, it is still a snow-capped volcano. It is interesting to speculate about the reasons for painting the model: Did that reflect the wish of adults to show younger generations what the mountain looked like, to remind themselves, or perhaps even to show the mountain herself their hope that she would return to her earlier state?

Elderly workshop participants recalled that during their youth earthquakes were more frequent than today and more water flowed in rivers and springs. Today, they say, the rivers are more like irrigation canals. A 49-year-old woman from the community of Turuco said, "Rivers such as Pichavi and Yanayacu were wider 40 years ago. Some people and animals died because they were swept away [by the river when it was at peak flow]." Another indigenous woman recalled that the Pichavi River was once about 8 m wide and would sometimes carry away people and houses. The river has been dry for about 20 years. Other participants pointed to springs where they once drew household water but that are dry today.

FIELD OBSERVATIONS AND WATER MONITORING

In an effort to cross-check the observations of local people, we conducted direct field observations and monitoring of selected water sources. Three rivers (the Pichavi, the Yanayacu, and the Pichanviche), several creeks, springs, and Lake Cuicocha were the main water sources at the beginning of the twentieth century. Today all these sources are experiencing declining water flows. For example, Chumavi Creek, which starts at Chumavi Falls 3,780 m above Lake Cuicocha, once carried water that flowed into the lake. Although the streambed is 30 m wide, it is dry today except for a small spring at approximately 3,600 m. This is the only source for the Chumavi and San Nicolas water systems, which supply nine communities, including 487 families (Mayorga 2004). The Alambi River, with a streambed 30 m wide, is located on the east side of Cotacachi directly under the now-vanished glacier. Thirty years ago, the Alambi ran full of water and was used to irrigate the northern part of the Cotacachi landscape, as is evidenced by an abandoned intake structure located at 2,800 m. Field observations reveal that the now dry Alambi and Chumavi riverbeds extend directly from the glacier.

Lake Cuicocha is also experiencing a lowering of the water level. Since the construction of a hotel on the lake's edge eight years ago, the water level is reported to have dropped about 5 m. Sensors in the lake show that from October 2003 until December 2005 the water level fell some 90 cm.

WATER AND CONFLICT IN COTACACHI

Water availability and use in the Cotacachi area are tied to dramatic social changes since the early 1960s, when a national agrarian reform program began to break up the area's large haciendas. Previously, indigenous people had no title to their lands. They had to work for the owners of the haciendas to obtain usufruct rights to small parcels and access to irrigation water. The agrarian reform divided up the large private estates but did not adequately address issues of water rights. The hacienda owners relinquished to indigenous people the land that had the worst conditions for agriculture, including poor topography and little access to water. When the indigenous communities obtained land, they asked for water concessions from the state, which has formal title to all water resources in Ecuador. Water concessions are obtained from the Consejo Nacional de Recursos Hídricos (National Council of Hydrological Resources [CNRH]). The decisions made often favor the economically and politically powerful. This injustice has angered indigenous communities, and they have filed formal complaints in Quito against the local authorities. As a form of protest, they have blockaded roads and even shut down the nation's major road, the Pan-American Highway (*Diario El Hoy*, July 5, 2005).

The difficulty of the farmers' situation is compounded by the fact that present-day water concessions are based on outdated flow figures. Despite a realization that less water is available, concession holders want to claim the same volume as before. At the same time, demand for water has increased dramatically. Development projects have installed new household potable water and irrigation systems in the growing indigenous communities. Moreover, the number of water users has increased as average plot size has declined as a result of the land fragmentation that follows the division of plots among many heirs in indigenous families.

New activities and users place additional demands on these shrinking water supplies. The owners of several large haciendas have planted water-demanding eucalyptus to demonstrate their active use of the land to government authorities who might expropriate it or otherwise challenge them and to rural activists who might otherwise encourage the poor and landless to occupy their lands (Carse 2006). Simultaneously, an expansion of the lower urban municipality and the growth of floriculture have increased the demand for water. While a partial solution would involve revision of the concessions and their more equitable distribution, the basic problem of insufficient water remains. Local authorities predict that conflict will increase dramatically in the coming years if nothing is done about it.

CONCLUSIONS

While social science researchers in tropical mountains may be disappointed by the scarcity of readily available data on glacier retreat, we have demonstrated that the use of multiple methods can generate significant information. We need to refine these methodologies further and to integrate the study of environmental phenomena with elicitation of human perceptions and responses. Our assumption is that by understanding local people's awareness of glacier retreat and its consequences, we can better understand their decision making and local adaptations to global change. One question that needs further investigation is the relation between present indigenous knowledge of the environment and global climate change as manifested locally. Logically, farmers' local knowledge forms the basis of their decision making and should be incorporated into any strategy meant to mitigate the impact of climate change (Vedwan and Rhoades 2001, 117). However, it should be recognized that the demise of the Cotacachi glacier and its consequences for water availability are entirely new phenomena for Cotacacheños. Local people—young or old—have no previous history with the disappearance of a physical and spiritual feature that has always dominated their cultural landscape. In their collective memories, the glacier has always been present and Mama Cotacachi has always supplied abundant water. The anthropological literature on indigenous knowledge largely draws on local people's long-term experiences with past and ongoing patterns in their environment (Bricker, Sillitoe, and Pottier 2004). On the basis of our study, we suggest that local knowledge is inadequate in the face of external global change that produces unprecedented events. Determining whether long-term adaptive strategies of mountain people (e.g., economic diversification and expansion, agricultural intensification, the development of new regulations) will help mitigate negative impacts will require more in-depth and comparative research at the local level (Netting 1976). While farmers are intensely aware that their climate is changing in ways they do not understand, the novelty of these changes—especially in relation to water availability—may signal that local indigenous knowledge of plant and animal behavior is less secure than in the past.

It is critical that we understand what is happening concretely in communities that have historically depended on glacier resources. This calls for a vigorous social science linked with the natural sciences and for cross-fertilization between local understanding and scientific analysis. Only with a strong interdisciplinary approach that involves the participation of the people directly affected can we hope to achieve solutions to what may become major disruptions of ancient cultures deeply rooted in glacier-fed mountain landscapes.

ACKNOWLEDGMENTS

This research was conducted under the Sustainable Agriculture and Natural Resource Management (SANREM) program, funded by the Office of Agriculture, Bureau for Economic Growth, Agriculture and Trade, USAID, under the terms of Cooperative Agreement Number PCE-A-00-98-00019-00. We thank Juana Camacho for assisting in the workshop in Cotacachi and Virginia Nazarea for providing input on the Cotacachi glacier and folktales.

REFERENCES CITED

BBC World News. 2003. Melting glaciers threaten Peru. World Edition Web site. http://news.bbc.co.uk/2/hi/americas/3172572.stm (accessed October 9, 2003).

Bermeo, J. 1987. *Ascensiones a las altas cumbres del Ecuador: Aventuras en las montañas.* Quito: Impresores MYL.

Bricker, A., P. Sillitoe, and J. Pottier. 2004. *Development and local knowledge.* London and New York: Routledge.

Byers, A. 1987. An assessment of landscape change in the Khumbu region of Nepal using repeat photography. *Mountain Research and Development* 7(1):77–81.

Cáceres, L., R. Mejía, and G. Ontaneda. 2001. Evidencias del cambio climático en Ecuador. INAMHI. http://www.unesco.org.uy/phi/libros/enso/caceres.html (accessed September 15, 2004).

Campbell, B. C. 2006. Why is the earth tired? A comparative analysis of agricultural change and intervention in northern Ecuador. In *Development with identity: Community, culture, and sustainability in the Andes,* ed. R. E. Rhoades, 255–70. Wallingford, UK: CABI International.

Carse, A. 2006. Trees and trade-offs: Perceptions of eucalyptus and native trees in Ecuadorian highland communities. In *Development with identity: Community, culture, and sustainability in the Andes,* ed. R. E. Rhoades, 103–22. Wallingford, UK: CABI International.

Cruikshank, J. 2005. *Do glaciers listen? Local knowledge, colonial encounters, and social imagination.* Vancouver: University of British Columbia Press.

Ekkehard, J., and S. Hastenrath. 1998. Glaciers of Ecuador. In *Satellite image atlas of glaciers of the world,* ed. R. S. Williams Jr. and J. Ferrigno. U.S Geological Survey Professional Paper 1386-I-3. http://pubs.usgs.gov/prof/p1386i/index.html (accessed September 30, 2005).

EPA. 2000. Global warming impacts, mountains. U.S. Environmental Protection Agency. http://yosemite.epa.gov/OAR/globalwarming.nsf/content/ImpactsMountains.html (accessed September 9, 2005).

FAO. 2002. Besieged mountain ecosystems start to turn off the tap. FAONEWSROOM. UN Food and Agriculture Organization. http://www.fao.org/english/newsroom/2002/9881~en.html (accessed October 12, 2005).

Gutiérrez, C. 2004. Ecuador: Sube el termómetro. *Revista Vistazo,* October, 26–27. Quito.

Hastenrath, S. 1981. *The glaciation of the Ecuadorian Andes.* Rotterdam: A. A. Balkema.

Hillenbrandt, C.G. von. 1989. Estudio geovulcanológico del complejo volcánico Cuicocha-Cotacachi y sus aplicaciones. M.Sc. thesis, Escuela Politécnica Nacional.

Humboldt, A. von. 1853. *Kleinere Schriften*. Vol. 1. Stuttgart and Tübingen: Cotta.

Mayorga, O. 2004. *Informe sobre el sistema de agua entubada Chumavi*. Quito: Pontificia Universidad Católica del Ecuador y Proyecto SANREM-Andes.

Mool, P.K., B.R. Bajracharya, and S.P. Joshi. 2001. *Inventory of glaciers, glacial lakes, and glacial lake outburst floods*. Kathmandu: ICIMOD.

Nazarea, V., R. Guitarra, M. Piniero, C. Guitarra, R. Rhoades, and R. Alarcón. 2004. *Cuentos de la creación y resistencia*. Quito: Abya Yala.

Netting, R. 1976. What Alpine peasants have in common: Observations on communal tenure in a Swiss village. *Human Ecology* 4:135–46.

Ontaneda, G., G. García, and A. Arteaga. 2002. *Evidencias del cambio climático en Ecuador*. Proyecto ECU/99/G31 Cambio Climático. Fase 2. Comité Nacional sobre el Clima GEF-PNUD, Ministerio del Ambiente.

Orton, J. 1870. *The Andes and the Amazon, or Across the continent of South America*. New York: Harper.

Price, M.F., and R.G. Barry. 1997. Climate change. In *Mountains of the world: A global priority*, ed. B. Messerli and J.D. Ives, 409–45. New York and London: Parthenon Publishing Group.

Rhoades, R., ed. 2006. *Development with identity: Community, culture, and sustainability in the Andes*. Wallingford, UK, and Cambridge, MA: CABI Publishing.

Troya, R. 1913. Photo from the historical archive of the Central Bank of Ecuador. Quito, Ecuador. Code 95.FO490.04 from the Photos Catalogue.

Vacas, H. 1978. Carchi e Imbabura. In *Maravilloso Ecuador*, ed. Edgar Bustamante, 152–61. Quito: Industria Gráfica Provenza.

Vedwan, N., and R. Rhoades. 2001. Climate change in the Western Himalayas of India: A study of local perception and response. *Climate Research* 19:109–17.

Wagner, M. 1870. *Naturwissenschafliche Reisen im tropischen Amerika*. Stuttgart: Cotta.

Whymper, E. 2001 (1892). *Viajes a través de los majestuosos Andes del Ecuador*. 3d ed. Quito: Ediciones Abya Yala.

Wolf, T. 1892. *Geología y geografía del Ecuador*. Leipzig: Brockhaus.

Zapata, X., R. Rhoades, M. Segovia, and F. Zehetner. 2006. Four decades of land use change in the Cotacachi Andes: 1963–2000. In *Development with identity: Community, culture, and sustainability in the Andes*, ed. R.E. Rhoades, 64–74. Wallingford, UK: CABI International.

PART FIVE

Responses

Adaptation and Accommodation

17

The Politics of Place

INHABITING AND DEFENDING GLACIER HAZARD ZONES IN PERU'S CORDILLERA BLANCA

Mark Carey

Glacier retreat during recent decades has threatened human populations worldwide. In the Peruvian Andes, the consequences of melting glaciers have been particularly dramatic and deadly. Glacier retreat in Peru's Cordillera Blanca has caused two dozen glacier avalanches and glacial lake outburst floods since the 1930s (Ames Marquez and Francou 1995; Zapata Luyo 2002). After a 1941 outburst flood killed 5,000 residents of Huaraz, the Peruvian government began monitoring glaciers and draining glacial lakes to prevent additional glacier disasters. Despite its efforts, in 1970 the Cordillera Blanca produced one of the world's most deadly glacier disasters when an earthquake triggered a massive glacier avalanche that buried Yungay and killed thousands (Figure 17.1). In response, the state expanded disaster mitigation programs to include hazard zoning. Experts agreed that hazard zoning would keep people out of potential avalanche and flood paths, thereby reducing their vulnerability to glacier disasters. But in every town where the authorities tried to implement hazard zoning, residents resisted government initiatives. Even though the region had just experienced a series of cataclysmic glacier disasters, people refused to move to safe areas; instead, they recolonized the areas that had been destroyed. Analysis of hazard zoning in the case of Yungay, the area most devastated by the 1970 glacier avalanche, helps to explain how hazard zoning was derailed.

Hazard zones and hazard zoning held widely divergent meanings for scientists, government officials, and local residents. To scientists, hazard zones represented paths that avalanches and outburst floods could follow. Hazard zoning was seen as the most prudent way to avoid future glacier disasters, and because the 1970 earthquake and avalanche had destroyed most of the region it did not entail moving structures or communities; rather, it required shifting reconstruction to new areas. To the avalanche survivors, however, the hazard zones were historically produced spaces with cultural, economic, social, and political meanings. Relocating to a safe place meant major compromises and significant risks to their livelihoods, connection with ancestors, material well-being, social status, and political power. Consequently,

FIGURE 17.1. The 1970 Yungay avalanche path from Glacier 511. (Photo ©George Plafker, courtesy of the U.S. Geological Survey.)

their decisions about whether to remain or to relocate involved the ranking of risks. For those who rejected hazard zoning, there were cultural, social, economic, and political risks associated with leaving that outweighed the risk of unknown and unpredictable glacier disasters. In other words, while scientists focused on a single risk, residents contended with a host of them. By analyzing the rationality of Yungay residents' risk perception—the historical forces informing that perception, the multiple meanings they assigned to the hazard zone, and their reasons for resisting relocation—it is possible to understand why experts, policy makers, and local residents clashed over the 1970s disaster mitigation policies.

Natural disasters occur not only because environmental processes inflict damage but also because vulnerable people, property, and infrastructure get in the way. Scholars acknowledge that some people choose to live in vulnerable locations because they are wealthy enough to afford insurance or powerful enough to be guaranteed government relief (Davis 1998; Steinberg 2000). More often, and especially for the developing world, researchers suggest that people inhabit vulnerable hazard zones because power imbalances, government neglect, poverty, racism, economic development, or social injustice force marginalized populations into areas prone to flooding, hurricanes, fire, drought, and other natural disasters. To protect people, many scholars call on governments and experts (scientists, engineers, planners, etc.) to develop policies that reduce human vulnerability to natural hazards (Maskrey 1993; Alexander 2000; Wisner et al. 2004).

While marginalized populations do suffer disproportionately from natural disasters, scholarship that deflects the *reasons* for people's vulnerability to forces beyond their control can yield two problematic interpretations. First, by overlooking people's own role—even the role of marginalized or developing world populations—in influencing their vulnerability to natural disasters, researchers can deny the historical agency of these groups. Second, by calling on governments and experts to implement policies that reduce people's vulnerability to natural disasters, scholars often assume that vulnerable populations will embrace these plans. The case of Yungay in the 1970s challenges these two views.

On the one hand, when Yungay residents rejected hazard zoning, they played a vital role in determining their vulnerability to future glacier disasters. They acted in multifaceted and complex ways that stemmed from historical, social, cultural, economic, political, and ideological factors (Johnston and Klandermans 1995; Rubin 2004; Chuang 2005). On the other hand, their views of hazard zones as space and hazard zoning as policy clashed markedly with the perspectives of scientists, planners, engineers, and national government officials. These local reactions to state plans demonstrate the intimate relationship between science and power (Arnold 1993; Scott 1998; Prakash 1999; Mitchell 2002). In short, this Yungay case study helps to clarify not only the local consequences of global warming and glacier retreat but also the conflicts that can arise when governments attempt to implement scientific policy, mitigate natural disasters, and adapt to climate change.

CORDILLERA BLANCA GLACIER MONITORING AND DISASTER MITIGATION

The Cordillera Blanca in Peru's Ancash Department contains approximately 600 glaciers that cover slightly less than 600 km^2 (Georges 2004). A half-million people inhabit the valleys and upland slopes surrounding this range. These residents consist of Quechua-speaking indigenous people, mestizos (of mixed Spanish-indigenous descent), and whites (of Spanish descent). The majority of them live in the Santa River valley, known as the Callejón de Huaylas, along the western base of the cordillera (Figure 17.2). More than 70% of Cordillera Blanca glacier meltwater drains into the Santa River, which therefore supports extensive agriculture and livestock as well as Peru's important Cañón del Pato hydroelectric station.

Peruvians living near the Cordillera Blanca have lived with repeated glacier disasters since the 1930s (Carey 2005*a*). As a result of rising global temperatures since the end of the Little Ice Age (~1350–1850), glacier melting in the cordillera has caused two types of glacier disasters (Portocarrero 1995; Ames 1998; Kaser and Osmaston 2002). First, glacial lake outburst floods have occurred when lakes grew in the space left by retreating ice. The number of glacial lakes has increased significantly, from 223 in 1953 to 374 in 1997 (Fernández Concha and Hoempler 1953; ElectroPerú 1997). Dammed

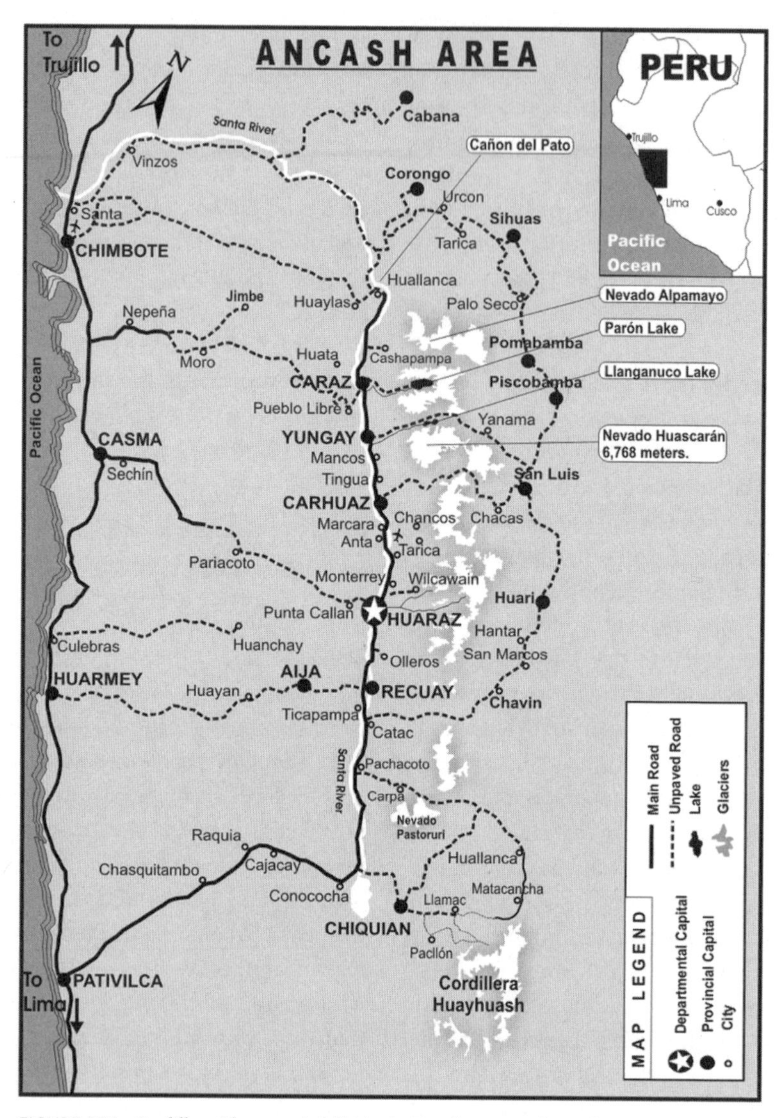

FIGURE 17.2. Cordillera Blanca and Callejón de Huaylas, Peru (drawn by Tito Olaza).

behind unstable moraines that sometimes ruptured, these glacial lakes produced 13 major outburst floods between 1932 and 1959 (Zapata Luyo 2002; Ames Marquez 2003). Three floods were particularly catastrophic: the 1941 Huaraz (5,000 deaths), the 1945 Chavín (500 deaths), and the 1950 Los Cedros (200 deaths).

A second type of glacier disaster has occurred when glacial ice thinned, fractured, and generated landslides. In 1962, for example, Glacier 511 on Mt. Huascarán caused an avalanche that destroyed the town of Ranrahirca, killing an estimated 4,000 inhabitants. The most deadly glacier disaster occurred on May 31, 1970, when an earthquake registering 7.7 on the Richter scale dislodged Glacier 511 again and the ensuing avalanche buried Yungay (Ericksen, Plafker, and Concha 1970). Most sources cite the avalanche death toll as 15,000 or 18,000 (e.g., Oficina Nacional de Información 1970; Oliver-Smith 1986; Ames Marquez and Francou 1995). Analysis of the most recent census data prior to 1970, however, indicates that in 1961 only 15,068 lived in the entire district of Yungay (República del Perú 1968). Given that many survived the earthquake and avalanche and even accounting for population growth between 1961 and 1970, the estimate of 15,000 deaths is perhaps inflated. Even so, the Yungay avalanche remains one of the world's most deadly glacier disasters.

To prevent glacial lake outburst floods and glacier avalanches, the Peruvian government has monitored glaciers and drained glacial lakes since the 1941 Huaraz flood. In 1951, after the Los Cedros outburst flood destroyed the nearly completed 50-megawatt Cañón del Pato hydroelectric station, the Peruvian government created the Comisión de Control de Las Lagunas de la Cordillera Blanca (CCLCB). The CCLCB conducted hundreds of studies and developed one of the world's first systematic glacial lake classification indexes to identify and rank glacial lake hazards. This classification system helped categorize glacial lakes so that unstable lakes could be drained and dammed before they inundated the Cañón del Pato hydroelectric station or the Callejón de Huaylas communities. The CCLCB partially drained and dammed 19 glacial lakes between 1952 and 1971, when the Peruvian Corporación Peruana del Santa (CPS), which ran Cañón del Pato, absorbed it into its own Division of Glaciology and Lakes Security (Gálvez Paredes 1970; Carey 2005*b*). Spurred partly by the 1962 Ranrahirca glacier avalanche, which pointed to ice—and not just glacial lake—hazards, and partly by its economic interest in determining the amount of water stored in glaciers, the CPS created the division in 1966 with the Peruvian glaciologist and Huaraz resident Benjamín Morales as its head (Morales 1969). By 1970, the commission and the company had carried out large-scale glacial lake control projects and extensive glacier monitoring. With the exception of declaring the 1941 Huaraz flood path off-limits to rebuilding, however, the glacier-disaster prevention agenda had not focused on reducing people's vulnerability to floods or avalanches. The magnitude of the 1970 Yungay avalanche expanded the government's disaster mitigation agenda to include hazard zoning.

DISASTER MITIGATION FROM ABOVE

Responses to Peru's 1970 earthquake and avalanche took place in two important contexts. First, the scale of the earthquake and Yungay avalanche stimulated increased state funding for glacier-disaster prevention projects. During the first half of the 1970s, government investments in glacier monitoring, lake control, and hazard zoning surpassed those of all previous eras. Second, Peru's political situation in 1970 was unique. In 1968, General Juan Velasco had seized Peru's presidency, and his "Revolutionary Government" had begun implementing far-reaching social, economic, political, and even cultural changes, including accelerated economic growth, redistribution of income and wealth, integration of the indigenous population into mainstream society, reduction of foreign dependence, and, essentially, the fabrication of a "new society" (Jaquette and Lowenthal 1987; Contreras and Cueto 2000). While the execution of these sweeping changes generally fell short of the goal, Velasco's reforms significantly altered the traditional social landscape.

In many ways, Velasco saw the 1970 earthquake and avalanche as an opportunity to rebuild the region according to his ideals. The near-complete obliteration of the Callejón de Huaylas offered a clean slate on which to create a "new and improved" society. He was determined to create a society less vulnerable to glacier avalanches and outburst floods, and hazard zoning—along with glacier monitoring and lake control—would reduce these long-term risks. To distribute relief aid and reconstruct the region, he established the Comisión de Reconstrucción y Rehabilitación de la Zona Afectada (CRYRZA). As a centralized, bureaucratic agency run from Lima, CRYRZA did not always operate efficiently or in cooperation with the Yungay survivors (Oficina Nacional de Información 1970; Oliver-Smith 1986; Bode 1990). Most scholars have criticized Velasco and

CRYRZA for imposing top-down reconstruction agendas and attempting social engineering. But to (mis)classify hazard zoning as just another aspect of Velasco's social experiment neglects the contribution of hazard zoning to long-term disaster mitigation—an approach that natural-disaster scholars advocate today (Maskrey 1993; Sarewitz, Pielke, and Mojdeh 2003; Wisner et al. 2004).

In June 1970, experts identified three principal problems that disaster prevention was subsequently to address: (1) unpredictable glaciers that required continuous monitoring, (2) unstable glacial lakes caused both by continued glacier retreat and by earthquake damage to previously constructed dams and drainage canals, and (3) human habitation of hazard zones, the riparian zones through which future avalanches and outburst floods could pass (Lliboutry 1970; Lliboutry et al. 1977).

First, given the unpredictability of glacier avalanches, the experts called for comprehensive glacier monitoring and research. The CPS (and its successor, ElectroPerú) maintained a distinct office for glacier research and hired dozens of new scientists and engineers after 1970. This research required access to remote Cordillera Blanca canyons in which researchers sometimes spent weeks at isolated camps above 4,000 m. A typical glacial lake was at least 20 km from and 1,500 m above the nearest road, and access to it involved climbing steep slopes and navigating boulder fields, dense brush, and marshy valleys, with mules and indigenous porters carrying the equipment. Glacier monitoring often benefited from the guidance of indigenous residents who were familiar with the high-elevation terrain from pasturing cattle and sheep or gathering plants and firewood. These locals also furnished information about recent glacier history and glacial lake formation. Glacier hazard monitoring took scientists to new heights and into unprecedented danger. In August 1973, for example, scientists spent a night at 6,400 m during an expedition to analyze a glacier crevasse on Mt. Huascarán Norte (Morales 1972; Zamora Cobos 1973).

Second, the scientists recommended effective lake control, especially at Lake Llanganuco above Yungay. Located in the province of Yungay, the upper and lower parts of Lake Llanganuco are two of the region's largest lakes. On June 5, 1970, scientists flying over the Cordillera Blanca noted that the upper lake outlet was blocked by a rock and ice dam formed by an earthquake-induced landslide that had covered part of the valley floor (*El Comercio* 1970*a*; Lliboutry 1970). By June 20, 30 men were digging frantically through long days and into the nights to open a drainage canal through the new dam (Ortega 1970). By July 1970 they had succeeded in reopening the drainage canal, thereby lowering the water level and averting catastrophe in the valley below. Through subsequent years, President Velasco allocated funds for several glacial lake control projects, including the drilling of four drainage tunnels and the digging of five drainage canals to remove water from unstable glacial lakes. Because access to these lakes was difficult, he also funded construction of 50 km of roads to glacial lakes (Morales 1970). From 1970 to 1978 the Division of Glaciology and Lakes Security drained, dammed, and secured a dozen glacial lakes (División de Glaciología y Seguridad de Lagunas 1972; ElectroPerú 1975, 1984; Lliboutry et al. 1977).

Finally, the experts advocated the relocation of towns that were vulnerable to avalanches and outburst floods. Because glacier disasters such as the 1962 and 1970 avalanches from Glacier 511 could not be predicted or avoided, scientists proposed hazard zoning and the relocation of several towns. As Morales observed, "The catastrophes produced by avalanches from hanging glaciers such as Huascarán will be periodic events, which is to say that they will occur again in the future. This type of phenomenon has no solution; there are no measures to avoid them or control them. Consequently, populations located below glaciers with these characteristics must be relocated" (Morales 1970, 71). Experts identified many hazard zones below Cordillera Blanca glaciers and glacial lakes, including areas of Huaraz, Carhuaz, and Yungay (Lliboutry et al.

1970; Oberti 1975). Because the earthquake and avalanche had destroyed most of these towns, relocation from hazard zones did *not* involve relocation of intact communities. Rather, it generally involved the rebuilding of previously demolished structures in different, safer places. It was proposed that Yungay be moved to a site called Tingua, 15 km away, which was believed to be protected from glacier disasters (*El Comercio* 1970*e*; CRYRZA 1970). By the time authorities announced relocation plans in November 1970, however, Yungay survivors had already begun reconstruction adjacent to the avalanche path in an area they named Yungay Norte. Scientists determined that the settlement would soon spread into the Yungay hazard zone, and therefore they made efforts to move it to Tingua.

LOCAL RESISTANCE FROM BELOW

Despite decades of glacier disasters, the people of Yungay and other Callejón de Huaylas residents were eager to resettle the hazard zone. The Yungay elite—property owners, businesspeople, politicians, entrepreneurs, and professionals—led this resistance to relocation and gained the support of other locals. Resistance to hazard zoning should not be taken to indicate that Yungay residents were unafraid or ignorant of glacier-related hazards. Rumors circulated widely in Yungay about the instability of the Huascarán glaciers. Some believed that the Cordillera Blanca was made up of water volcanoes that could erupt at any point and send deadly floods or landslides into populated valleys below (Oliver-Smith 1986; Walter 2003). Others fled in fear of future disasters and the horrifying postdisaster reality; of the approximately 27,000 earthquake and avalanche survivors from Yungay Province, an estimated 4,000–5,000 moved to Lima immediately (Angeles 1970*a*; *El Comercio* 1970*d*). At the same time, fatalism sometimes led to inaction among residents who believed that natural disasters were beyond their control. Instead of recognizing their vulnerable location beneath unstable glaciers, they implicated others for causing natural disasters, among them God, who took revenge on sinners and controlled nature, or the French, who had detonated underwater atomic bombs in May 1970 that many believed had triggered the earthquake (Ramírez Gamarra 1971; Miano Pique 1972).

Still others decided to rebuild their community in its previous location. Trauma, grief, and a sense of place helped motivate this choice (Oliver-Smith 1982). The avalanche, after all, had buried thousands of people, and their survivors wanted to remain close to them. One of them said, "This is where we want to be. We are accustomed to dying, to losing family.... We want to be here, to die where we were born" (quoted in Bode 1990, 201). Beyond this, despite fear of disaster and hatred of the peak for causing catastrophe, residents felt attached to Mt. Huascarán and Lake Llanganuco (Flores Vásquez 1972). As one observer explained, "Most struggle to live and die on these prodigal lands because, although Huascarán and the lakes threaten us with their fractured bases, they have spirits that attract and captivate everyone who lives here and sees them" (Zavaleta Figueroa 1970, 16).

The Yungay hazard zone was also vital for economic productivity. Relocation to Tingua, residents argued, would undermine their capacity to maintain commercial relations within the province and exploit the rural labor force. As an urban area that possessed both the geographical location and the gravitational pull to attract people, trade, labor, and resources from surrounding areas, Yungay had become a "central place" within the Callejón de Huaylas (Oliver-Smith 1977*b*). Before 1970 it was the economic center of the province and attracted markets, transportation, labor, food products, agriculture, and natural resources. Yungay leaders feared that moving the town to Tingua would diminish its importance and cripple its economy.

The hazard zone held great potential for tourism as well. Yungay was the access point for the ascent of Mt. Huascarán (6,768 m a. s. l.), Peru's highest peak, and for recreation at Lake Llanganuco, a stunning turquoise glacial lake. Prior to 1970 Yungay Province had been a principal tourist destination in Peru. After the 1970 disaster, Yungay survivors believed that tourists would come not only for traditional recreation at Huascarán and Llanganuco but also to see the avalanche site (Angeles 1970*c*; *El Comercio* 1970*b*). Tourism was so important that the Yungay authorities reopened the road to Lake Llanganuco as quickly as they built schools and hospitals (Ángeles Asín 2002). The safe zone at Tingua simply did not possess the proximity to the Cordillera Blanca that made Yungay "a worldwide tourist attraction" (Angeles 1970*b*).

Beyond its economic potential, the hazard zone also helped determine social status. Yungay's town boundaries had historically divided rural and urban, upland and valley, and indigenous and mestizo populations. Throughout Peruvian history, the country's dominant social classes, including the Yungay elite, had defined the indigenous sphere as rural, highland, and close to nature (Orlove 1993, 1998). With the conflation of race, class, and geographical location in Peru, habitation of Yungay's urban area signified superior social status over rural inhabitants labeled as indigenous (Stein 1974; Walton 1974). Although historical migration, miscegenation, and shifting identities had blurred their boundaries, these categories remained prominent in Peruvians' minds from the colonial era through the twentieth century (de la Cadena 2000). The Yungay elite maintained their superior status over the rural population (67% of the population in 1967) in part because they inhabited an urban area. Further, elite anxiety about their privileged social standing was acute after the disaster because, while rural survivors descended to Yungay for aid, many urban residents fled to Lima (Walton 1974). In the first year after the disaster, the Yungay population rose from several hundred survivors to 2,000 because of an influx of rural residents (Oliver-Smith 1982). Maintenance of political positions, professional jobs, property, and control of the urban space hinged partly on clear markers to distinguish rural from urban, lower from upper class (Oliver-Smith 1977*a*; Ángeles Asín 2002), and the 1970 avalanche had destroyed the identifiable urban-rural division that was one such marker. According to urban survivors, whereas Yungay possessed remnants of these markers and historical memories of the division, Tingua had no recognizable boundaries to demarcate social standing. They therefore rejected relocation because improved safety from avalanches and outburst floods threatened to reduce their social status and control (Cabel 1973).

Rural Yungay districts did not necessarily oppose relocation to Tingua, and Yungay leaders used political incentives to gain allies in their struggle. For example, in the district of Yanama, they persuaded Mayor Isidro Obregón to reject relocation by promising to allocate funds to the building of a long-awaited road across the Cordillera Blanca (Ángeles Asín 2002). Further, to persuade the new indigenous immigrants in Yungay Norte to resist relocation, they made sure that those in the relief camp received sufficient food and shelter (Oliver-Smith 1982). Meeting their immediate needs linked residents to Yungay Norte and demonstrated the leaders' effectiveness, thereby generating political support for them and creating a popular base to resist hazard zoning.

Hazard zones became sites not only for regional power struggles but also for national politics. Even though it was scientists who had identified the hazard zones, urban survivors believed that relocation signified their subordination and loss of power to Lima. Several factors help explain why Yungay survivors saw hazard zones as political battlegrounds. First, it was CRYRZA planners and engineers, rather than glaciologists and hydrologists, who attempted to implement the Tingua relocation plan. The scientists generally worked near the glaciers, where they did not appear in survivors' daily lives, while CRYRZA officials set up their offices in survivor camps and interacted with local residents daily.

Second, the CRYRZA representatives sought to distribute aid equally among all survivors, but for the urban elite accustomed to advantages over the indigenous population equality was an insidious proposition. A Yungay woman captured the rift when she grumbled about CRYRZA's aid distribution: "The people of the heights, the Indians, never had anything, so why should they get help? On the other hand, we, the real Yungainos, have lost everything, so we should get more" (quoted in Oliver-Smith 1977*a*: 8). By helping rural and indigenous people in the Callejón de Huaylas, CRYRZA planned to eliminate elite social status and political control—an agenda the Yungay elite emphatically opposed (Walton 1974).

Third, President Velasco was threatening Yungay landowners with a vast agrarian reform program. He and his successor eventually redistributed land to 375,000 families (25% of all farm families) throughout Peru (Klarén 2000). As elsewhere, agrarian reform in the Callejón de Huaylas turned landowners against Velasco because land expropriation diminished their regional power and economic base (Barker 1980; Kay 1982). In Yungay, where the elite already felt threatened by CRYRZA's support for the downward-migrating indigenous population, the risk of government land expropriation posed another threat. Hazard zoning and government relocation plans resembled agrarian reform in that the state determined who would inhabit (or leave vacant) a specific plot of land. Yungay survivors thus interpreted the scientists' well-intentioned efforts to relocate them to safe areas as a national-government affront (Cabel 1973; Walton 1974).

Finally, Yungay residents were alienated by CRYRZA's failure to connect relocation plans with local needs. Discussions about building Tingua did not involve local people, and lack of faith in Velasco, combined with historical mistrust of the state, led many of them to doubt that the government would support the move. In November 1970, residents demanded that government planners "give more attention to future considerations, such as population, production, design of the city, and communication and transportation" (*El Comercio* 1970*b*). In subsequent weeks, the local authorities from Yungay insisted that CRYRZA present concrete plans for the development of Tingua (*El Comercio* 1970*c*; *La Prensa* 1970). People complained because official plans ignored them.

To the Yungay elite, hazard zones were literal and figurative spaces that signified much more than potential avalanches or floods. The Yungay hazard zone and the proposed relocation site at Tingua had complex meanings tied to cultural beliefs, livelihoods, material well-being, social status, identity, and power structures. The glacier avalanche had thus been much more than a natural disaster; it had also triggered cultural, economic, social, and political catastrophes. To many, and especially to the Yungay elite, recovery from these multiple disasters meant rebuilding their lives and their societies in the hazard zone. The risks of further losses of social status, economic security, political power, and cultural beliefs were far more pressing and important than the risk of a glacier avalanche or an outburst flood. The occupation of hazard zones and resistance to relocation thus reflected Yungay residents' ranking of risks.

CONCLUSION

Peruvian responses to the 1970 Yungay avalanche illustrate the complexities of contending with glacier retreat and mitigating natural disasters. Challenges to hazard zoning below the Cordillera Blanca glaciers emerged because distinct groups attributed different meanings to the hazard zone. Scientists saw hazard zoning as a way to reduce the risk of glacier disasters. The government recognized hazard zoning as a political opportunity to remake Peruvian society according to President Velasco's ideals. And many Yungay residents believed that hazard zoning was less important than resolution of cultural, social, economic, and political disruptions. While all three groups recognized the hazard, their perceptions of risk involved a host of interconnected historical, environmental, and human factors.

Clearly, people can influence not only their vulnerability to natural disasters but also the implementation of policy. No major Cordillera

Blanca avalanche has occurred since 1970, so Yungay residents may have made the right choice. However, if global warming persists, glaciers will likely continue to retreat, thin, fracture, and possibly produce more avalanches and outburst floods. If, in contrast, cooling occurs, a new set of problems may emerge as glacier tongues advance into existing glacial lakes and threaten to displace lake water. Ironically, local people's having chosen to resettle vulnerable zones, in part to defy government authority, has made them more dependent than ever on the government's use of science and technology to protect them from glacier disasters.

ACKNOWLEDGMENTS

I thank Ben Orlove, Charles Walker, Ellen Wiegandt, anonymous reviewers, and participants in the Wengen Workshop 2004. I thank the Mabelle McLeod Lewis Memorial Fund and the S.V. Ciriacy-Wantrup Postdoctoral Fellowship at the University of California, Berkeley, for supporting the writing of this essay. Field research was funded by the American Meteorological Society, the Pacific Rim Research Program, and the International Dissertation Field Research Fellowship Program of the Social Science Research Council, with funds provided by the Andrew W. Mellon Foundation.

REFERENCES CITED

Alexander, D. 2000. *Confronting catastrophe: New perspectives on natural disasters.* New York: Oxford University Press.

Ames, A. 1998. A documentation of glacier tongue variations and lake development in the Cordillera Blanca, Peru. *Zeitschrift für Gletscherkunde und Glazialgeologie* 34:1–36.

Ames Marquez, A. 2003. Chronology of ice avalanches and floods occurring in the Cordilleras Blanca and Huayhuash since the beginning of the 18th century. MS, Huaraz.

Ames Marquez, A., and B. Francou. 1995. Cordillera Blanca glaciares en la historia. *Bulletin de l'Institut Français d'Études Andines* 24:37–64.

Ángeles Asín, L. 2002. Mientras haya un yungaíno con vida, Yungay no desaparecerá. In *Vida, muerte y resurrección: Testimonios sobre el sismo-alud 1970*, ed. R. Pajuelo Prieto, 187–92. Yungay: Ediciones Elinca.

Angeles, P.M. 1970*a*. Acuerdo no. 9: Reubicación y recomendaciones sobre las ciudades y centros poblados en Yungay, 15 de diciembre. In *Lo mejor de nuestra juventud al servicio de Yungay y de los yungainos por una vida mejor a través del Centro Unión Yungay*, ed. Centro Unión Yungay y la Junta Directiva. Lima: Centro Unión Yungay.

———. 1970*b*. Carta al decano del colegio de arquitectos del Perú, Lima, 7 de setiembre. In *Lo mejor de nuestra juventud al servicio de Yungay y de los yungainos por una vida mejor a través del Centro Unión Yungay*, ed. Centro Unión Yungay y la Junta Directiva. Lima: Centro Unión Yungay.

———. 1970*c*. Carta al sr. Director del diario El Comercio, Lima, 13 de noviembre. In *Lo mejor de nuestra juventud al servicio de Yungay y de los yungainos por una vida mejor a través del Centro Unión Yungay*, ed. Centro Unión Yungay y la Junta Directiva. Lima: Centro Unión Yungay.

Arnold, D. 1993. *Colonizing the body: State medicine and epidemic disease in nineteenth-century India.* Berkeley: University of California Press.

Barker, M.L. 1980. National parks, conservation, and agrarian reform in Peru. *Geographical Review* 70:1–18.

Bode, B. 1990. *No bells to toll: Destruction and creation in the Andes.* New York: Paragon House.

Cabel. 1973. *Literatura del sismo: Reportaje a Ancash.* Lima: Juan Mejia Baca.

Carey, M. 2005*a*. Living and dying with glaciers: People's historical vulnerability to avalanches and outburst floods in Peru. *Global and Planetary Change* 47:122–34.

———. 2005*b*. People and glaciers in the Peruvian Andes: A history of climate change and natural disasters, 1941–1980. Ph.D. diss., University of California, Davis.

Chuang, Y.-C. 2005. Place, identity, and social movements: *Shequ* and neighborhood organizing in Taipei City. *Positions* 13:379–410.

Contreras, C., and M. Cueto. 2000. *Historia del Perú contemporáneo.* Lima: Instituto de Estudios Peruanos.

CRYRZA (Comisión de Reconstrucción y Rehabilitación de la Zona Afectada). 1970. La reubicación de las ciudades del Callejón de Huaylas. *Revista Peruana de Andinismo y Glaciología* 19(9):28–29.

Davis, M. 1998. *Ecology of fear: Los Angeles and the imagination of disaster.* New York: Vintage Books.

de la Cadena, M. 2000. *Indigenous mestizos: The politics of race and culture in Cuzco, Peru, 1919–1991.* Durham: Duke University Press.

División de Glaciología y Seguridad de Lagunas, Corporación Peruana del Santa. 1972. *Estudios glaciológicos, bienio 1971–1972*. Biblioteca, Unidad de Glaciología y Recursos Hídricos, Huaraz, Doc #. I-GLACIO-015.

El Comercio. 1970*a*. Hacen estudios para que laguna de Llanganuco no siga siendo una amenaza. *El Comercio*, June 5.

———. 1970*b*. Piden que el gobierno profundice estudio sobre reubicación de la ciudad de Yungay. *El Comercio*, November 15.

———. 1970*c*. Sugieron profundizar los estudios de reubicación de pueblos en prov. Yungay. *El Comercio*, December 21.

———. 1970*d*. Yungaínos acuerdan formar colonias hogares en Lima. *El Comercio*, October 8.

———. 1970*e*. Yungay, Carhuaz, Mancos y Ranrahirca tendrán otra ubicación, señala CRYRZA. *El Comercio*, November 12.

ElectroPerú. 1975. *Memoría bienal del programa de glaciología y seguridad de lagunas*. Huaraz, febrero. Biblioteca, Unidad de Glaciología y Recursos Hídricos, Huaraz, Doc #. I-MEMORIAS-008.

———. 1984. *Información básica de la labor realizada por la unidad de glaciología y seguridad de lagunas entre los años 1973 y 1984*. Huaraz, agosto. Biblioteca, Unidad de Glaciología y Recursos Hídricos, Huaraz, Doc #. I-MEM-002.

———. 1997. *Mapa indice de lagunas de la cordillera blanca*. Huaraz, octubre. Biblioteca de ElectroPerú, Lima, Doc #. Caja 060902, No. H-10.

Ericksen, G. E., G. Plafker, and J. F. Concha. 1970. Preliminary report on the geological events associated with the May 31, 1970, Peru earthquake. *United States Geological Survey Circular* 639:1–25.

Fernández, C. J., and A. Hoempler. 1953. *Indice de lagunas y glaciares de la Cordillera Blanca*. Comisión de Control de Las Lagunas de la Cordillera Blanca, Ministerio de Fomento, Lima, mayo. Biblioteca, Unidad de Glaciología y Recursos Hídricos, Huaraz, Doc #. I-INVEN-011.

Flores Vásquez, A. 1972. Discurso del Prof. Alejandro Flores Vásquez, pronunciado al conmemorarse el primer aniversario de la catástrofe. *Forjando Ancash (Organo del Club Ancash)* 21:30–31.

Gálvez Paredes, H. 1970. *Breve información sobre lagunas de la Cordillera Blanca*. Huaraz, abril. Biblioteca de ElectroPerú, Lima, Doc #. 701 27.725.

Georges, C. 2004. 20th-century glacier fluctuations in the tropical Cordillera Blanca, Peru. *Arctic, Antarctic, and Alpine Research* 36:100–107.

Jaquette, J. S., and A. F. Lowenthal. 1987. The Peruvian experiment in retrospect. *World Politics* 39:280–96.

Johnston, H., and B. Klandermans, eds. 1995. *Social movements and culture*. Minneapolis: University of Minnesota Press.

Kaser, G., and H. Osmaston. 2002. *Tropical glaciers*. New York: Cambridge University Press.

Kay, C. 1982. Achievements and contradictions of the Peruvian agrarian reform. *Journal of Development Studies* 18:141–70.

Klarén, P. F. 2000. *Peru: Society and nationhood in the Andes*. New York: Oxford University Press.

La Prensa. 1970. Solicitan estudios adecuados para ubicar pueblos de Yungay. *La Prensa*, December 29.

Lliboutry, L. 1970. Informe preliminar sobre los fenómenos glaciológicos que acompañaron el terremoto y sobre los peligros presentes. *Revista Peruana de Andinismo y Glaciología* 19(9):20–26.

Lliboutry, L., V. Mencl, E. Schneider, and M. Vallon. 1970. *Evaluación de los riesgos telúricos en el Callejón de Huaylas, con vista a la reubicación de poblaciones y obras públicas*. Paris: UNESCO.

Lliboutry, L., A. B. Morales, A. Pautre, and B. Schneider. 1977. Glaciological problems set by the control of dangerous lakes in Cordillera Blanca, Peru. 1. Historical failures of morainic dams, their causes and prevention. *Journal of Glaciology* 18:239–54.

Maskrey, A., ed. 1993. *Los desastres no son naturales*. Bogotá: La Red de Estudios Sociales en Prevención de Desastres en América Latina.

Miano Pique, C. 1972. *¡¡Basta!! La bomba atómica francesa, la contaminación atmosférica y los terremotos*. Lima: Tangrat.

Mitchell, T. 2002. *Rule of experts: Egypt, techno-politics, and modernity*. Berkeley: University of California Press.

Morales, A. B. 1969. Las lagunas y glaciares de la Cordillera Blanca y su control. *Boletín del Instituto Nacional de Glaciología* (Peru) 1:14–17.

———. 1970. El día más largo en el hemisferio sur. *Revista Peruana de Andinismo y Glaciología* 19(9):63–71.

———. 1972. Comentarios sobre el memorandum del Dr. Leonidas Castro B. en el caso Huascarán. Lima, 20 diciembre. Biblioteca, Unidad de Glaciología y Recursos Hídricos, Huaraz, Doc #. I-GLACIO-005.

Oberti, I. L. 1975. Estudio glaciológico del cono aluviónico de Huaraz. Biblioteca, Unidad de Glaciología y Recursos Hídricos, Huaraz, Doc #. I-GLACIO-014.

Oficina Nacional de Información. 1970. *¡Cataclismo en el Perú!* Lima.

Oliver-Smith, A. 1977*a*. Disaster rehabilitation and social change in Yungay, Peru. *Human Organization* 36:5–13.

———. 1977*b*. Traditional agriculture, central places, and postdisaster urban relocation in Peru. *American Ethnologist* 4:102–16.

———. 1982. Here there is life: The social and cultural dynamics of successful resistance to resettlement in postdisaster Peru. In *Involuntary migration and resettlement: The problems and responses of dislocated people,* ed. A. Hansen and A. Oliver-Smith, 85–103. Boulder: Westview Press.

———. 1986. *The martyred city: Death and rebirth in the Andes.* Albuquerque: University of New Mexico Press.

Orlove, B. 1993. Putting race in its place: Order in colonial and postcolonial Peruvian geography. *Social Research* 60:301–36.

———. 1998. Down to earth: Race and substance in the Andes. *Bulletin of Latin American Research* 17:207–22.

Ortega, J. 1970. Enviado especial comprobó desaparición de siete pueblos. *El Comercio,* June 20.

Portocarrero, C. 1995. Retroceso de glaciares en el Perú: Consecuencias sobre los recursos hídricos y los riesgos geodinámicos. *Bulletin de l'Institut Français d'Études Andines* 24:697–706.

Prakash, G. 1999. *Another reason: Science and the imagination of modern India.* Princeton: Princeton University Press.

Ramírez Gamarra, H. 1971. *Ancash: Vida y pasión.* Lima: Editorial Universo.

República del Perú. 1968. *Censos nacionales de población, vivienda y agropecuario, 1961.* Vol. 2. *Departamento de Ancash.* Lima: Dirección Nacional de Estadística y Censos.

Rubin, J. W. 2004. Meanings and mobilizations: A cultural politics approach to social movements and states. *Latin American Research Review* 39:106–42.

Sarewitz, D., R. Pielke Jr., and K. Mojdeh. 2003. Vulnerability and risk: Some thoughts from a political and policy perspective. *Risk Analysis* 23:805–10.

Scott, J. 1998. *Seeing like a state: How certain schemes to improve the human condition have failed.* New Haven, CT: Yale University Press.

Stein, W. W. 1974. *Countrymen and townsmen in the Callejón de Huaylas, Peru: Two views of Andean social structure.* Buffalo: Council on International Studies, State University of New York at Buffalo.

Steinberg, T. 2000. *Acts of God: The unnatural history of natural disaster in America.* New York: Oxford University Press.

Walter, D. 2003. *La domestication de la nature dans les Andes Péruviennes: L'alpiniste, le paysay et le parc national du Huascarán.* Paris: L'Harmattan.

Walton, N. K. 1974. Human spatial organization in an Andean valley: The Callejón de Huaylas. Ph.D. diss., University of Georgia, Athens.

Wisner, B., B. Piers, T. Cannon, and I. Davis. 2004. *At risk: Natural hazards, people's vulnerability, and disasters.* New York: Routledge.

Zamora Cobos, M. 1973. Informe sobre la ascención al pico norte del nevado "Huascarán." Huaraz, octubre. Biblioteca, Unidad de Glaciología y Recursos Hídricos, Huaraz, Doc #. I-GLACIO-010.

Zapata Luyo, M. 2002. La dinámica glaciar en lagunas de la Cordillera Blanca. *Acta Montana* (Czech Republic) 19(123):37–60.

Zavaleta Figueroa, I. 1970. *El Callejón de Huaylas antes y después del terremoto del 31 de mayo de 1970.* Caraz: Ediciones Parón.

18

Responses to Glacier Retreat in the Context of Development Planning in Nepal

Shardul Agrawala

The potential and actual responses to climate change and glacial retreat in the small Himalayan kingdom of Nepal are shaped by both geographical and socioeconomic conditions. Nepal borders on the two most populous nations in the world, China and India. It is famed for its mountains, since it contains eight of the world's ten highest, among them Mt. Everest, at 8,848 m a.s.l. Less well-known is its extraordinary range of elevation. Some of its lowest regions are less than 100 m a.s.l., and the variation occurs within a very narrow range, only 200 km—a distance so short that it would barely fill a grid-box of most climate models. This range is responsible for the country's enormous variation in climate, from arctic to tropical regimes, and also creates the potential for hydropower.

Socioeconomic variables are also important. Nepal's population of 23 million is much smaller than that of its large neighbors. The Nepalese are among the poorest people in the world; over 80% possess an income below the international poverty line of US$2 per day, and nearly 80% somehow cope with an income of less than US$1 per day (World Bank 2003). Rural people, who make up 88% of the population, live at an average density of 686 per km^2. The association of poverty, rural residence, and agriculture is shown by the fact that agriculture is the main source of income for 81% of the population but contributes only 40% of the gross domestic product. Most farmers depend on irrigation rather than rainfall for water supplies, tapping the flow of the numerous streams that descend from the Himalayas. These streams are fed by rainfall during the monsoon season and by the meltwater from snowfields and glaciers during the dry season. The Himalayas contribute to Nepal's economy in other ways. They are one of the major attractions that draw the numerous tourists who flock to the country, and they support the hydropower on which the nation relies for 91% of its electricity generation. As this brief sketch suggests, Nepal's water resources, including glaciers, are central to the country's economic development.

Note: This paper draws upon S. Agrawala et al., "Development and Climate Change in Nepal: Focus on Water Resources and Hydropower," Environment Directorate, Development Co-Operation Directorate, Working Party on Global and Structural Policies, Working Party on Development Co-Operation and Environment, © OECD 2003. The views expressed in this paper are the author's own and not necessarily those of the OECD or its member countries.

IMPLICATIONS OF CLIMATE CHANGE

Recent climate trends and anticipated climate change have significant implications for Nepal's critical natural resources. The rivers that are fed by rains and by melting snows and glaciers provide water for the agriculture on which the bulk of the population relies and for hydropower generation as well. Even tourism rests on the white peaks that reach higher than anywhere else in the world. Since the late 1970s, Nepal has experienced significant increases in temperature, a trend that has been particularly marked at higher elevations (Shrestha et al. 1999) and has been noted also in adjacent areas such as Tibet (Liu and Chen 2000). The greater degree of warming at higher elevations poses a significant threat to Nepal's glaciers, whose retreat it may accelerate. The effects of climate change on rainfall and on river flows are not as sharply defined as the increasing temperature and the shrinking of glaciers, though researchers have documented an increased frequency of extreme precipitation events and longer runs of consecutive days of flooding.

The various climate change scenarios that have been constructed for Nepal through the use of different climate models all concur that warming will continue. Averaging these models yields a projected increase of 1.2 °C by 2050 and 3 °C by 2100 (Agrawala et al. 2003). A number of scientists suggest an intensification of the summer monsoon, a trend that is consistent with the higher frequency of intense rainfall events in that season.

IMPACTS ON GLACIER RETREAT AND RIVER FLOW

The Himalayan region of Nepal has already experienced significant impacts of these observed changes in climate. Virtually all the glaciers in Nepal are retreating, at rates that often range from 3 to 6 m per year (Shrestha and Shrestha 2004) and are projected to continue or accelerate. Melting is proceeding more quickly at lower elevations (Fujita et al. 2006) and is likely to increase at higher elevations in the future. The flow of rivers in the dry season is likely to decline as a result of this retreat and of the decrease in winter precipitation that is also forecast in climate change scenarios. This reduction is projected to take place across the entire Himalayan region, with minor differences in the projected timing that reflect specific basin geomorphologies and local variations in the distribution of precipitation (Rees and Collins 2006). These lower levels of river flow will present difficulties for irrigated agriculture and for the generation of hydropower.

The observed intensification of the summer monsoon, projected to increase under climate change, will also have severe impacts on human life and on infrastructure in the form of geohazards, particularly flooding and landslides. The danger of these monsoon-related floods is shown by an event in the Bhotekoshi Valley in central Nepal in July 1996. A massive landslide caused by heavy precipitation and runoff in an already weakened valley blocked the Bhairab Kunda Stream, which breached it and carried over 100,000 m^3 of debris downstream. When the debris rushed into the village of Larcha, over 70% of the houses were swept away and 54 people were killed (Adhikari and Koshimizu 2005). In addition to their immediate impacts, landslides have longer-term effects through the increase of sediment loads in rivers, which reduces the useful life span of reservoirs and other water-supply infrastructure. Taken together, these dry- and wet-season impacts will make river flow more variable and less reliable (Shakya 2003).

GLACIAL LAKE OUTBURST FLOODS

Nepal possesses more than 2,000 glacial lakes. These lakes may be formed behind ice or moraine dams, and some are located beneath or on the surface of glaciers. Glacier retreat and ice melt have increased the area and volume of a number of these lakes. Many of them drain themselves of excess water periodically when an increasing lake level reaches natural outlets. Because of their topography, many glacial lakes can release water catastrophically by breaking

TABLE 18.1

Glacial Lake Outburst Floods Recorded in Nepal (Shrestha and Shrestha 2004)

DATE	RIVER BASIN	NAME AND LOCATION OF LAKE
450 years ago	Seti Khola	Machhapuchhare, Nepal
August 1935	Sun Koshi	Taraco, Tibet
September 21, 1964	Arun	Gelaipco, Tibet
1964	Sun Koshi	Zhangzangbo, Tibet
1964	Trishuli	Longda, Tibet
1968	Arun	Ayaco, Tibet
1969	Arun	Ayaco, Tibet
1970	Arun	Ayaco, Tibet
September 3, 1977	Dudh Koshi	Nare, Tibet
June 23, 1980	Tamur	Nagmapokhari, Nepal
July 11, 1981	Sun Koshi	Zhangzangbo, Tibet
August 27, 1982	Arun	Jinco, Tibet
August 4, 1985	Dudh Koshi	Dig Tsho, Nepal
July 12, 1991	Tama Koshi	Chubung, Nepal
September 3, 1998	Dudh Koshi	Sabai Tsho, Nepal

through their dams or by responding to large calving events, producing devastating glacial lake outburst floods (GLOFs).

Also known by their Icelandic name *jökulhlaups* (literally "glacier-leap"), these outburst floods may discharge enormous volumes of water, destroying entire settlements and infrastructure. A number of factors can trigger them: avalanches into lakes, breakage of glacial ice into lakes, earthquakes, and spontaneous weakening and collapse of the moraine dams (Ives 1986). Since higher temperatures have promoted the melting of glaciers and snow, increasing the area and volume of many of Nepal's glacial lakes, the risk of outburst floods has increased. Table 18.1 lists the better documented of such events, many of which in fact originated in lakes that are located north of Nepal in Tibet. During one of them, floodwaters swept a Chinese military truck from Tibet into Nepal.

The greatest recorded damage associated with an outburst flood in Nepal took place in 1985. In this event, a debris-laden wall of water 10–15 m high swept 90 km down the gorges of the Bhote Koshi and Dudh Koshi Rivers, destroying 14 bridges, including new suspension bridges. The peak flow of this event, 2,000 m^3/sec, was several times greater than the maximum monsoon flood levels. The event devastated the nearly completed Namche Small Hydro Project, in which over US$1 million had been invested. The damage was so extensive that no restoration of the project was possible. The canals that led water into the plant were destroyed, and topsoil was stripped from many fields. A major trail linking an airstrip at Lukla to the Mt. Everest base camp was heavily eroded, and the prices of staples increased by about half when it was reopened (Ives 1986). Fortunately, the flood did not take place during the main trekking season, and therefore the loss of human life was lower than otherwise would have been the case.

An inventory of glaciers and glacier lakes in Nepal was jointly conducted in 2001 by the United Nations Environment Program (UNEP) and International Center for Integrated Mountain Development (ICIMOD). These groups registered 3,252 glaciers, 2,323 glacial lakes, and 20 potential glacial lake outburst flood sites (ICIMOD 2002). This inventory provides a view of risk at one point in time. In addition, site-based monitoring of specific glacial lakes documents increases in lake volumes over time, a finding that is consistent with the previously mentioned trends in temperature increase at high altitudes in the Himalayas. These lines of evidence, taken as a whole, indicate the serious nature of the hazards associated with climate change.

ADAPTATION OPTIONS FOR OUTBURST FLOOD RISKS AND STREAM-FLOW VARIATION

A variety of adaptation strategies provide means of coping with outburst flood risks and variation in stream flow. Some of these responses are already in varying stages of implementation in the context of development projects. The emphasis thus far, however, has been more on engineering solutions and considerably less on social measures to reduce vulnerability to such risks (Kattlemann 2003).

SITING OF HYDROPOWER FACILITIES IN LOW-RISK LOCATIONS

To protect hydropower infrastructure from the risk of outburst-flood-related damage, such facilities can be sited in locations at low levels of hazard. One noted geographer, reviewing the documentation from the Namche Project, comments that there is no evidence that any "special attention was paid to the possible occurrence of catastrophic geomorphic events, despite the fact that the project was being sited in one of the highest and most precipitous mountain regions in the world" (Ives 1986, 18). After the disaster, however, the Austrian government built its plant in another location where research suggests that the risk of an outburst flood is low.

In many cases, though, the high cost of relocation of existing facilities may preclude this option. For this reason, siting in low-risk locations may be possible only for new facilities. Costs may be incurred in such cases; there is the question of whether generating capacity would be reduced or transmission costs increased in moving to the alternative site. Indeed, planners may face a trade-off between risk reduction and profitability: sites that may offer greater hydropower potential and lower generating costs may also be ones that are at higher risk. Another concern is that, given the general uncertainty of outburst-flood risks, investors and energy planners may be reluctant to choose alternative locations when such risks must be integrated with many other factors in choosing a site. Two key barriers to effective incorporation of these risks in project siting are the lack of detailed and reliable spatial mapping of lakes indicating which of them are at risk of breaching and the interpretation of such maps, given the fact that these floods can travel as far as 200 km downstream. Catchmentwide analyses would be required to determine vulnerability to flooding downstream of hazardous glacial lakes. Furthermore, secondary damming resulting from the initial outburst-flood events can lead to the formation of large temporary reservoirs that are prone to bursting, creating additional serious risks for hydropower plants. In fact, these secondary flood risks may be even greater than the direct ones, since the reservoirs are typically at lower elevations and closer to hydropower plants. An integrated risk-management approach is therefore a necessary complement to satellite-based risk mapping of the glacial lakes themselves.

EARLY WARNING SYSTEMS

Since the impacts of glacial lake outburst floods are so severe, it is very important to establish early warning systems downstream from glacial lakes that are considered at risk of such events. In contrast to the situation with many other disaster warning systems, there is no forecasting capability for outburst floods. Nevertheless, timely warning once an event has begun can still save lives downstream. The speed with which floodwaters and debris move means that the effect of early warning systems is likely to be significant only in places rather far downstream from the point of origin. Further, while early warning can save lives, it cannot save infrastructure such as the hydropower facilities, bridges, and roads that are critical to the livelihoods of local populations. Another major constraint is the high setup and maintenance costs of automated early warning systems, estimated at around US$1 million for just one river basin. Such amounts represent a burdensome investment for a poor country like Nepal, particularly given that several of its river basins are at risk of glacial lake outburst floods and that such risks will increase with rising temperatures.

MICRO- VERSUS STORAGE HYDROPOWER

Planners face difficult decisions in choosing the forms of hydropower best suited to the risky setting of Nepal, since climate change has two different kinds of impacts on hydropower generation. One of these would favor the creation of many small power plants and the other the concentration of power generation in a few large ones. On the one hand, increased risk of outburst flooding would argue for numerous dispersed micro-hydropower facilities in small river basins. This would spread the risk and limit the overall loss of infrastructure and generation capacity. Micro-hydropower has the potential to fulfill a significant portion of the rural demand for energy. Water wheels, known locally as *ghatta*, have been used in Nepal for hundreds of years to process agricultural products. Nepal has 6,000 rivers and rivulets and 25,000 *ghatta* in use. Current micro-hydropower plants range in capacity from 1 to 25 kW, and there are 924 units in the country, producing approximately 10 MW of electricity. The development of micro- and small hydropower facilities is in line with Nepal's development priorities and is being encouraged by both the government and international donors. In other words, climate change may be an additional reason for promoting a strategy that is already being implemented for reasons of economic development.

At the same time, climate change is likely to lead to a situation in which interseasonal variation in river flows increases and dry-season river flows are reduced. Hydropower generation requires a certain level of reliability of river flows and certain baseline minimum flows during the dry season. Since smaller rivers are expected to show greater variability, adaptation to these impacts would require that hydropower facilities be built on larger rivers. Moreover, as dry-season flows decrease, planners will increasingly consider the construction of storage hydropower facilities. However, having large volumes of stored water might in fact be a maladaptation to the risk of glacial lake outburst floods, as they might cause a severe added risk if the storage facility were breached. This discussion shows a significant difficulty in adaptation responses: They may not always be internally consistent or, for that matter, coherent with other social or development priorities. Indeed, there could also be conflicts that might require careful consideration of the trade-offs involved.

INCORPORATION OF FLOOD RISK AND STREAM-FLOW VARIATION INTO PROJECT DESIGN

Though it is impossible to eliminate all the impacts of glacial lake outburst floods on downstream hydropower infrastructure, some design measures can nevertheless limit some of the damage. Building powerhouses underground will protect them from flood damage. Tailraces can be designed so that they are protected from floods and debris flows, and the water storage itself can be designed in a way that addresses the risk of excess sediment deposition due to flood-transported debris (Shrestha and Shrestha 2004).

Other less certain impacts of climate change can be incorporated through greater flexibility in the design. Hydropower generation already has several mechanisms in place to cope with stream-flow fluctuations as a result of current seasonal as well as climate variability. For example, a plant may have three intake channels and turbines to generate electricity during peak runoff in the monsoon season, one or more of which can be shut off during the dry season. This flexibility allows the plant to generate electricity more efficiently and without incurring losses for excess capacity. This option should be investigated to analyze the economic benefit of designing hydropower plants with the possibility of reduced capacity in future decades. Small hydropower plants are currently being designed with an average expected useful life span of 50 years, with most investors expecting a return on their investment within 7 years. One hydropower expert commented that reduced runoff and electricity generation in coming decades would affect his decision making only if two conditions were met: increased likelihood of significantly reduced runoff and the occurrence of this reduction within 20 to 25 years. His statement shows the challenges to planning on medium- to long-term time scales caused by the uncertainties in the magnitude and timing of many climate change impacts.

DIRECT REDUCTION OF RISKS

Other adaptive responses center on the physical reduction of the flooding risks of glacial lakes. Such measures include draining dangerous glacial lakes with siphons or pumps, cutting drainage channels that allow the lake to drain periodically, and constructing flood control measures downstream (Rana et al. 2000). A side benefit of such mitigation measures is that the "methods of remediation can be harnessed to facilitate safe management of the water resource for hydroelectric power at a local scale and for export" (Reynolds and Richardson 1999, 53). In addition to hydropower, the siphoned water could be used to supplement dry-season flows, maintain adequate water levels in downstream ecosystems to protect valuable fish stocks, supply water for local usage, and even provide recreational facilities.

However, all of these direct risk reduction measures have significant disadvantages. It is very expensive to pump water out of the glacial lakes. Their remote location at high altitudes makes it necessary to fly heavy equipment to the site by helicopter. Nepal's topography, with its steep elevation gradients, makes the downward movement of flood waters rapid and unpredictable, and therefore flood control measures are very difficult to design and operate. Moreover, these measures do nothing to prevent outburst floods from happening in the first place. Early warning systems tend to be expensive to set up and maintain and benefit only populations far enough downstream to have sufficient lead time. These disadvantages notwithstanding, there is one instance in Nepal—the Tsho Rolpa Risk Reduction Project—in which such responses have in fact been implemented in an integrated manner (Agrawala et al. 2003).

The Tsho Rolpa lake, at an altitude of about 5,000 m, increased in size from 0.23 km^2 in 1957–58 to 1.65 km^2 by 1997. It was estimated to be storing approximately 90–100 million m^3 in that year and considered a hazard that called for urgent attention. A 150-m-high moraine dam held the lake and, if breached, could have allowed a third or more of it to flood downstream. The likelihood of an outburst flood and the risks it posed to the 60-MW Khimti hydropower plant that was under construction downstream were sufficient to spur the government to initiate a project in 1998, with the support of the Netherlands Development Agency, to drain the lake. An expert group recommended lowering the lake 3 m by cutting a channel in the moraine and constructing a gate to allow water to be released as necessary. While the lake draining was in progress, an early warning system was established in 19 villages downstream of the Rolwaling Khola on the Bhote/Tama Koshi River. Local villagers have been actively involved in the design of this system, and drills are carried out periodically. The World Bank provided a loan to construct the system. The four-year project ended in December 2002, with a total cost of US$2.98 million from The Netherlands and an additional US$231,000 provided by the government of Nepal. The goal of lowering the lake level was achieved by June 2002, reducing the risk of an outburst flood by 20%. The prevention of an outburst flood would require further lowering of the lake level, perhaps by as much as 17 m. Expert groups are now undertaking further studies, but it is obvious that the cost would be substantial. Lake lowering might need to be repeated at Tsho Rolpa as lake levels rise again with further melting of surrounding ice. Furthermore, there are perhaps a dozen or more other dangerous lakes in Nepal, and similar response measures are beyond the capacity of the government and international donors. The cost, however, might be much less than the potential damage that would be caused by an actual event in terms of lost lives, destroyed communities, development setbacks, and reduction in energy generation.

CONCLUDING REMARKS

This analysis reveals the acute need for integrating considerations of climate change and development activity in Nepal. Rising temperatures will exacerbate the problems of glacier retreat and glacial lake outburst flooding that already impact human well-being, economic livelihoods, and development infrastructure. Nepal must therefore adapt to these risks and to projected climate changes such as reduced dry-season rainfall and stream flow. A consideration of initiatives that are already under way shows a complex mosaic of choices that frequently require high up-front costs and involve decision making under significant uncertainty. Moreover, these choices require citizens and planners to address complex trade-offs between risk reduction and profitability and between adaptation to one set of risks and other adaptation measures or development priorities. Since many of these adaptation measures are very expensive, they have required the involvement of the government and international donors. This involvement of two very different sets of actors creates difficulties. To be sure, they have collaborated on occasion, as shown by the Tsho Rolpa project. However, donors often mention the lack of coordination across various national government agencies, while government agencies point to a similar lack of coordination across donors. Donor-supported programs have on occasion also favored academic institutions from the donor country, with instances in which graduate students from such institutions were placed in expert roles. In another instance, computer software and project materials were available only in the donor language, limiting their utility to the government and its experts.

Another obstacle to the implementation of projects that address climate change risks and development priorities is the difficulties that host agencies and institutions face as they attempt to field simultaneous requests from various donors. The amount, continuity, and scope of project funding remain a continuing concern. Funding in the hydropower sector has also traditionally been more readily available for infrastructure for risk reduction as opposed to training and capacity-building efforts that might contribute to vulnerability reduction. Further, generally only current risks are incorporated into project planning. The evidence on whether plans and projects incorporate the increase in risks that is projected with a changing climate is at best mixed. This projected increase might offer opportunities for climate change funds and projects to be used to complement existing development funding by focusing on training and capacity building and on longer-term risk and vulnerability reduction.

Even if all these obstacles to planning in Nepal were resolved, an additional matter would remain to be addressed: the regional dimension of both climate change impacts and responses. Many catastrophic glacial lake outburst floods in Nepal in fact originated in Tibet. Moreover, decisions about water resource management and hydropower generation in Nepal affect neighboring countries downstream, particularly India and Bangladesh. These decisions therefore not only involve national discourses on the linkages between climate change and development but must take place in the context of regional discussions that can formulate coordinated strategies. As consideration of climate change so often does, the discussion of Nepalese responses to glacial retreat shows the interconnected nature of human life in the twenty-first century, with all its attendant difficulties and opportunities.

REFERENCES CITED

Adhikari, D. P., and S. Koshimizu. 2005. Debris flow disaster at Larcha, upper Bhotekoshi Valley, central Nepal. *The Island Arc* 14:410–23.

Agrawala, S., V. Raksakulthai, V., M. van Aalst, P. Larsen, J. Smith, and J. Reynolds. 2003. *Development and climate change in Nepal: Focus on water resources and hydropower.* COM/ENV/EPOC/DCD/DAC(2003)1/Final. Paris: OECD.

Fujita, K., L.G. Thompson, Y. Ageta, T. Yasunari, Y. Kajikawa, A. Sakai, and N. Takeuchi. 2006. Thirty-year history of glacier melting in the Nepal Himalayas, *Journal of Geophysical Research* 111, D03109, doi:10.1029/2005JD005894.

ICIMOD (International Center for Integrated Mountain Development). 2002. *Inventory of glaciers, glacial lakes, and glacial lake outburst floods: Monitoring and early warning systems the Hindu Kush–Himalayan Region, Nepal.* Kathmandu.

Ives, J.D. 1986. *Glacial lake outburst floods and risk engineering in the Himalaya.* International Center for Integrated Mountain Development Occasional Paper 5. Kathmandu: ICIMOD.

Kattelmann, R. 2003. Glacial lake outburst floods in the Nepal Himalaya: A manageable hazard? *Natural Hazards* 28:145–54.

Liu, X., and B. Chen. 2000. Climatic warming in the Tibetan Plateau during recent decades. *International Journal of Climatology* 20:1729–42.

Rana, B., A.B. Shrestha, J.M. Reynolds, R. Aryal, A.P. Pokhrel, and K.P. Budhathoki. 2000. Hazard assessment of the Tsho Rolpa glacier lake and ongoing remediation measures. *Journal of Nepal Geological Society* 22:563–70.

Rees, H.G., and D.N. Collins. 2006. Regional differences in response of flow in glacier-fed Himalayan rivers to climatic warming. *Hydrological Processes* 20:2157–69.

Reynolds, J., and S. Richardson. 1999. Geological hazards: Glacial. In *Natural disaster management: A presentation to commemorate the International Decade for Natural Disaster Reduction,* ed. J. Ingleton. Leicester: Tudor Rose.

Shakya, N.M. 2003. Hydrological changes assessment and its impact on hydro power projects of Nepal. Paper presented to the Consultative Workshop on Climate Change Impacts and Adaptation Options in Nepal's Hydropower Sector with a Focus on Hydrological Regime Changes including GLOF, Department of Hydrology and Meteorology and Asian Disaster Preparedness Center, March 5–6, 2003, Kathmandu.

Shrestha, M.L., and A.B. Shrestha. 2004. Recent trends and potential impacts of climate change on glacier retreat/glacial lakes in Nepal. Paper presented at OECD Global Forum on Sustainable Development, November 11–12. ENV/EPOC/GF/SD/RD(2004) 6. Paris: OECD.

Shrestha, A.B., C.P. Wake, P.A. Mayewski, and J.E. Dibb. 1999. Maximum temperature trends in the Himalaya and its vicinity: An analysis based on temperature records from Nepal for the period 1971–94. *Journal of Climate* 12:2775–89.

World Bank. 2003. *World development report 2003: Sustainable development in a dynamic world.* New York: Oxford University Press.

19

Glaciers and Efficient Water Use in Central Asia

Urs Luterbacher, Valerii Kuzmichenok, Gulnara Shalpykova, and Ellen Wiegandt

Among their many roles, glaciers serve as important reserves of freshwater. Water in the form of glaciers is particularly valuable because it constitutes a stock that tends to be released in periods when water in other forms is scarcest. These general characteristics of water create complex management problems. In addition, the current retreat of many glaciers alerts us that supplies may be changing in the future. When the inherent qualities of water and of glaciers are superimposed on evolving political and economic circumstances, the potential for conflict is substantially enhanced. Central Asia constitutes just such a case. Here, a political structure in which internal relations were managed by a strong central authority under the Soviet system has become one of international relations among upstream and downstream states, in which other asymmetries also play an important role. Formerly interdependent regions vie as autonomous entities for water for irrigation (under conditions of low precipitation) and for energy production. Currently, tensions are high, but there is no overt conflict. Avoiding future outbreaks of violence among the republics, especially between those upstream, where water resources originate, and downstream, where the level of use is highest, will require complex negotiations among the parties. Achieving agreements will require understanding the fundamental issues of water allocation, forecasting the evolution of glaciers, and balancing the needs and wants of the various populations. Our research into these interacting aspects leads to some concrete proposals for solutions for the water management problems in the region.

GLACIERS AND WATER RESOURCES IN CENTRAL ASIA

The Central Asian states of Kyrgyzstan, Tajikistan, Turkmenistan, Uzbekistan, and Kazakhstan draw their water from the Amu Darya and Syr Darya Rivers. The latter originates in the Tien-Shan and Pamir-Alay Mountains of eastern Kyrgyzstan, at altitudes of 7,000 m, and then flows through Kyrgyzstan, Tajikistan, and Uzbekistan to Kazakhstan and into the Aral Sea. The Amu Darya has its source in the Pamir region of Tajikistan and then forms the border between Tajikistan and Afghanistan before

FIGURE 19.1. The geographical and geopolitical situation.

becoming the border between Uzbekistan and Turkmenistan and then flowing into the Aral Sea (Figure 19.1). Kyrgyzstan is thus a vital link in the hydrological system of the region and, in addition, contains 47% of the glaciers of the Central Asian republics of the former Union of Soviet Socialist Republics (USSR) (Figure 19.2).

All Central Asian states are dependent to varying degrees on irrigated crops for survival and draw their water from the Amu Darya and Syr Darya river systems fed by the upland glaciers. Intensive production has led to overuse of water resources and to the drying of the Aral Sea. The resulting water shortages have created tensions with the upstream republics of Kyrgyzstan and Tajikistan, which rely on glacier meltwater for the production of electricity from hydropower. Glacier meltwater provides for irrigation but also serves to fill high-altitude retention lakes constituted by dams placed in certain river basins. How these resources are exploited and distributed depends largely on the entitlements attached to the water and the accompanying rights, privileges, obligations, and limitations related to its use. These constitute property rights, and they act in effect as quotas that control access and limit quantities used. How effectively these rules are designed and followed determines whether use of particular resources will be sustainable.

The prevailing view in economics has long been that private property rights lead to efficiency of resource use. As Dasgupta and Heal (1979, 48–52) show, however, clear definition and effective enforcement of property rights are essential, and flowing water is a resource for which property rights are difficult both to define and to enforce. It cannot be treated as a separate commodity because it is in a "constant state of diffusion" (p. 49) or movement, and a precise unit therefore cannot be allocated to a single individual. In the case of pools of underground water, property rights can be defined as an area of land above them, but it will not be clear whether the water extracted really comes from the area below this surface because of the fluid nature of the resource. In the case of glaciers and the water resources associated with them, similar problems arise. Glaciers are usually situated in relatively remote areas that cannot easily be controlled. What flows from up high is therefore difficult to attribute to a single owner and sometimes even to a single country. Principles for sharing the resource must be designed, a fact recognized by most legal systems even at the international level. There is, however, no international water regime in place.

The interrelation between the inherent characteristics of water and the technologies used to exploit it has frequently led to common-property arrangements and collective structures for allocating and managing water resources to prevent some groups or individuals from overusing them at the expense of others. Without socially imposed limits, the end result is overexploitation and dissipation of the resource (Figure 19.3). The use of a natural resource such

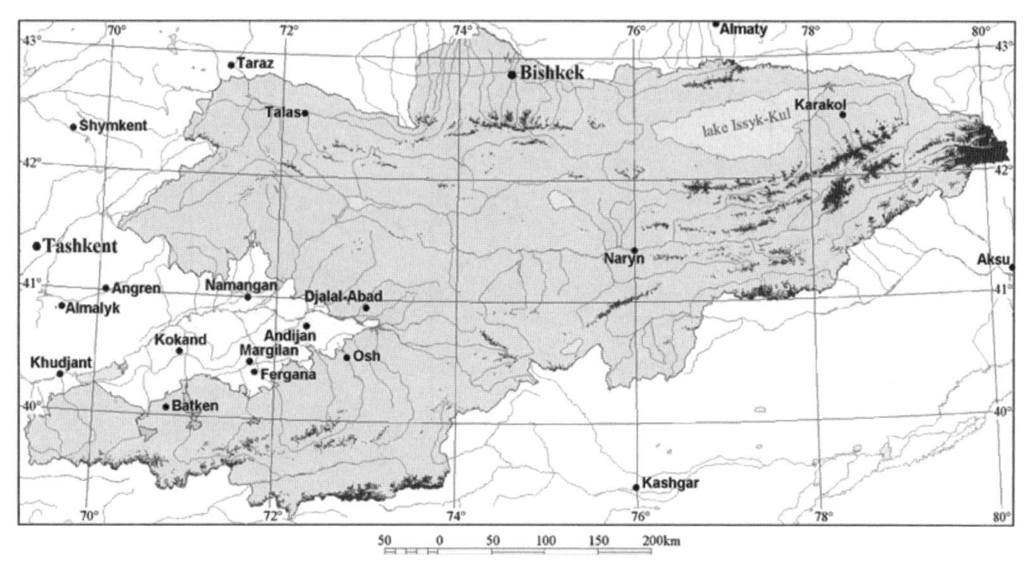

FIGURE 19.2. The glaciers of Kyrgyzstan (black dots).

as water for irrigation, depends on the number of producers willing to share in its supply. The operation of the production system (taking into account initial investments) requires a relatively large number of producers, leading initially to increasing returns to scale. As more and more producers join, however, diminishing returns set in. This is illustrated by the S shape of the production curve. The fixed cost per unit of production for each producer crosses the production curve at two points: at A, where the number of producers is sufficient to initiate profitability, and at C, where an excess of producers has eliminated profits or surpluses. The optimum lies at B, which maximizes surplus production over costs. This optimal situation can be maintained by limiting the number of producers entering the process either by exclusion or by taxes, which increase the fixed costs for each entrant. A solution to the problem involving taxes is represented by the thin straight line parallel to the cost line and tangent to the production curve at B, showing that correct taxation of entrants limits their number and ensures maintenance of the scarcity rent.

Inefficiency from overexploitation of a common resource can always be corrected by an appropriate taxation policy. Establishing such a tax is relatively easy in the general case because profit seeking leads to a single maximum. The tax limits the number of users entering production. Dasgupta and Heal (1979) show, however, that such an outcome usually does not obtain when asymmetries are present, as is the case with upstream-downstream relations under differentially defined property rights. In the case, for example, of a downstream firm that must base its production on water that is already being used by an upstream firm, there are two important considerations. One is the degree of use due to the first firm, which may significantly reduce the possible profits for the second firm. The other is the property rights (or legal rights and obligations) that the firms have with respect to each other in terms of clean water. If the first firm has an unlimited right to the water and the second rights only to what remains of it, the second will eventually be driven out of business in the absence of some negotiated arrangement between the two. Conversely, if the first firm is constrained to limit its use in consideration of the needs of second, it may have to cease its activities. Assigning property rights to one side or the other changes the production possibilities and the relations between the two firms. The firms' production possibilities are no

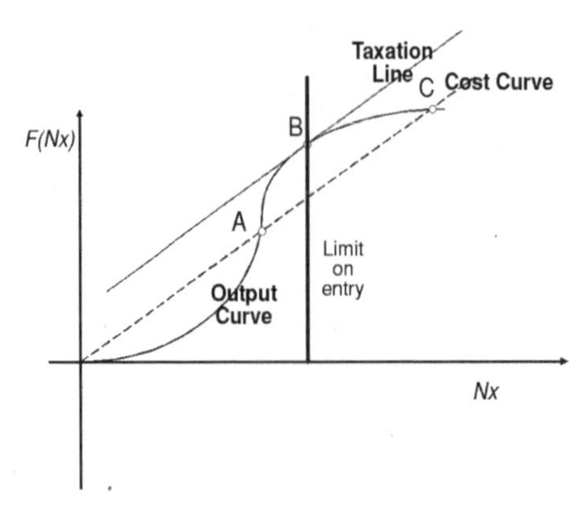

FIGURE 19.3. Limiting natural resource use. $F(Nx)$, use of the resource; Nx, number of producers willing to share in its supply.

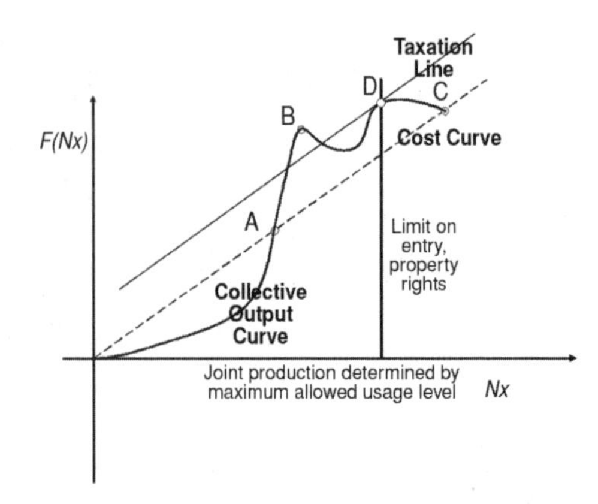

FIGURE 19.4. Multiple taxation equilibrium case due to preestablished property rights.

longer convex, and neither taxation nor buying or selling of rights will bring a uniquely defined equilibrium solution.

In Figure 19.4, B would represent a higher level of profit than D, but joint production will remain at D because property rights permit one firm to keep the other from using the resource more efficiently by restricting usage for the first. If the two firms can agree on a solution that, under redistribution of profits arising from higher efficiency, allows them both to be better off, then they should adopt it. In reality, however, the problem can be more complex, because giving up property rights ultimately raises a credibility problem. If the first firm abandons some of its property rights to the second firm in exchange for a share of higher profits, it has to have a strong guarantee that the second firm will indeed share its profits. Whereas at the domestic level this problem can usually be solved by contract, nothing like this occurs in an international setting between sovereign states. In Central Asia, the situation might be better overall if upstream countries were allowed to produce electricity with enough water for their use (at B) but had to release it to less efficient downstream users (at D), but the credibility problem we have mentioned keeps downstream republics from relinquishing their water rights.

In cases where glaciers are present, the most common confrontations among users revolve around the positions that various users occupy between the high-altitude source of the water (the glacier) and the place where the river system enters a sea. A cursory analysis would conclude that upstream users have control and power over downstream users. However, this is not necessarily the case, because downstream users may have more economic and power resources with which they can retaliate against any upstream attempt to cut off the supply of water. Only when an upstream nation has substantial economic and power resources in addition to its location can it impose its goals on downstream users.

CENTRAL ASIA: THE PROBLEM

Reports from Central Asia regularly alert the international community to worsening ecological conditions, the dire social and economic status of its population, and the ensuing potential for serious civil and interstate conflict. The recent riots and upheavals in Uzbekistan and Kyrgyzstan are an illustration of latent hostilities. The situation is particularly complex and delicate because familiar problems of overextensive irrigation agriculture and population increase have become mixed with interstate politics as a result of the collapse of the USSR.

Fluvial water resources play essential roles in the economy and society of the Central Asian states of Kyrgyzstan, Tajikistan, Turkmenistan, Uzbekistan, and Kazakhstan, all dependent in varying degrees on irrigated crops. Cotton, the most important irrigation crop, is the major source of income and employment in Turkmenistan and Uzbekistan. Its production was encouraged under the Soviet system as a source of hard currency. Overuse of water resources for irrigation, however, is responsible for the drying of the Aral Sea, whose surface area and volume have declined by 35% and 58%, respectively, since the mid-1980s. This has aggravated local climate conditions, reduced the amount of water available for agriculture, and deprived the upstream countries of Kyrgyzstan and Tajikistan of the amounts necessary for their hydropower production, which accounts for 50% of their electricity production. These competing water uses are only made worse by demographic patterns that have led to an increase in population of 140% between 1959 and 1989 (Horsman 2001, 71) and a further projected increase of 35–50% in most of the states between now and 2050 (Population Reference Bureau 2002).

Although tensions over water allocation are not new, they have taken on new significance since the collapse of the USSR. Previously managed from Moscow by a centralized administration, water systems have suddenly come under the control of separate sovereign states that have no history of agreements or coordination structures. This poses important allocation problems. Upstream states Kyrgyzstan and Tajikistan need water for hydroelectric production as well as irrigation, while Kazakhstan and Uzbekistan use water mostly for irrigation. The upstream republics have held up the release of water or threatened to charge for delivery downstream to pressure downstream users to compensate them for the energy production forgone when water is released for downstream irrigation. Despite their control over the source of water, upstream states are implicated in allocation schemes that oblige them to provide water downstream. Soviet patterns of allocation were reaffirmed in the Almaty agreement of 1992 but have never been perceived to be equitable.[1] They continue to reflect the favored status that downstream countries had achieved. Kyrgyzstan and Tajikistan would like to expand irrigation agriculture as well as electricity production. However, their dominant upstream position does not permit them to achieve these goals because of their political weakness in the face of the downstream users' control over coal and gas (Horsman 2001, 74–75). Indeed, after independence Uzbekistan and Kazakhstan introduced market prices for gas and coal. Kyrgyzstan could not pay these higher prices, and its response was to increase electricity production to augment revenues. This meant that the amount of water available for irrigation in Uzbekistan and Kazakhstan was reduced. As a consequence, agreements were not respected. In breaching or threatening to breach those agreements, upstream states became vulnerable to reprisals from downstream states.

The case of Central Asia illustrates two important aspects of the way in which water is distributed and used. One is that upstream republics can control the quantities of water sent downstream but are subject to reprisals because they do not control other critical resources such as gas and coal. The other is that, when users of a natural resource have both unclear property rights and crosscutting powers or access, there is no obvious solution that is both equitable and efficient. Management schemes must therefore be negotiated.

A potential solution for Central Asia would be to compensate for the asymmetry in access to water by industrial and agricultural developments that would benefit all the countries concerned. Two countries or regions could share in the advantages created by the development of water resources in one of them by specializing in complementary activities, for example, concentrating industry in the area less suitable for agriculture, taking advantage of the cheaper electric power made available by dam construction upstream. High-altitude countries could develop high-pressure dams,[2] like those in the Alps, that are not very harmful to countries downstream. These dams also have the advantage of providing large amounts of electric power under peak load conditions. This allows them to serve the industrial needs of firms located relatively far away and thus be useful beyond the boundaries of a given country. The high-altitude but relatively poor countries of Kyrgyzstan and Tajikistan could benefit from such schemes. International organizations such as the World Bank could devise policies that favor such positive spillovers and thus help to resolve otherwise intractable water disputes, in part by offering to be guarantors of future profit sharing.

Under the present system, Kyrgyzstan receives only 10% of the waters of the Syr Darya basin, the remainder being used for irrigation downstream. There may be ways of producing gains in hydropower development that would more than make up for the potential losses in agriculture from slightly diminished irrigation. Of course, such gains will be of interest only if all the countries in the region benefit from them, and this will be the case only if they cover significant regional needs and if all the countries have some institutional control over hydropower use. We will suggest that both possibilities exist. In particular, we will show that the development of hydropower can enhance industrial efficiency for the whole region and that all the countries can profit from it. Moreover, given the legal obligations of the Almaty agreement, a transnational institutional setup for this development can also be established.

GLACIERS AND ENHANCED WATER RESOURCES IN CENTRAL ASIA

To examine future water resource potential, we studied the current and projected extent and water content of key glaciers. Glaciers within Kyrgyzstan make up 47% of the total glacial area in the region. The largest glaciers are situated at an altitude of about 4,000 m. Climate change and the resulting glacier melt and loss of ice mass over time have consequences for the hydrological system of the country, notably by increasing annual water flows of several major rivers. Moreover, water supply will be enhanced not only by additional glacier melt but also by an expected increase in precipitation.

Translating all these characteristics into potential water flows for Kyrgyzstan, we find that they amount to about 51 km^3. Under climate change, these flows could actually increase by 10% because of increased precipitation and glacier melt, making available more than 56 km^3. According to the estimates made by the Institute of Water Problems and Hydropower, this additional water could be used to produce 150 billion kWh of electricity, more than the present combined electricity production of Kazakhstan, Uzbekistan, and Kyrgyzstan, if its potential were fully used. There is therefore considerable potential for growth, and it could be nearly doubled if similar resources in Tajikistan were also used. To examine the impact of this capacity on the country's economy and that of its neighbors, we carried out empirical analyses of present conditions and constructed socioeconomic models to explore future developments in the various countries of the region.

Economic trends for the countries of Central Asia tend to confirm the tendencies noticed for other transitional states. When they abandoned Soviet-style economies in the 1990s, these countries all experienced a period of decline in their gross domestic products. Then, from around 1995 on, expansion resumed. Measurements of value added,[3] both in agriculture and in industry, show the same. For agriculture, however, the rebound was slower for Kazakhstan than for Kyrgyzstan and Uzbekistan, and Kyrgyzstan, originally the lowest agricultural producer of the three, showed the highest growth rates after 1995 (World Bank 2001).

To calculate the effects of specific capacity improvement schemes, we constructed value-added country models involving important sectors of the economy and their water resources input, either direct (as in the case of irrigation) or indirect (as in the case of electricity). We derived the gross domestic product (GDP), the most commonly used measure in national accounts statistics for economic growth, as a sum of gross value

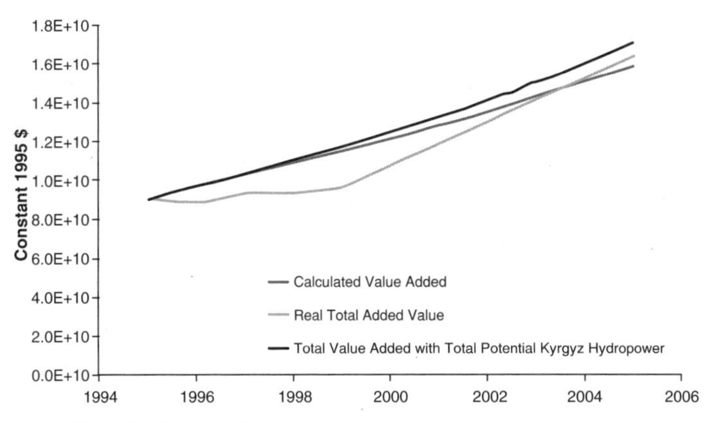

FIGURE 19.5. Total value added.

added (GVA) by resident producers net of taxes (taxes − subsidies). Typically, to arrive at the GDP for the economy, economic activity is divided on a sectoral basis and then summed again. Thus, the GDP consists of the GVA generated in the agricultural, industrial, and services sectors. These quantities can then be represented in term of their monetary values by a vector of input (p_i) and output prices (p_o): GVA = Value of output ($p_o.q_{output}$) − Value of intermediate inputs ($p_i.q_{input}$) + Net Taxes. Water enters as an input in the measurement of GVA in each of the three sectors. Thus the total water available in the system must be distributed across these three sets of activities as well as consumption for domestic activities and by the environment, which act as a constraint on the water available for productive activities. While it would be reasonable to assume that consumption by the environment is more or less a static amount, the same is not true for domestic consumption. The latter may increase or decrease depending on the rate of population growth and the assumptions made about per capita water consumption. Thus each scenario, whether baseline, increasing, or decreasing, has implications for the water available for agriculture, industry, and services. Here our data consist of the value added in agriculture and industry for the three countries considered. We then try to determine the evolution of gross value added for each sector. The evolution of gross value in agriculture is represented as the difference between the desired level, which depends on income and available water for irrigation, and the actual level, which dampens the increase and depends on the current level of gross value added. In addition, labor, which can be approximated by the proportion of the total population involved in agricultural activities, also influences the evolution of gross value added (for a complete representation of the dynamic equations corresponding to this verbal description, see the appendix). Industrial value-added evolution depends on income (GDP), electricity production, and the proportion of the population involved in industry. Such equations, along with others representing population evolution, can now be written for each Central Asian country.

Once values are calculated for the levels of gross value added in agriculture and industry for Kyrgyzstan, Kazakhstan, and Uzbekistan, we can compare them with real data. We can also add these series together to get total value-added figures (minus gross value added in services) for each country and then a total gross value-added number for the whole region. Our strategy then allows us to simulate the impact on the whole region of a significant increase in the electricity production of Kyrgyzstan in terms of total value added.

From Figure 19.5 we can see that calculated gross value added for the region as a whole tracks real value added for the three countries, particularly in the later period, and that potential total value added reaches levels significantly higher than the current real ones. Thus, an increase in Kyrgyzstan's hydroelectric capacity would ultimately benefit all three countries. Moreover, this result could be achieved by constructing hydroelectric dams in mountainous regions where the population density is relatively low. A simple geographic analysis of population densities shows that they are highest in the northern and southwestern regions of Kyrgyzstan, relatively far from elevated areas. The mountains are in the south and southeast, where dam construction and operation would have relatively little impact on resident populations. The crucial factor in achieving added productive capacity of electricity is the negotiation of agreements with neighboring countries regulating the seasonal use of the dams. Kyrgyzstan would need water especially during the winter and would also need to modulate river runoff somewhat during the summer when the demand for irrigation water was high. This adjustment of summer water would seem to involve relatively insignificant quantities and could easily be achieved as dams filled up because of glacier melt.

CONCLUSIONS

We have suggested that freshwater use generates many complex interactions among various users. Often such situations lead to tensions, which in turn can escalate into water-based conflicts as they become tied to other oppositions. The importance of water as the source of life often imbues these conflicts with heavy symbolism. The link between water and agriculture reinforces these symbolic aspects, and, since it is by far the largest user of freshwater, agricultural use is often at the heart of water conflicts. In certain regions, agriculture is impossible without a substantial amount of irrigation. It prevails for reasons of national sovereignty and autonomy of food production and because of the strength of agricultural lobbies. Reliance on irrigation can occur because water is subsidized or because costs are improperly calculated and therefore artificially low. Agriculture's privileged access to water is increasingly being contested, however, as competition among sectors and regions increases. In recent years, traditional agricultural water use has been in competition with requirements for power generation. Upstream-downstream relations tend to accentuate this competition, which can lead to shortages and thus, increasingly, to cross-boundary conflicts over resources. Reducing them requires a new organization of production processes with redistribution of benefits resulting from better use of the comparative advantages of regions.

It is important to understand current water use and the factors that influence it, develop predictors of future availability under conditions of climate change, and evaluate existing legal frameworks regulating cross-boundary use. These can become the basis for developing circumscribed initiatives to foster cooperation. Applying this approach to the Central Asian case, we have come to the following conclusions: The development of Kyrgyzstan's hydropower could lead to significant improvement in the welfare of the whole region by creating higher potential total gross value added. Such a policy would have two additional advantages. It would lessen regional dependence on fossil fuels and even allow their exportation to Western Europe or China, thereby providing outside income to the region. It would also diminish the importance of the largely inefficient agricultural production of the downstream countries, where industrialization based on the availability of cheap electricity from upstream hydroelectric plants could be further developed.

APPENDIX

1. The structuring of value-added relations: $GVA_{economy} = GVA_{agriculture} + GVA_{industry} + GVA_{services}$.
2. The evolution of value-added sectors:[4]
   2.1 Agriculture: GVAAG.=(A*GDP + B*WAT)*(1-ALF*GVAAG) + H*POP, where GVAAG is gross value added in agriculture, GDP is gross domestic product, WAT is water input for agriculture, and POP is population. A, B, ALF, and H are parameters, the

latter representing the proportion of the population involved in agriculture.

2.2 Industry: GVAIN. = C*GDP + K*EL + D*POP, where GVAIN is gross value added in industry, GDP is gross domestic product, and EL is electricity inputs to industry. C, K, and D are parameters, the latter representing the proportion of the population involved in industry.

2.2.1 Electricity: EL. = GAM*WAT (electricity produced by other sources is just an extrapolation from current trends).

2.3. Services: GVAS. = E*GVAS, where E is a parameter.

3. Total value added: GVAG + GVAIN + GVAS.

REFERENCES CITED

Dasgupta, P., and G. Heal. 1979. *Economic theory and exhaustible resources*. Cambridge, UK: Cambridge University Press.

Horsman, S. 2001. Water in Central Asia: Regional cooperation or conflict? In *Central Asian security*, ed. R. Allison and L. Jonson, 70–94. London: Royal Institute of International Affairs.

Luterbacher, U., T. M. Clarke, P. Allan, and N. Kessler. 1987. Simulating the response of a small open politico-economic system to international crises. *Management Science* 33:270–87.

Population Reference Bureau. 2002. *World population data sheet*. http://www.prb.org/pdf/ WorldPopulationDS02_Eng.pdf (accessed August 2007).

Shalpykova, G. 2002. Water disputes in Central Asia: The Syr Darya River Basin. Master's thesis, International University of Japan.

World Bank. 2001. *World development indicators*. Washington, DC.

20

Research and Development in Mountain Glaciers

AN OVERVIEW OF INSTITUTIONS AND ORGANIZATIONS

Denis Knubel, Greg Greenwood, and Ellen Wiegandt

Fear, reverence, fascination, observation, and scientific analysis encompass the different ways glaciers have been assimilated into our views of nature-society interactions. Early perceptions inspired local legends and shaped historical cultural practices regulating the use of glacier water and the response to glacier-induced disasters. Progressively, scientific inquiries addressed glacier dynamics and sought to understand, predict, and respond to the potential dangers and opportunities presented by glaciers. These often individual endeavors are sometimes linked together in university centers, networks, or consortia. In parallel, more institutionalized efforts have been made since the nineteenth century to systematize and coordinate information. The history of these individual and collective efforts is a window into a particular domain of the history of science but also into the politics of scientific inquiry and the relation between science and policy at the national and the international level. Given the close linkage between glacier dynamics and climate change, the role of glacier science in alerting society to the potential impacts of glacier melting acquires new urgency. Developing a consensus about what we know and identifying gaps in our knowledge have implications for our understanding of what the consequences of glacier retreat may mean not just for our vision of our surroundings but also for our energy use, our agriculture, and our vulnerability to catastrophes. The inventory of glacier organizations that follows provides an overview, albeit incomplete, of sources of information about glaciers and thus points to what we know and what we still need to know about glaciers' dynamics, their links to climate change, and the impacts their shrinkage will have on the ways of life not only of people living among them but of the far greater numbers of people who value them in their landscapes and depend on them, if only indirectly, for energy and food.

The science of glaciology is generally agreed to have originated in Switzerland. Early fieldwork carried out by Ignace Venetz identified glacier moraines downslope from existing glacier limits. He concluded in an 1821 article, presented to the Swiss Society of Natural Sciences in 1822, that this could be explained only by previous glacier advance, thus launching the theory of

Ice Ages (Mariétan 1959).[1] Though first received with skepticism, this view was soon accepted by other scholars, including Louis Agassiz, whose international reputation encouraged further research and theoretical developments.

To confirm these theories and to confront conflicting religious interpretations of geologic history, there was a strong impetus for field observation and monitoring. Much of the development of monitoring efforts and early scientific understanding began in Switzerland, where the International Glacier Commission was founded at the Sixth International Geological Congress in Zurich in 1894. This group carried out regular and rigorous surveys at selected glaciers but also relied on mountain inhabitants to provide supplementary observations (see Haeberli, Steiner et al., and Wiegandt and Lugon, this volume). Other individual countries, especially those where large areas are covered by glaciers, such as Norway or Russia (and the former Soviet Union), developed their own academic departments and monitoring programs. In the details of this early history we can discern the role that organizations can play in ensuring the collection of continuous and intercomparable data and even in identifying or defining critical research areas, thus potentially influencing policy responses. Glacier monitoring proceeded in fits and starts, succumbing to scientific debates and fashion and to broader political contexts that frequently intervene to slow or encourage research in particular fields. Haeberli (this volume) identifies the different phases of international glacier observation and the focus on particular aspects of glacier dynamics and the corresponding methodologies. The importance of recognizing these different phases is that they explain why our knowledge of historical glacier patterns may have missing pieces or suddenly include new kinds of information. The earliest observations provided maps of such accuracy that they can be used to identify long-term volume and mass changes. Between the two world wars, interest and capacity to maintain these data were wanting, and, according to Haeberli (this volume), we therefore lack comprehensive scientific assessment of the climate warming and glacier shrinkage of the 1930s and 1940s. New measurement techniques for calculating mass balance reenergized research by the late 1950s. Subsequent scientific quarrels again put systematic monitoring in peril, and it was not until the 1980s that efforts picked up again, stimulated by theoretical developments and the growing concern about the consequences of anthropogenic climate warming for glacier extent.

Our survey of organizations currently focusing on glaciers is testimony to the geographic breadth of interest and also implicitly reflects the historical trajectory of scientific concerns. In presenting such an inventory and overview of key groups and organizations, we at once document this history and signal present concerns. We thus highlight the state of knowledge but also point to potential gaps in regional coverage and scientific understandings. These become all the more relevant in the context of the international political negotiations to address the consequences of climate change embodied in the United Nations Framework Convention on Climate Change and its Kyoto Protocol.

BUILDING AN INVENTORY OF ORGANIZATIONS DEVOTED TO GLACIERS

The organizations described here do not constitute a systematic inventory of organizations devoted to glaciers. We have tried to identify some key groups and to achieve a breadth of coverage not only geographically but also in terms of the nature of the organizations' interest in glaciers. Most that we have selected have scientific concerns ranging from observation to advocacy. The World Wide Web has been a major source, and while this has introduced significant limitations on coverage, it allows us to present the reader with a mechanism for exploring information and deriving other links. We have also arbitrarily organized the information according to a set of criteria that we first defined and then used to summarize key findings.[2] We adopted a two-stage approach. First, we searched the World Wide Web with phrases such as "glaciology," "glaciers, protection," and "glaciers, communities" to find relevant organizations. Our main research was done in English and French, with additional searches in German, Russian, and Spanish. Even with this restriction on language, search results were plentiful and interesting. Second, we explored additional links found on the numerous Web sites visited. In this way, we gained insight into existing and partly overlapping networks, mostly but not exclusively in the English-speaking community. (Some of the most important links were found on a Norwegian and a Chinese Web site.)

This search is by definition incomplete. Reliance on the World Wide Web can produce only organizations that choose to present themselves there. Thus it is probable that there are organizations for which mountain glaciers are extremely important but that nonetheless do not appear in this list simply because they lack a presence on the Web. Moreover, it is certain that important organizations are present on the Web but publish only in Chinese, Japanese, or some other language not covered in our search, and they also would have been missed.

Given that hundreds of entries may result from a single Google search, it was necessary to define a point beyond which we would not go. When the same Web sites started to appear over and over again or when the libraries of explored links did not lead to anything new, we decided that we had passed the point where marginal cost equaled marginal return and therefore stopped the search.

Since polar institutions represent important members of the community interested in glaciology, we have also included them in our list. Mountain glaciers as a distinct topic are often inextricably associated with polar glaciers and ice caps. Sites that subsume mountain glaciers under the polar cryosphere are presented separately in the database.

RESULTS

We created three categories to impose some structure on the search results: archives and monitoring, research and academic institutions, and institutions for which glaciers have special significance. This structure is entirely ad hoc but nevertheless provides a framework for discussion.

ARCHIVES AND MONITORING

Data archives are essential for long-term and cumulative research. There are several parallel structures that fulfill different but overlapping objectives. We located three World Data Centers (WDCs) for cryospheric data, located in Boulder (Colorado, United States), Cambridge (UK), and Lanzhou (China). These include data on mountain glaciers but have a broader focus on the cryosphere. The World Data Center system was established as part of the International Geophysical Year (1957–58). These centers undertake international data exchanges in terms of the principles established by the International Council of Scientific Unions (ICSU). As one of more than 40 WDCs around the world that are collecting, archiving, and distributing geophysical data, the World Data Center for Glaciology provides a focus for snow and ice information services. Data sets cover the subject areas of glaciers, avalanches, snow cover, polar ice masses, ice cores, sea ice, and freshwater ice.

Among the other institutions characterized as "data centers" are organizations that process, archive, and redistribute data captured by NASA or the European Space Agency (ESA) satellites. While some organizations are specifically linked to a single "provider" (e.g., Centre ERS [European Remote Sensing Satellite] d'Archivage et de Traitement [French ERS Processing and Archiving Facility], CERSAT, which acts as a "node of the ESA ground segment for the ERS-1 and ERS-2 Earth observation satellites"), others accumulate data from different sources. Located in Fairbanks and considered one of the NASA-sponsored Distributed Active Archive Centers, for example, the Alaska Satellite Facility (ASF) processes data coming from ESA, the Japanese NASDA, and the Canadian Space Agency satellites. Some of these institutions specialize in archiving data related to polar regions, among them the Arctic System Science (ARCSS) Data Coordination Center (ADCC) and the Arctic and Antarctic Research Center (AARC) of La Jolla (California, United States).

Closely related to data archives are glacier-monitoring programs. The most comprehensive is the World Glacier Monitoring Service (WGMS), which provides time-series data on specific glaciers worldwide. Its organizational linkages reflect a set of historical antecedents different from those of the WDCs. It operates within the context of the United Nations system through the Global Environment Monitoring System (GEMS/GTOS) of the United Nations Environment Program (UNEP) and the International Hydrological Program (IHP) of the United Nations Educational, Scientific, and Cultural Organization (UNESCO). Tied also to the Commission on Cryospheric Sciences of the International Union of Geodesy and Geophysics (CCS/IUGG) and the Federation of Astronomical and Geophysical Data Analysis Services (FAGS/ICSU), the WGMS today collects and publishes worldwide standardized glacier data. Its main tasks are to complete and maintain an inventory of the world's glaciers, periodically assess ongoing changes, publish results of mass balance measurements from selected reference glaciers at two-year intervals, continue collecting and publishing standardized data on glacier fluctuations at five-year intervals, and include satellite observations of remote glaciers to approach global coverage.

While WGMS is largely tabular in format, several other projects have existed in the past (e.g., Landsat 7 Glacier Inventory and International Work Group on Geospatial Analysis of Glaciated Environments [GAGE]) or currently exist to promote the use of remotely sensed data on glaciers. The Global Land Ice Measurements from Space (GLIMS) program monitors the world's glaciers using primarily data from the Advanced Spaceborne Thermal Emission and Reflection (ASTER) radiometer aboard the Eos Terra spacecraft, launched in December 1999. Its objectives are to develop a set of software tools that can be applied to the tracking of characteristics of glaciers such as areal extent, location of snow line at the end of the melt season, velocity field, and location of terminus. It also seeks to develop a network of centers around the world that will monitor the glaciers in their regions and create the infrastructure for storing and manipulating approximately 80 gigabytes of additional data per year, thus contributing to a global glacier database of derived glaciological parameters.

In addition, there are many regional or national monitoring projects that use both point and remotely sensed data (see examples for Switzerland, Italy, and Europe in the database). These projects span the whole world, from the Scandinavian Svartisen or Ostkindan regions to the Mexican tropical glaciers. Some of these initiatives have yet to be started, among them the Cryosat program of satellite data collected by the European Space Agency, which has been in preparation since 1999 and is due to be relaunched in 2009. Others date back to the 1950s.

Finally, nongovernmental organizations (NGOs) and even some individuals have developed glacial monitoring sites, if only using retrospective photography from different dates (including the Greenpeace photoarchives at http://www.greenpeace.org and individual sites on glaciers in the Pyrenees, as well as the Gletscherarchiv Project at http://www.gletscherarchiv.de and the Opération Glaciers site at http://www.unifr.ch/geoscience/geographie/glaciers).

RESEARCH AND ACADEMIC INSTITUTIONS

In addition to these umbrella organizations with international and/or governmental support, there are numerous institutions with research and academic interests in mountain glaciers. Several international academic societies or consortia serve to link many scholars into specific communities and organizations. The International Union of Geology and Geophysics/Union Commission for Cryospheric Sciences, for example, creates an international framework of glacial specialists. Heir to the

International Commission on Snow and Ice, founded in 1894, this institution is one of the oldest organizations dealing specifically with glaciological issues. The International Commission on Snow and Ice has made efforts to promote all cryospheric sciences and make them more prominent within the International Union of Geology and Geophysics.

Another pioneer is the International Glaciological Society, which focuses largely on publishing scientific research on glaciers. It was founded in 1936 to provide a focus for individuals interested in the practical and scientific aspects of snow and ice. It seeks to stimulate interest in and encourage research on the scientific and technical problems of snow and ice in all countries, facilitate and increase the flow of glaciological ideas and information by publishing scientific journals, and sponsor scientific meetings.

Finally, a number of national professional associations are active in the field of glaciology. In the United States we can identify the Cryosphere Focus Group of the American Geophysical Union, in Russia the Glaciological Commission of the Russian Geographic Society, and in Switzerland the Working Group for the Protection of Geotopes.

Similarly, individual national scientific organizations often have commissions that address glaciers. Some are specifically dedicated to the study of polar regions—for example, the Italian National Research Council's Polarnet and the Belgian Scientific Research Programme on the Antarctic—but others are larger in scope, such as the Commission for Glaciology of the Bavarian Academy of Sciences (http://www.glaziologie.de) and the different organizations sponsored by the Russian and Chinese Academies of Sciences.

As one might expect, national governments themselves often have entities that address glaciers. For instance, the National Snow and Ice Data Center (NSIDC), which manages one of the WDCs noted previously, is affiliated with the U.S. National Oceanic and Atmospheric Administration (NOAA). Funded by the National Aeronautics and Space Administration, it archives and distributes digital and analog snow and ice data, making use of NASA's past and current satellites and field measurement programs. It maintains information about snow cover, avalanches, glaciers, ice sheets, freshwater ice, sea ice, ground ice, permafrost, atmospheric ice, paleoglaciology, and ice cores. NSIDC also supports the National Science Foundation through the Arctic System Science Data Coordination Center and the Antarctic Glaciological Data Center.

Institutions such as the Northern Rocky Mountains Science Center of the U.S. Geological Survey and the National Glaciology Program affiliated with the Geological Survey of Canada tend to be interested in specific issues such as resources availability. Other national governments have even more focused concerns and channel them through line rather than research agencies. Nepal's Department of Hydrology and Meteorology, for example, has a strong interest in glacial lake outburst floods.

National governments are also very active in the field of polar studies, particularly of the Antarctic, since major national interests are perceived to be at stake in these regions. From Brazil to New Zealand, from France to South Africa, one can find numerous relevant organizations generally affiliated with a department of the national government.

National governments and international consortia have tended to focus on the crucial tasks of collecting and archiving data with some level of spatial and temporal breadth and depth. There are also a vast number of university departments and laboratories that focus on a wide range of basic scientific issues related to glaciers, although they are also frequently the entities that manage the data archives or participate in specific projects funded by government or international donor agencies.

Some of these arrangements are quite complex. The French Laboratory for Glaciology and Environmental Geophysics (LGGE), for example, is a mixed research unit under the joint responsibility of the National Center for Scientific Research and the Joseph Fourier University in Grenoble and receives technical support from the Paul-Emile Victor French Polar Institute (IPEV) for all activities undertaken in polar regions. In addition, it houses researchers from the Institute for Research and Development (IRD) who work on tropical glaciers. Research and academic structures within this organization are often devoted to specific regional interests. The Great Ice unit, for example, studies the evolution of glaciers in the Andes and Himalayas to evaluate the future of nivo-glacier water resources and to describe recent climate variability by referring to the Little Ice Age (fifteenth to nineteenth centuries), thus allowing for converging analyses of numerous decadal data series. Research carried out since 1991 with South American, European, and, more recently, Indian partners has led to the establishment of a high-quality data bank including meteorological, hydrological, and glaciological data. In the Andes the unit is responsible for the Research and Development Observatory GlacioClim. It brings together instrumental measures from recent periods, during which human forcings must be taken into account, and indirect indicators from past eras that are found in environmental and historical archives. Great Ice carries out ice-core analysis and undertakes high-altitude observations and observations along trajectories of atmospheric fluxes. In parallel, it develops methods for spatial imaging, dating, and use of environmental tracers. Taken together, these data allow the calibration of hydro-glaciological models using or generating climate simulations.

Certain departments have even taken on individual glaciers as their central interest and developed a specific tradition about them. Here we can mention the Juneau Icefield Research Program (JIRP) of the Universities of Alaska and Idaho, which mounts an eight-week field research program every year. Others, such as the Worthington Glacier Project of the Universities of Wyoming and Colorado, aim to integrate field-based measurements of glacier movement and numerical models that describe and simulate glacier flow. Some of these projects focused on individual glaciers were started many years ago. The Svartisen Research Project, now a joint program of the Department of Geography of the University of Manchester, England, and the Department of Earth Sciences of the University of Aarhus, Denmark, began in 1957, when there were no accurate maps of the Svartisen region in northern Norway. The project established photographic stations in front of many of the region's glaciers in the late 1950s and early 1960s. Photographs taken from these stations in subsequent years provide a record of the changes of the lower parts of the glaciers and underlie the program's focus on observations of glacier change in relation to climatic conditions.

Taken as a whole, the category of university departments is the largest. Found in cities as far apart as Sapporo, Japan, and Porto Alegre, Brazil, these glaciology groups are quite numerous, although the permanent staff of any one group may vary from 100 people to a single individual.

On top of all these institutions—professional associations, national governments, and university departments—there are international research projects that bring together different actors on an ad-hoc basis. One of them, Climate and the Cryosphere (CliC), includes mountain glaciers, although its objective is much broader. It was initiated in 2000 by the World Climate Research Programme (WCRP), one of the major international research programs on climate change. The CliC project addresses the entire cryosphere (i.e., snow cover, sea, lake and river ice, glaciers, ice sheets, ice caps and ice shelves, and frozen ground including permafrost) and its relation to climate. Its principal goal is to assess and quantify the impacts of climate variability and change on components of the cryosphere and the consequences of these impacts for the climate system. An additional goal is to determine the stability of the global cryosphere.

Some international projects have far more specific goals and are sometimes very short-lived, such as the Greenland Ice Core Project

(GRIP, 1989–95) and the North Greenland Ice Core Project (NGRIP, 1996–2003), which aimed at drilling and retrieving 3-km-long ice cores. It is striking that most of these international projects deal with polar issues and not glaciers.

INSTITUTIONS FOR WHICH GLACIERS HAVE SPECIAL SIGNIFICANCE

In addition to institutions centered on glacier monitoring or on physical science research, a number of organizations have specific policy or action-oriented agendas that grow out of concern with the human dimensions of glacier processes. Knowing how glaciers behave and monitoring their dynamics are important scientific issues, but this information acquires added significance from the strong impacts of glaciers on society. As the source of catastrophes, glaciers can exact a deadly toll. Yet they are also vital to vast numbers of people who depend directly or indirectly on their water for drinking water, irrigation, hydropower, and other purposes. Unique landscapes are defined by the presence of glaciers, and these shape the identities of populations living among them and provide pleasure and recreation for those who visit. This realization underlies many efforts to gather information on specific thematic or regional issues.

Glaciers as hazards are a particularly compelling theme because of their potential to devastate landscapes and property and cause human deaths, often in the thousands (Carey, Wiegandt and Lugon, this volume). According to a recent inventory undertaken by the International Center for Integrated Mountain Development (ICIMOD) in Nepal, there are more than 20 potentially dangerous lakes in Nepal and in Bhutan alone. While glacial lake outburst floods are nothing new to the Himalayan countries, including China and India, the glacier melting induced by climate change makes the situation more acute. As noted earlier, the Nepal Department of Hydrology and Meteorology has a major project related to glacial lake outburst floods, an interest that it shares with ICIMOD and the UNEP Regional Resource Centre for Asia and Pacific (RRCAP). Among others, these institutions try to develop early-warning systems and to support mitigation measures to reduce the incidence of catastrophic events.

Even in less severe situations, organizations are interested in glacier hazards. In Switzerland, for example, the University of Zurich maintains a database on glacier hazards. There is recognition that hazards related to glaciers are common in most glacierized high-mountain areas in the world but that their impact on society depends on the density of human structures and settlements in those regions. The Swiss Alps are particularly affected by glacier hazards because of their steep slopes and the proximity of installations and villages to glacial environments. Thus, experience and data on glacier disasters, including historical cases, in Switzerland have been systematically collected over the past two decades. From these a digital database has been developed that includes all documented catastrophic events related to glacier hazards in the Swiss Alps.

The European Union has launched a similar project, engaging a different Swiss partner, the Federal Polytechnic Institute in Zurich (ETH). Called Glaciorisk, it is partly funded by the European Union and involves 11 institutes from six European countries, including both EU members and nonmembers (France, Norway, Switzerland, Italy, Iceland, and Austria). It aims to identify, survey, and prevent catastrophic events that may occur as a result of glacial hazards. Because of the recent increase in both occupation and economic activities in some European mountainous areas, a growing concern about security for people and infrastructure led to the awareness of potential danger. Recognizing the importance of past catastrophic events as indicators of the hazards that might be expected for any given glacier, the project's researchers compiled an inventory of some 350 known flood outburst events. While acknowledging that extrapolations to the future remain hypothetical, they hope to predict future hazards by considering how varying environmental and climatic conditions might lead to changes in the catastrophic event itself. Moreover, they seek to assess how new patterns in population and economic distributions could affect damages from glacial catastrophic events.

Closely related to the theme of glacier hazards is the focus on glaciers as significant indicators of climate change. Data and evidence presented throughout this volume demonstrate that understanding of glacier dynamics is critical to identifying patterns of glacier advance and retreat over time. All of the monitoring organizations document current shrinkage of most glaciers throughout the world (with the notable exception of some in New Zealand, Norway, and California, which are nevertheless expected to retreat under the strong warming conditions that are predicted for the future). The stark evidence of their shrinkage has made them icons in political discussions about global warming. The threat of their eventual disappearance figures importantly in debates over energy policy because of their role as reservoirs of water for the production of hydroelectricity. The reports and campaigns of institutions involved in these discussions have made glacier issues more visible to the general public. The World Wide Fund for Nature forcefully sounds the alarm by pointing to the potential sea-level rise, flooding, and, paradoxically, water shortage that will occur as a result of warming-induced glacier melting.

In addition to the international efforts and concern about global impacts of glacier retreat, there is the third core theme of regionalism, which unites institutions focusing on the specificity and importance of glaciers to particular areas of the world. This approach seems particularly characteristic of academic research and has been presented previously here, but it is also a feature of some institutions and research consortia. This regional focus often overlaps with attention to hazards, as the Glaciorisk project demonstrates. While some regionally based organizations are uniquely devoted to glacier issues, others include them as part of a more general concern for fragile mountain environments and cultures. In both instances, these specialized organizations are testimony to the vulnerability of mountain regions and the concern for their future that is manifested either by local populations or by international organizations that have designated them for special attention. The International Center for Integrated Mountain Development (ICIMOD), mentioned previously, is an example of the latter type of organization. It is a large and well-known international center for mountain learning and knowledge bringing together eight member countries of the Hindu Kush–Himalayan area (Afghanistan, Bangladesh, Bhutan, China, India, Myanmar, Nepal, and Pakistan). It is committed to improving sustainable livelihoods and securing a better future for the people of the extended Himalayan region. Its range of activities includes biodiversity, landscapes, forests, water, gender questions, and indigenous knowledge, but it includes a program area devoted to mountain risks and hazards and has collaborated on an atlas that documents natural hazards in the region (Zurick et al. 2006).

A much smaller regional organization is L'Association Moraine, a consortium of research institutions, regional governments, and water agencies dedicated to improving understanding of glaciers in the French Pyrenees. These small regional glaciers are not the focus of any other organization or network. In fact, they are simply not well studied, and thus basic features such as their number, location, and characteristics are not well known. Systematic observation efforts began only at the end of the 1980s. In contrast, on the Spanish side annual measurements are carried out and therefore glacier characteristics are far better known. Recognition of these gaps led to the creation of the association in 2001. Its mission is to study the glaciers and their environmental context in the French Pyrenees. It undertakes annual monitoring and disseminates this knowledge to a broad public. Collaboration takes place with Spanish colleagues to ensure complete information on glaciers throughout the region.

Other examples of regional institutions include various NGOs, such as Pro Natura Centre Aletsch, at the edge of the UNESCO World Heritage site Jungfrau-Aletsch-Bietschhorn in Switzerland, and museums such as the Gletschergarten in Lucerne, Switzerland. Some organizations focus on the spiritual value of particular mountains, such as Mt. Kailash (Tibet) or Mt. Shasta (California), and advocate for the mountains and their glaciers.

Beyond the research institutions and the political activist organizations there are also corporations that draw on themes of nature conservation to illustrate their environmental benevolence. The large manufacturing firm Unilever, specializing in food, cosmetics, and domestic appliances and products, organized a campaign in 2005 that emphasized not only the beauty of glaciers but their role in water storage. To counter effects of pollution, it set up a program to remove litter and construction debris from glacier sites in the Italian Alps.

On the opposite side of the battle lines, nature conservationist organizations and movements often coalesce to oppose regional development projects, and glaciers become dramatic evidence of the impact of various projects on landscapes. Many of these involve skiing developments. For example, the Jumbo Creek Conservation Society opposes a proposed ski resort in British Columbia that would construct ski lifts directly across several glaciers. The Canadian international mining company Barrick Gold's plans to relocate three glaciers in the mountain range between Argentina and Chile to gain access to gold and silver deposits have met with opposition from farmers and residents of the surrounding Huasco Valley, who use the glaciers for irrigation and who fear degradation of the ecological balance and agricultural production of the fertile river valley.

Finally, some researchers are focusing on the significance of glaciers to indigenous communities. ICIMOD has emphasized the approximately 70 large glaciers in the Himalayas that feed rivers such as the Indus, the Ganga, and the Brahmaputra, on which millions of people depend. As has been noted elsewhere in this volume, societies located close to glaciated areas are dependent on meltwater for primary production and are therefore very vulnerable to changes in glacier dynamics. The consequences for the local populations are particularly severe in the central and eastern Himalayas because, compared with the rest of the world, the population density near glaciers in this region is very high and most people living in such areas are still dependent on the meltwater for both agriculture and pasture. Moreover, the effects will extend beyond the glacier sites to areas downstream that are also heavily dependent on water for crop irrigation. Perhaps surprisingly, such livelihood issues occupy a rather small place on glacier-related Web sites. This contrasts with the substantial evidence brought by various articles in this volume of their crucial role for irrigation systems, particularly in the developing world, for energy production (witness the 60% of Swiss energy production furnished by glacier-fed dams), and for tourism, which constitutes a major source of revenue for all mountain countries in the developed world.

GLACIER ORGANIZATIONS PRESENT AND FUTURE

This inventory of groups, organizations, and institutions dedicated to observation, study, and action relating to glaciers does not claim to be exhaustive, but it nevertheless permits reflection. Rather than a coordinated and systematic repository of knowledge, it shows disparate perspectives and competing activities. The important strengths that emerge are the long history of data collection, the geographic breadth of coverage, and the variety of themes that have been documented. These comprehensive efforts create the basis on which research can go forward. Implicit in many descriptions of glacier-related organizations is the importance of the continuing study of glaciers, which are identified as crucial indicators of global warming. The organizations listed here have emphasized that continuity of glacial monitoring and of the analysis of glacier dynamics is critical to our ability to understand the processes that regulate glacier behavior and that understanding of glacial processes is a precondition for predicting what will happen under conditions of climate change. The underlying but implicit concern is that the expected changes will have significant impacts on society. Making clear this connection between processes and their outcomes is essential for developing the effective policy responses that will be one of the cornerstones of adaptation.

In many scientific circles, there is great concern that support for continuous and consistent data collection is waning in favor of short-term projects that are attractive in their novelty but do not permit the accumulation of long-term data sets. One of the original and fundamental lessons of early glacier research was that the earth has seen dramatic shifts in climate and conditions of life. If we do not continue to monitor and measure glaciers, we will not be able to learn the second lesson—that our human activities are fundamentally altering our current environment and that glaciers can provide us with clues about some of the consequences. Sea-level rise is one of the most obvious, but there will also likely be changes in the patterning of hazards and, potentially, in their intensity as well. The implications for the critical role of glaciers in providing water for hydroelectricity and irrigation and snow for winter sports are much less well understood. Indeed, much less attention is being given to changing landscapes, perceptions, and impacts on resource availability than to hazards that are more directly linked to physical changes. Yet the transformation of resource, energy, and cultural domains as a result of glacier melting will in all likelihood affect greater numbers of people than glacier-related catastrophes will. International efforts to design appropriate policy to counter impacts and devise strategies of adaptation depend, of course, on accurate knowledge of glacier dynamics. We come away from our study, however, with the sense that not only must basic scientific monitoring and analysis continue but also more attention must be paid to the themes raised in this volume—the perceptions of glaciers that have shaped the cultures of people living near them and the social and institutional patterns that have been developed to meet their hazards and benefit from their resources, now under threat from rapid glacier retreat. These factors, as well as the international will to confront climate change, will be critical to success in adapting to inevitable glacier retreat in the next decades. Basic research carried out by individuals such as the contributions to this volume will be the foundation for this understanding. In addition, the collective and coordinated efforts in this article will help the partially overlapping sets of researchers locate and communicate with each other, thereby enhancing our capacity to respond to the challenges of the future.

REFERENCES CITED

Mariétan, I. 1959. La vie et l'oeuvre d'Ignace Venetz. *Bulletin de la Murithienne* 76:2–51.

Pálsson, S. 2004 (1795). *Draft of a physical, geographical, and historical description of Icelandic ice mountains on the basis of a journey to the most prominent of them in 1792–1794.* Trans. R. S. Williams, Jr., and O. Sigupsson. Reykjavík: Icelandic Literary Society.

Zurick, D., J. Pacheco, B. R. Shrestha, and B. Bajracharya. 2006. *Atlas of the Himalaya.* Lexington: University of Kentucky Press.

NOTES

CHAPTER 3

1. "Glacial hazards" are critical glacier fluctuations (advancing or retreating glaciers), glacier floods, or ice avalanches. In Switzerland, the main concentration of historical events of glacier catastrophes is in the Canton Valais because of its large glacierized areas and particularly steep terrain (Huggel, Kääb, and Haeberli 2002).

2. There is an extensive body of local history that describes the intricate irrigation networks and their management in the Valais. See, for example, the papers from an international symposium published by the Société d'Histoire du Valais Romand in 1995. Wiegandt (1980, n.d.) has studied *bisses* organization as part of the overall pattern of collective risk sharing.

3. This perspective challenges the standard view developed by Kahnemann and Tversky (2000) that in situations with low probability (rare events), people tend to be risk-averse for gains and risk-preferring for losses.

4. For the most part, these are local people already responsible for security questions in their local communities. They have received training from glaciologists about glacier-related risks and in effect constitute an intermediary level between the scientific expert and the lay population.

5. Recent melting on the Blanche de Perroc (the "White" peak) is sufficient to warrant the readoption of the mountain's old name "Lex blava" ("Grey Wall" in the old local dialect) (Delaloye and Haeberli 2004). This face was still partially glaciated during the 1970s (Reynald Delaloye, personal communication) but now has completely lost its ice cover.

CHAPTER 4

1. The U.S. Geological Survey officially names two more: Mud Creek and Watkins (Miesse 2005).

2. We used 2 × 2 chi-square tests for each of the five values, with the cells containing the four possible combinations of birthplace (local vs. nonlocal) and value (mentioned vs. not mentioned).

3. We asked directly about glacial change only at the end of the interview. The number of people who said that they were "very concerned" or "somewhat concerned" about glacial retreat was 29, larger than the 26 who stated, early in the interview, that there were glaciers on the mountain. It is possible that the people who did not know that there were glaciers but expressed concern about retreat were surprised to learn from our question that there were glaciers and hence were concerned or that they sensed our involvement with environmental issues and felt that these answers were the ones we preferred to "not concerned."

4. We used the Mann-Whitney test for the first comparison and the Kruskal-Wallis test for the second.

CHAPTER 10

1. Previous studies (e.g., Walters and Meier 1989; Hodge et al. 1998; McCabe, Fountain, and Dyurgerov 2000) have noted inverse correlations between mass balance records from Alaskan glaciers and records from glaciers in the U.S. Pacific Northwest and southwestern Canada. These relationships are related to the strength and position of the Aleutian Low and its impact on the movement of weather systems and snowfall totals in the two regions.

2. Snow water equivalent is for April 1 (within seven days) for the Canadian stations and May 1 for the stations in the United States. A few missing values were replaced in the Canadian records using simple linear regression with the most highly correlated station. Canadian data were obtained from the 2004 version of the Canadian Snow Data CD-ROM (Meteorological Service of Canada 2000). The records for Montana were obtained from the National Resources Conservation Service (NRCS) anonymous ftp server (ftp.wcc.nrcs.usda.gov).

3. Principal components analysis (using a varimax rotation and three retained principal components, not shown) of all stations' maximum temperature records suggests that there may be a clustering of stations in relation to location with regard to the Continental Divide and latitude. However, missing data and potential inhomogeneities in non-AHCCD temperature records would suggest cautious interpretation of such findings.

4. A similar winter mass balance reconstruction for Glacier National Park was developed using Biondi, Gershunow, and Cayan's (2001) reconstruction, which uses data from California (not shown).

CHAPTER 11

1. In this chapter *Alps* or *Alpine* refers explicitly to the European Alps; the terms *alps* and *alpine* are purely generic.

CHAPTER 19

1. For a thorough examination of the legal situation, see Shalpykova (2002).

2. High-pressure dams take advantage of an altitude differential (hence pressure increase) between the point where water is gathered and the point where it is processed into electric power through a turbine. Water can be released only at appropriate moments. This is very different from a turbine that is moved by a steady flow of water such as an installation on a river.

3. Value added is the net output of a sector after adding up all outputs and subtracting intermediate inputs. It is calculated without making deductions for depreciation of fabricated assets or depletion and degradation of natural resources.

4. The coding of these expressions is carried out in terms of the SPARE simulation language developed at the Graduate Institute of International Studies (see Luterbacher et al. 1987 for an overview of the workings of such a dynamic simulation language). SPARE codes a differential equation in terms of FORTRAN coding rules. A dot after a variable represents a derivative with respect to time, so that dx/dt is coded *x.*; an asterisk corresponds to the multiplication sign.

CHAPTER 20

1. The history of scientific ideas is often the chase after the earliest reference to some germ of a theory elaborated at a later date. Thus the reference to Venetz might be contested. Indeed, other scholars point to the work of Icelandic explorer Sveinn Pálsson (2004 [1795]), whose treatise contained the first elements of a theory of glaciers but was forgotten for nearly a century.

2. What is summarized here is available in raw form at http://mri.scnatweb.ch.

INDEX